T0201502

Plant Cells and their Organelles

Plant Cells and their Organelles

EDITED BY

William V. Dashek

and

Gurbachan S. Miglani

Library of Congress Cataloging-in-Publication Data

Names: Dashek, William V., editor. | Miglani, Gurbachan S., editor.
Title: Plant cells and their organelles / by William V. Dashek, Gurbachan S. Miglani.
Description: [Hoboken, N.J.] : Wiley, 2017. | Includes bibliographical references and index.
Identifiers: LCCN 2016024724 (print) | LCCN 2016026083 (ebook) | ISBN 9780470976869 (cloth) |
  ISBN 9781118924761 (pdf) | ISBN 9781118924754 (epub)
Subjects: LCSH: Plant cells and tissues–Textbooks.
Classification: LCC QK725 .D36 2017 (print) | LCC QK725 (ebook) | DDC 581.3–dc23
LC record available at https://lccn.loc.gov/2016024724

A catalogue record for this book is available from the British Library.

Contents

Contributors

Milee Agarwal
Scientist
Pharmacology and Toxicology Department
B.V. Patel PERD Centre
Ahmedabad, Gujarat, India

James E. Bidlack
Professor of Biology and CURE-STEM
Scholar
Department of Biology
University of Central Oklahoma
Edmond, OK, USA

Amy M. Clore
Professor of Biology
Division of Natural Sciences
New College of Florida
Sarasota, FL, USA

William V. Dashek
Retired Faculty
Adult Degree Program
Mary Baldwin College
Staunton, VA, USA

D. Davis
Graduate Student
Hellman Fellow Plant Sciences
Department of Plant Sciences
University of California Davis
Davis, CA, USA

P. Desai
Scientist
Cellular and Molecular Biology Department
B.V. Patel PERD Centre
Ahmedabad, Gujarat, India

Robert Donaldson
Professor
Department of Biological Sciences
George Washington University
Washington, DC, USA

Georgia Drakakaki
Associate Professor
Hellman Fellow Plant Sciences
Department of Plant Sciences
University of California Davis
Davis, CA, USA

Rajdeep Kaur Grewal
Senior Research Fellow
Department of Physics
Bose Institute
Kolkata, India

J. Kenneth Hoober
Professor Emeritus
School of Life Sciences
Center for Photosynthesis
Arizona State University
Tempe, AZ, USA

Dasmeet Kaur
Research Assistant
School of Agricultural Biotechnology
Punjab Agricultural University
Ludhiana, India

Gurbachan S. Miglani
Visiting Professor
School of Agricultural Biotechnology
Punjab Agricultural University
Ludhiana, India

Harish Padh
Vice-Chancellor
Sardar Patel University
Vallabh Vidyanagar, Anand
Gujarat, India

Ray J. Rose
Emeritus Professor
Center of Excellence for Integrative
Legume Research
School of Environmental and Life Sciences
The University of Newcastle
Callaghan, New South Wales, Australia

Soumen Roy
Associate Professor
Department of Physics
Bose Institute
Kolkata, India

Saptarshi Sinha
Senior Research Fellow
Department of Physics
Bose Institute
Kolkata, India

Terence W.-Y. Tiew
Graduate Student
Center of Excellence for Integrative
Legume Research
School of Environmental and Life Sciences
The University of Newcastle
Callaghan, New South Wales, Australia

Yogesh Vikal
Senior Geneticist
School of Agricultural Biotechnology
Punjab Agricultural University
Ludhiana, India

T.E. Wilkop
Senior Project Scientist
Hellman Fellow Plant Sciences
Department of Plant Sciences
University of California Davis
Davis, CA, USA

Preface

Plant Cells and their Organelles is an advanced textbook to enhance the plant biology student's knowledge of the structure and function of plant cells and their organelles. The book assumes that the student has had introductory courses in plant science and chemistry. The book emphasizes the research literature in plant cell biology concerning cell and organellar structure. However, the literature from plant physiology, molecular genetics, and biochemistry has been utilized to augment the discussions of cell and organellar function.

Acknowledgments

Dashek is grateful to Drs. W.G. Rosen, W.F. Millington, and D.T.A. Lamport for training enabling a career in teaching and research in plant biology. Dashek appreciates the grant support of the USA's NIH, NSF, DOE, and USDA Forest Service. Dashek thanks Ms. Katherine Mumford, Ms. Retha Howard, and Ms. Abigail M. Johnson for technical assistance in the preparation of the manuscript.

Miglani wishes to record his appreciation for Dr. Darshan S. Brar, Honorary Adjunct Professor, School of Agricultural Biotechnology, Punjab Agricultural University, Ludhiana, India for his valuable technical suggestions. Miglani thanks Dr. (Mrs.) Parveen Chhuneja, Director, School of Agricultural Biotechnology, Punjab Agricultural University, Ludhiana, India for motivating me to prepare this volume, and the Punjab Agricultural University for providing facilities for this work.

We thank the Wiley editorial staff members for their attention to detail.

CHAPTER 1

An introduction to cells and their organelles

William V. Dashek

Retired Faculty, Adult Degree Program, Mary Baldwin College, Staunton, VA, USA

Cells

Parenchyma, chlorenchyma, collenchyma, and sclerenchyma are the four main plant cell types (Figure 1.1, Evert, 2006). Meristematic cells, which occur in shoot and root meristems, are parenchyma cells. Chlorenchyma cells contain chloroplasts and lack the cell wall thickening layers of collenchyma and sclerenchyma. Certain epidermal cells can be specialized as stomata that are important in gas exchange (Bergmann and Sack, 2007). The diverse cell types (Zhang *et al.*, 2001; Yang and Liu, 2007) are shown in Table 1.1. Photomicrographs of certain of these cell types can be found in Evert (2006), Fahn (1990), Beck (2005), Rudall (2007), Gunning (2009), MacAdam (2009), Wayne (2009), Beck (2009), Assmann and Liu (2014) and Noguchi *et al.* (2014).

How do cells arise?

Cells arise by cell divisions (see Chapter 8 for mitosis and meiosis) in shoot and root (Figures 1.2 and 1.3) meristems (Table 1.2, Lyndon, 1998; McManus and Veit, 2001; Murray, 2012). The shoot apex is characterized by a tunica–corpus organization (Steeves and Sussex, 1989). The tunica gives rise to the protoderm and its derivative, the epidermis. In contrast, the corpus provides the procambium which yields the primary xylem and phloem. In addition, the ground tissue derives from the corpus originating the pith and cortex. Following divisions, cells can differentiate into tissues (Table 1.3) and organs of the mature plant body (Leyser and Day, 2003; Sachs, 2005; Dashek and Harrison, 2006). The leaf primodium arises on the apex (Micol and Hake, 2003). The mature angiosperm leaf consists of palisade cells and spongy mesophyll cells sandwiched between the upper and the lower epidermis (Figure 1.4). The epidermis possesses guard cells with associated stomata that function in gas exchange. *KNOX* genes affect meristem maintenance and suitable patterning of organ formation (Hake *et al.*, 2004). In dissected leaves, *KNOX* genes are expressed in leaf primordia (Hake *et al.*, 2004). Hake *et al.* (2004) suggest that

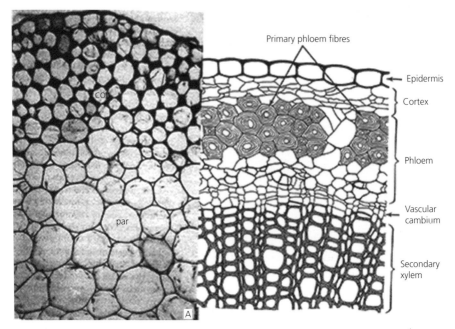

Figure 1.1 Plant cell types: Left: parenchyma (par) and collenchyma (co). Right: sclerenchyma. Source: Evert (2006). Reproduced with permission of John Wiley & Sons.

KNOX genes may be important in the diversity of leaf form. Extensive discussions of leaf development occur in Sinha (1999), Micol and Hake (2003) and Efroni *et al.* (2010). Under appropriate stimuli the vegetative apex can be converted to a floral apex (Figure 1.5). Photoperiod (Mazumdar, 2013), such as short days and long days and combinations of the two, is one such stimulus (Glover, 2007; Kinmonth-Schultz *et al.*, 2013). This induction results in the production of florigen (Turck *et al.*, 2008), the flowering hormone (Zeevaart, 2006). While early reports suggest that florigen is an mRNA species (Huang *et al.*, 2005), a more recent investigation indicates that florigen is a protein complex (Yang *et al.*, 2007; Taoka *et al.*, 2013). Taoka *et al.* state that florigen protein is encoded by the gene, Flowering Locus T, in *Arabidopsis* species (Shresth *et al.*, 2014). It is believed that florigen is induced in leaves and that it moves through the phloem to the shoot apex. Plant hormones (see Appendix A) can influence floral development (Howell, 1998). Gibberellins (Blázquez *et al.*, 1998), auxins, and jasmonic acid can affect petal development. In contrast, auxin can influence gynoecium development. The ABC model has been proposed for regulating the development of floral parts (Soltis *et al.*, 2006). The *A* gene expression is responsible for sepals, while the petals are the result of co-expression of *A* and *B* genes. The *B* and *C* genes are responsible for stamen development and carpels require *C* genes. In certain plants, vernalization (low temperature) can induce flowering in certain plants (Kemi *et al.*, 2013). A diagram of the mature angiosperm plant body is presented in Figure 1.6. Plant

Cell types	Characteristics	References
Epidermal cells	Unspecialized cells; one layer of cells in thickness; outer covering of various plant parts; variable in shape but often tabular	Evert (2006)
Examples		
Guard cells	Specialized epidermal cells; crescent shaped; contain chloroplasts; form defines stomatal pore	Wille and Lucas (1984)
Subsidiary cells	Cells which subtend the stomatal guard cells	http://anubis.ru.ac.za/Main/ANATOMY/guardcells.html
Trichomes	An outgrowth of an epidermal cell; can be unicellular or multicellular	Callow (2000)
Parenchyma cells	Isodiametric, thin-walled primary cell wall; in some instances may have secondary walls; not highly differentiated; function in photosynthesis, secretion, organic nutrient and water storage; regeneration in wound healing	Evert (2006) and Sajeva and Mauseth (1991)
Examples		
Transfer cells	Specialized parenchyma cells; plasmalemma greatly expanded; irregular extensions of cell wall into protoplasm; transfer dissolved substances between adjacent cell; occur in pith and cortex of stems and roots; photosynthetic tissues of leaves; flesh of succulent fruits; endosperm of seeds	Dashek et al. (1971) and Offler et al. (2003)
Collenchyma cells	Lamellar or plate collenchyma, with thickenings on the tangential walls Angular collenchyma, with thickenings around the cell walls Present in aerial portions of the plant body	Evert (2006)
Vascular cells		
Phloem		
Sieve cells		
Sieve elements		
Companion cells	Specialized parenchyma cells; possess numerous plasmodesmatal connections	Oparka and Turgeon (1999)
Albuminous cells in gymnosperms	Absence of starch; cytoplasmic bridges with sieve cells; dense protoplasm, abundance of polysomes, highly condensed euchromatin and abundant mitochondria	Alosi and Alfieri (1972) and Sauter et al. (1976)
Xylem		
Tracheids	Long tapering cell with lignified secondary wall thickenings; can have pits in walls; devoid of protoplasm at maturity; not as specialized as vessels; widespread	Tyree and Zimmerman (2002)
Vessels		Fukuda (2004) and Evert (2006)

(Continued)

Table 1.1 (Continued)

Cell types	Characteristics	References
Specialized cells – Hydathodes (modified parts of leaves and leaf tips or margins)	Consist of terminal tracheids epithem, thin-walled chloroplast-deficient cells, a sheath with water pores; guttation discharge of liquid containing various dissolved solutes from a leaf's interior	Lersten and Curtis (1996), https://www.biosci.utexas.edu/ and Maeda and Maeda (1988)
Laticifer cells	Cells or a series of cells which produce latex	Fahn (1990), Pickard (2008) and Botweb.uwsp.Edu
Simple	Single-celled	
Compound and articulated	Union of cells compound in origin and consist of longitudinal chains of cells; wall separating cells remain intact, can become perforated or entirely removed	
Salt glands	Modified trichomes, two-celled and positioned flat on the surface in rows parallel to the leaf surface; occur in *Poaceae*;	Evert (2006), Tan *et al*. (2010), Oross *et al*. (1985) and Thomson *et al*. (1988)
	Cap cell – large nucleus and expanded cuticle	Naidoo and Naidoo (1998)
	Basal cell – numerous and large extensive partitioning invaginations of plasmalemma	
Nectaries	Found in nectarines; produce nectar, usually at the base of a flower	Fahn (1990), Nicolson and Nepi (2005) and Paiva (2009)
Idioblasts	Crystal-containing cells	Lersten and Horner (2005)
Example		
Raphides	Produce needle-shaped crystals	
Mucilage cell	Occur in a large number of dicots, common in certain cacti; slimy mucilage prevents evaporation of water by binding to water; a parenchyma cell whose dictyosomes produce mucilage as in seed coats; cell walls are cellulosic and unlignified	http://www.sbs.utexas.edu/masuet/weblab/webchap9secretory/9.1-2.html, Western *et al*. (2000) and Arsovskia *et al*. (2010)
Oil cells	Specialized cells appear like large parenchyma cells; can occur in vascular and ground tissues of stem, and leaf cell wall has three distinct layers; cavity is formed after the inner wall layer has been deposited	Rodelas *et al*. (2008), http:brittanica.com and Lersten *et al*. (2006)
Druses	Spherical aggregates of prismatic crystals	Lersten and Horner (2005)
Cells in non-angiosperms		
Bryophytes		
Gemmae	One to many cells	http://buildingthepride.com/faculty/pgdavison/bryology_links.htm
Hydroids	Water-conducting cells	http://www.Biology-online.org

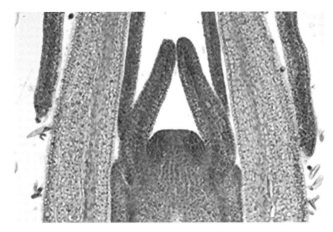

Figure 1.2 Angiosperm shoot meristem section. Source: Alison Roberts. Reproduced with permission of University of Rhode Island.

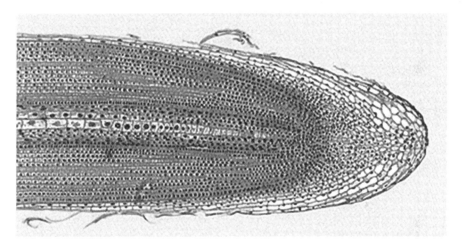

Figure 1.3 Angiosperm root meristem section. Source: Alison Roberts. Reproduced with permission of University of Rhode Island.

development is discussed in Fosket (1999), Moore and Clark (1995), Greenland (2003), Leyser and Day (2003) and Rudall (2007).

What is the composition of cells?

Certain plant components exhibit polar growth, for example, the tip growth of pollen tubes (Hepler *et al.*, 2001). The tubes elongate via the fusion of Golgi-derived vesicles with the plasmalemma and subsequent deposition of the vesicles' contents into the cell wall (Taylor and Hepler, 1997; Parton *et al.*, 2001 and others as reviewed in Malho (2006a, 2006b)). In 2007, Dalgic and Dane (2005) published a diagram depicting the now known tube-tip structural elements and physiological processes that facilitate tube elongation. The diagram represents a

Table 1.2 Meristems and their derivatives.*

Meristems	Derivatives
Primary	
Protoderm	Epidermis
From tunica (Evert, 2006)	
Procambium (provascular)	Primary xylem and phloem
From corpus (Evert, 2006)	Vascular cambium
Ground	Ground tissue: pith and cortex
Lateral	
Vascular cambium	
Fusiform initials	Secondary xylem
	Secondary phloem
Ray initials (Evert, 2006)	Ray cells
Cork cambium	
Phellogen	Replaces the epidermis when cork cambium initiates stem girth
Periderm (Evert, 2006)	increase; composed of 'boxlike' cork cells which are dead at
	maturity; protoplasm secretes suberin; some cork cells that are
	loosely packed give rise to lenticels which function in gas exchange
	between the air and the stem's interior. http://www.Biology-online.
	org, Evert (2006), http://www.vebrio.Sceince.vu.nl.en/virut
Phelloderm	Parenchyma cells produced on the inside by the cork cambium

*Meristems are discussed by Steeves and Sussex (1989).

Table 1.3 Plant tissues.

Tissue system			
Meristematic	Ground	Vascular	Dermal

significant advance over the early studies of pollen tubes as it assigns function to ultrastructural components, for example, signalling molecules, the Rho family of GTPases and phosphatidylinositol 4,5 bisphosphate appear to be localized in the apical plasma membrane. Besides pollen tubes, root hairs exhibit polar growth.

Cell organelles – an introduction

Organelles are required for plant growth, development and function (Sadava, 1993; Gillham, 1994; Herrmann, 1994, Agrawal, 2011). These organelles (Figure 1.7) are the loci for a myriad of physiological and biochemical processes (Tobin, 1992; Daniell and Chase, 2004 – see individual chapters).

There are many diagrams of a generalized plant cell. Some of these are available at www.explorebiology.com, http://www.daviddarling.info/images/plant_cell.jpg

Figure 1.4 SEM of a pecan leaf. Diagram of a leaf's interior is available at http://pics4learning. com. Source: Reproduced with permission of Asaf Gal.

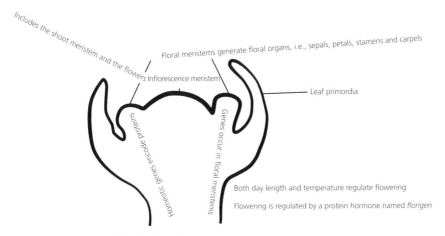

Figure 1.5 Schematic of the floral meristem.

and http://micromagnet.fsu.edu. The organelle contents of plant and animal cells in common and those unique to plant cells are depicted in Table 1.4. The dimensions of plant organelles are presented in Table 1.5. A plant organelle database (PODB) has been reviewed by Mano *et al.* (2008).

To enter a plant cell, molecules must traverse both the cell wall and the fluid mosaic plasmalemma (Singer and Nicolson, 1972; Leshem *et al.*, 1991; Larsson and Miller, 1990). In contrast to the fluid mosaic model (Figure 1.8) of the plasmalemma,

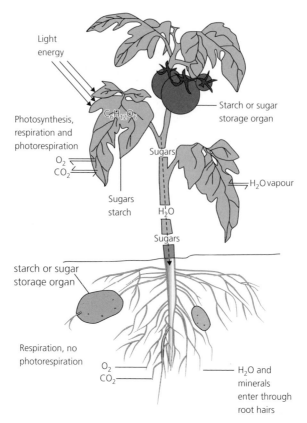

Figure 1.6 Diagram of angiosperm plant body. Source: From http://www.msu.edu/course/te/8021/science08plants/foods.html.

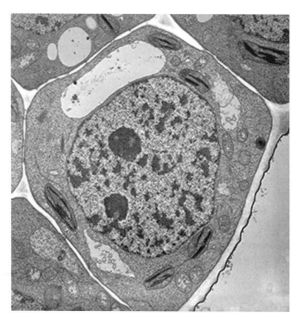

Figure 1.7 Electron micrograph of a plant cell and its organelles. Source: Reproduced with permission of H.J. Horner.

Table 1.4 Comparison of organelle contents of plant and animal cells.*

Organelle	Animal cell	Plant cell
Cell wall	Absent	Present
Centrioles	Present	Absent
Endoplasmic reticulum	Present	Present
Glyoxysomes	Absent	Present
Golgi apparatus	Present	Present
Microfilaments	Present	Present
Mitochondrion	Present	Present
Nucleus	Present	Present
Peroxisomes	Present	Present
Plastids	Absent	Present
Protein bodies	Absent	Present
Spindle	Present	Present
Vacuoles	Sometimes small	Present (mature cell – large central)

* Early discussions of plant cell organelles occur in Hongladarom *et al.* (1964), Pridham (1968), Reid and Leech (1980) and Tobin (1992).

Table 1.5 Dimensions of subcellular organelles.

Organelles	Dimension
Chloroplast	4–6 μm in diameter
Golgi apparatus	Individual cisternae, 0.9 μm
	Coated vesicles 50–280 μm in diameter
Microbodies	0.1–2.0 μm in diameter
Microtubules	0.5–1.0 μm in diameter
Mitochondria	1–10 μm
Nuclear envelope pores	30–100 μm in diameter
Nucleus	5–10 μm in diameter
Peroxisome	0.2–0.7 μm
Plasmodesmata	2–40 μm in diameter
Primary wall	1–3 μm
Protein bodies	2–5 μm in diameter
Vacuoles	30–90% of cell volume

the picket–fence model proposes the accumulation of membrane protein anchored in an actin network beneath the membrane (Kusumi *et al.*, 2012).

The plasmalemma is composed of water, protein and lipids. There are both integral and peripheral proteins (Leshem *et al.*, 1991). The integral proteins may be simple (classical α-helical structure that traverses the membrane only once) or complex (globular – composed of several α-helical loops which may span the membrane several times). Peripheral proteins can be easily isolated by altering

Fluid mosaic model of the plasmalemma

Consists of a lipid bilayer in which globular proteins are embedded; There are two types of proteins: integral and peripheral. Oliogsaccharides (2–20 monosaccharides) can be attached to the integral proteins. Phospholipids from the bilayer with a polar head on the outside and non-polar tails on the inside.

Fence model of the plasmalemma

There is a membrane skeleton with skeleton-anchored proteins and transmembrane proteins projected outwards into the cytoplasm. Cytoplasmic domains of proteins collide with the actin skeleton, yielding temporary confinement of the transmembrane proteins. The membrane can contain lipid rafts and related caveolae invaginations. The rafts are combinations of proteins and the lipids which may function in signalling. sphingolipids are prevalent in the rafts.

Picket model of the plasmalemma

Phospholipids can also be confined by the membrane skeleton. Some investigators combine the fence and picket models.

Figure 1.8 Top: Fluid mosaic model of the plasmalemma. Middle: Fence model of the plasmalemma. Bottom: picket model of the membrane.

the ionic strength or pH of the encasing medium. The transport proteins are pumps, carriers or chemicals (see section on membrane transport). The lipids are electro-negative and anionic phospholipids, sphingolipids (Figure 1.9), chloroplast-specific glycerolipids and sterols (Table 1.6).

Lipid rafts are specialized phase domains containing sterols and sphingolipids which may be important in signal transitions (Gray, 2004; Furt *et al.*, 2007; Grennan, 2007; Mongrand *et al.*, 2004). Caveolae, which give rise to clathrin-coated vesicles (Brodsky *et al.*, 2001), are anchored multifunctional platforms in lipids (Van Deurs *et al.*, 2003; Patel and Insel, 2009).

The organization of the caveolae (Bastani and Parton, 2010) in the plasma-lemma and clathrin-coated vesicles (Samaj *et al.*, 2005) is presented in Figure 1.10. The current discussion focuses on membrane transport mechanism. Plants can internalize certain molecules by endocytosis via invaginations of the plasmalemma yielding clathrin-coated vesicles (Figure 1.11, Holstein, 2003) which become the endosome (Low and Chandra, 1994; Battey *et al.*, 1999; Šamaj *et al.*, 2006). Proteins involved in clathrin-dependent endocytosis appear to be clathrin, adaptor proteins and two adaptins (Pearse and Robinson, 1990; Šamaj *et al.*, 2006). Plant endocytosis and endosomes (Contento and Bassham, 2012) seem to be significant in auxin-mediated cell–cell communication, gravity responses, stomatal movements, cytokinesis and cell wall morphogenesis (Šamaj *et al.*, 2006).

Ion channels

Plasma membranes contain potassium (K^+), calcium (Ca^{++}) and anion channels (Roberts, 2006). Voltage-gated ion channels are transmembrane ion channels activated by changes in electrical potential. Gating is the precise control of ion channel opening (Krol and Trebacz, 2000). An example of an ion channel is the K^+ the

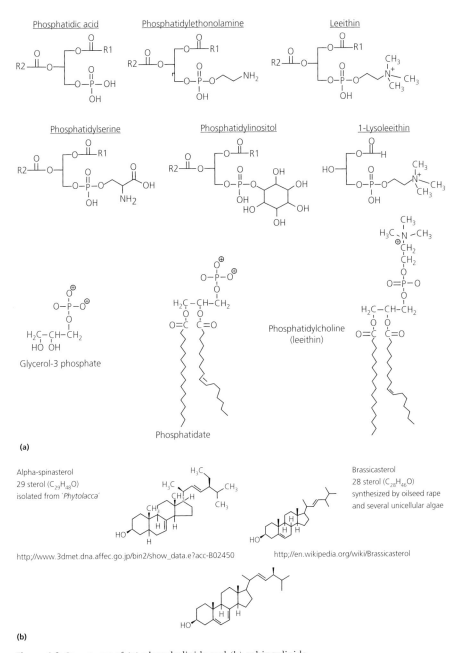

Figure 1.9 Structures of (a) phospholipids and (b) sphingolipids.

inwardly potassium channel. This type of channel possesses a positive charge in the cell. Stomatal pore movements are mediated by a rise in intracellular K^+ and anion contents of guard cells (Schroeder and Hagiwara, 1989). Another example is the adenosine triphosphate (ATP) binding cassette transporter or ABC transporter. These transport toxic substances from the cell or into the vacuole. These

Table 1.6 Composition of certain cellular membranes.

Chemical composition	
Fatty acyl groups in membrane lipids 16:0, 16:1, t-16:1, 16:3, 18:0, 18:1, 18:2, α18:3, δ18:3, 18:4, 22:0, 22:1, 24:0, 24:1	
Electroneutral phospholipids	Phosphatidylcholine, phosphatidylethanol, phosphatidylethanolamine
Anionic phospholipids	Phosphatidylserine, phosphatidylglycerol, phosphatidylinositides
Lyo-phospholipids	Cerebrosides
Sphingolipids	Galactolipids, sulpholipids
Chloroplast-specific glycerolipids	Diphosphatidylglycerol and monophosphatidylglycerol
Mitochondrial phospholipids	
Sterols	Sitosterol
	Campesterol
	Stigmasterol
	Unusual sterols
	Cycloartenol
	Cholesterol, minute quantities
Sterol glycosides	
Lanosterol	Pathogenic fungal membranes
Water	
Extramembrane water	Membrane is a bilayer sandwiched between two layers of water
	Water located within the bilayer which is attached to or in approximate contact with the expanses of membrane constituents
Proteins	May cross the membrane once or several times and are
Integral proteins	linked either electrostatically or by means of biophysical lipophilicity to the inner domains of the bilayer
Simple integral proteins	Classic α-helical structure that traverses the membrane only once
Complex integral proteins	Globular – comprised of several α-helical loops that may span the membrane several times
Peripheral proteins	Associated with only leaflet–easily isolated by altering ionic strength or pH of the encasing medium
Transport proteins	Pumps, carrier and channel

Source: From Leshem et al. (1991).

transporters are composed of four core domains, two cytosolic nucleotide-binding proteins and two transmembrane domains (Malmstrom, 2006).

Besides cation channels there are anion channels regulated by voltage, but their activity is also influenced by Ca^{++}, ATP, phosphorylation or membrane stretching (Tyerman, 1992). Anion plasma membrane channels function as efflux channels when they are open.

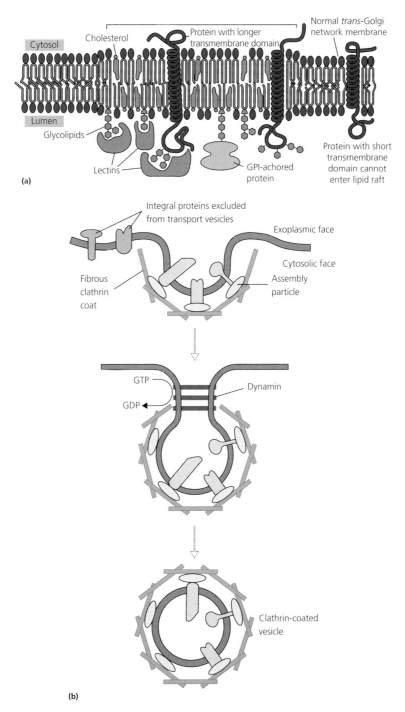

Figure 1.10 Depictions of a (a) lipid raft, (b) caveolae and a clathrin-coated vesicle. Source: Reproduced with permission of Caveolae and Clathrin Vesicle.

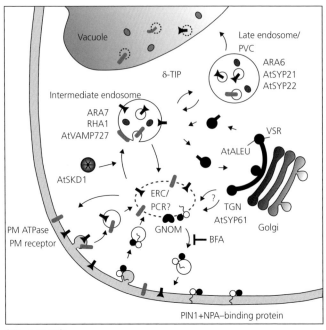

SVP – a syntaxin
GNOM – Plant-specific protein that participates in ADP-ribosylation
ESCRT – protein endosomal sorting complex
RHA – a member of the Rab GTPases function in trafficking pathways
ARA6 – a member of the Rab GTPases
SYP – a SNARE component of the late endosome
VSR – vacuolar sorting receptor
SKD – vacuolar protein suppressor
Ubiquitylation – signal that regulates the cell surface expression

Figure 1.11 Diagram of plant endocytosis. Source: Reproduced with permission of M. Otegui, University of Wisconsin.

Proton pumps

The transport of a substance against its electro channel gradient requires energy generated by ATP-proton pumps (Briskin and Hanson, 1992; Evert, 2006). One such pump is the V-ATPase found in both the plasmalemma and the tonoplast (Barkla and Pantoja, 1996; Vinay *et al.*, 2009). The H$^+$-ATPase in the plasmalemma is the P-ATPase which forms electrochemical gradients (Elmore and Coaker, 2011). Mitochondria and chloroplast membranes possess F-ATPases.

Water channels

Aquaporins are channel proteins which exist in the plasmalemma in intracellular spaces (Maurel *et al.*, 2008). These proteins permit water to move freely but exclude ions and metabolites (Chrispeels and Maruel, 1994; Muller *et al.*, 2007), providing for buffering osmotic fluctuations in the cytosol. Aquaporins are major intrinsic membrane proteins which are composed of four subunits, each of which comprises six transmembrane-spanning helices. Aquaporins are encoded by multiple gene families (Johansson *et al.*, 1998).

Carriers

Carriers are unitransporters and co-transporters (Evert, 2006). Unitransporters transport only one solute from one side of the membrane to the other. On the contrary, co-transporters transfer one solute with the simultaneous or sequential transfer of another solute. A thorough discussion of membrane transport processes occurs in Malmstrom (2006).

Organelle structure and function can be influenced by a variety of environmental parameters which affect plant growth. A discussion of parameters is presented because of the increasing pollution of the earth's atmosphere and ecosystem. In addition, global climate change is a current issue of urgent concern (Dashek and McMillin, 2009).

Both major and minor elements are required for growth and development (Table 1.7). Metals and metalloids at elevated levels can result from mining (Dashek and McMillin, 2009). What effects do these levels have on the structure and function of cellular organelles? (See Lepp, 1981; Medioini *et al.*, 2008; Yusuf *et al.*, 2011; see also Table 1.8.)

Elevated levels of SO_2, CO_2, NO_2 and O_3 (Treshow and Anderson, 1989) can occur in the atmosphere as a result of industrial and contemporary activities. Table 1.9 presents the effects of certain gases (Bell and Treshow, 2002) on the structure and function of organelles. Of special interests are the increasing levels

Table 1.7 Major and minor elements required for plant growth and development.

Element		mg/kg	Minor or major
Nitrogen	N	15000	Major
Potassium	K	10000	Major
Calcium	Ca	5000	Major
Magnesium	Mg	2000	Major
Phosphorus	P	2000	Major
Sulfur	S	1000	Major
Chlorine	Cl	100	Minor
Iron	Fe	100	Minor
Boron	B	20	Minor
Manganese	Mn	50	Minor
Zinc	Zn	20	Minor
Copper	Cu	6	Minor
Molybdenum	Mo	0.1	Minor

Table 1.8 Toxic metals and metalloids.

Metal or metalloid	Toxic level effects	References
Aluminium	Affects root cells of plasmalemma	Mossor-Pietraszewska (2001)
Arsenic	Pale green to yellow lesions on leaves and necrosis of leaves	Treshow and Anderson (1989)
	Defoliation	
	Impaired nitrogen metabolism	
	Needle abscission	
Cadmium	General chlorosis	Treshow and Anderson (1989),
	Reduced photosynthesis	Saadati et al. (2012) and Khateeb
	Reduced transpiration; toxic effects – changes in proline levels; changes in lipid peroxidation and seed germination	(2014)
Copper	Interference with normal metabolic reactions	Treshow and Anderson (1989)
	Blocks specific enzymatic reactions	and Shah et al. (2001)
Chromium	Contamination	Treshow and Anderson (1989)
	Can promote white dead patches on leaves	and Antonovics et al. (1971)
Lead	Condensation of nuclear chromatin; decrease in germination of two Brassica cultivars	Rout and Das (2003) and Hosseini et al. (2007)
Nickel	Dilution of nuclear membrane	Seregin and Kozhernikova (2006)
Zinc	Disruption of cortical cell	Rout and Das (2009)

Table 1.9 Effects of environmental pollutants on organelles.

Elevated CO_2		
	Stomatal openings reduce as CO_2 increases	Woodward et al. (1991)
	Affects both primary and secondary meristems of shoots and roots; alternation of leaf size and anatomy; increased branching and stem diameter	Pritchard et al. (1999)
	Increase in the number of mitochondria and amount of chloroplast stroma thylakoid membranes	Griffin et al. (2001)
	Stomatal densities decrease in two species of Spartia	Lammertsmaa et al. (2011)
Acid rain	Leaching of nutrients on tree needles; damages surfaces of needles and leaves and reduces a tree's ability to withstand cold	Godbold and Hüttermann (1994), Schulze et al. (2000) and White and Terninko (2003)
Nitric oxide	Necrotic lesions, marginal chlorosis	Lamattina and Polacco (2007)
Ozone and its derivatives	Changes in metabolism	Roshchina and Roshchina (2003)

of CO_2 in the atmosphere, which many scientists believe causes global warming (Dashek and McMillin, 2009). Table 1.10 offers the effects of sublethal and lethal temperatures on organelles. Franklin and Wigge (2014) discuss the effects of temperature on plant development. Other environmental parameters which can

Table 1.10 Effects of temperature or subcellular organelles.*

Temperature	Effect	Reference
Sublethal	Swollen chloroplasts and loss of chlorophyll in *Elodea* leaves	Quinn (1988)
Lethal	Plasmolysis of *Elodea* and soybean leaves; disintegration of cellular membranes	Daniell *et al.* (1969)

*Other effects of elevated temperature are on photosynthetic activities (Weis and Berry, 1988) and the plant immune response (Franklin and Wigge, 2014).

Biotic stress – chemical, humidity, mechanical, radiations, temperature and water

Abiotic stress – competition, herbivory, infection

ROS levels increase
superoxide, hydrogen peroxide, hydroxyl radical, singlet oxygen, Nitric oxide

involving cell wall, chloroplasts, mitochondria, phagosomes

Oxidative damage

Programmed cell death

Figure 1.12 Reactive oxygen species (ROS) and plant cell death.

affect organelle structure and function are radiation (Parida *et al.*, 2002; Mokobia and Anomohanran, 2005; Borzouei *et al.*, 2012) and salinity (Bennici and Tani, 2009; Kumar *et al.*, 2013).

Cell death

In certain mammalian systems, there appear to be two apoptotic pathways: extrinsic and intrinsic. Whereas the extrinsic pathway involves death receptor liquids, in the intrinsic pathway a variety of factors act upon mitochondria to promote loss of mitochondrial membrane potential. Whether these two pathways are significant in plant apoptosis remains to be established with certitude. Programmed cell death in plants (Bryant *et al.*, 2000) is viewed as a normal phase of development (Gray, 2004; Lam, 2008). These authors state that little is known about how plant cell death occurs and is regulated. However, reactive oxygen species (ROS) seem to be involved (Karuppanapandian *et al.*, 2011 – Figure 1.12). Fragmentation of nuclear DNA, involvement of Ca^{++}, alterations in protein phosphorylation, increases in nuclear heterochromatin and involvement of ROS seem to occur. Beers *et al.* (2000) conclude that proteases may possess a role in programmed cell death. In animals, caspases are significant components of programmed cell death. Although caspase attributes have been detected in plants, a role for these proteases in plant cell death is unclear (Lam, 2006). Lastly, van Doorn (2011) distinguished between ureolytic and non-ureolytic cell death. Whereas the former involves tonoplast rupture and subsequent destruction of the cytoplasm, the latter includes tonoplast rupture but not cytoplasmic destruction.

Finally, aspects of plant cells can be found in the following general plant cell biology textbooks: Batra (2009), Dashek and Harrison (2006), Gupta (2004), Pandian (2008), Pickett-Heaps and Pickett-Heaps (1994) and Wayne (2009).

References

Agrawal, G.K. (2011) Plant organelle proteomics: collaborating for optimal cell function. *Mass Spectrometry Reviews*, **30**, 772–853.

Alosi, M.C. and Alfieri, F.J. (1972) Ontogeny and structure of the secondary phloem in Ephedra. *American Journal of Botany*, **59** (8), 818–827.

Antonovics, J., Bradshaw, A.D. and Turner, R.G. (1971) Heavy metal tolerance in plants. *Advances in Ecological Research*, **7**, 1–85.

Arsovskia, A.A., Haughn, G.W. and Western, T.L. (2010) Seed coat mucilage cells of *Arabidopsis thaliana* as a model for plant cell wall research. *Plant Signaling and Behavior*, **5** (7), 796–801.

Assmann, S. and Liu, B. (2014) *Cell Biology (The Plant Sciences)*, Springer-Verlag, New York, NY.

Barkla, B.J. and Pantoja, O. (1996) Physiology of ion transport across the tonoplast of higher plants. *Annual Review Plant Physiology and Plant Molecular Biology*, **47**, 159–184.

Bastani, M. and Parton, R.G. (2010) Caveolae at a glance. *Journal of Cell Science*, **123**, 3831–3836.

Batra, V. (2009) *Plant Cell Biology*, Oxford Book Company, Oxford, UK/New York, NY.

Battey, N.H., James, N.C., Greenland, A.J. *et al.* (1999) Exocytosis and endocytosis. *The Plant Cell*, **11**, 643–660.

Beck, C.B. (2005) *An Introduction to Plant Structure and Development*, Cambridge University Press, Cambridge, UK.

Beck, C.B. (2009) *An Introduction to Plant Structure and Development: Plant Anatomy for the Twenty-First Century*, Cambridge University Press, Cambridge, UK.

Beers, E.P., Woofenden, B.J. and Zhao, C. (2000) Plant proteolytic enzymes: possible roles during programmed cell death. *Plant Molecular Biology*, **44**, 399–415.

Bell, J.N.B. and Treshow, M. (2002) *Air Pollution and Plant Life*, John Wiley & Sons, Inc., Chichester, UK.

Bennici, A. and Tani, C. (2009) Ultrastructural effects of salinity in Nicotiana bigelovii var. bigelovii callus cells and Allium cepa roots. *Caryologia*, **62** (2), 124–133.

Bergmann, D.C. and Sack, F.D. (2007) Stomatal development. *Annual Review of Plant Biology*, **58**, 163–168.

Blázquez, M.A., Green, R., Nilsson, O., Sussman, M.R. and Weigela, D. (1998) Gibberellins promote flowering of Arabidopsis by Activating the *LEAFY* promoter. *The Plant Cell*, **10**, 791–800.

Borzouei, A., Kafi, M., Akbari-Ghogdi, E. and Mousavi-Shalmani, M. (2012). Longterm salinity stress in relation to lipid peroxidation, super oxide dismutaseactivity and proline content of salt sensitive and salt-tolerant wheat cultivars. *Chilean Journal of Agricultural Research*, **72**, 476–482.

Briskin, D.P. and Hanson, J.B. (1992) How does the plant plasma membrane H^+-ATPase pump protons? *Journal of Experimental Botany*, **43** (3), 269–289.

Brodsky, F.M., Chen, C.Y., Knuehl, C. *et al.* (2001) Biological basket weaving: formation and function of clathrin-coated vesicles. *Annual Review of Cell and Developmental Biology*, **17**, 517–568.

Bryant, J.A., Hughes, S.F. and Garland, J. (2000) *Programmed Cell Death in Animals and Plants*, Taylor & Francis, Abingdon, UK.

Callow, J.A. (2000) *Plant Trichomes*, 1st ed, Academic Press, London, UK.

Chrispeels, M.J. and Maurel, C. (1994) Aquaporins: the molecular basis of facilitated water movement through living plant cells? *Plant Physiology*, **105**, 1, 9–13.

Contento, A.L. and Bassham, D.G. (2012) Structure and function of endosomes and plant cells. *Journal of Cell Science*, **125**, 1–8.

Dalgic, O. and Dane, F. (2005) Some of the molecular components of pollen tube growth and guidance. *Asian Journal of Plant Science*, **4**, 702–710.

Daniell, H. and Chase, C. (2004) *Molecular Biology and Biotechnology of Plant Organelles*, Springer, New York, NY.

Daniell, J.W., Chapell, W.E. and Couch, H.B. (1969) Effect of sublethal and lethal temperature on plant cells. *Plant Physiology*, **44**, 1684–1689.

Dashek, W.V. and Harrison, M. (2006) *Plant Cell Biology*, Science Publishers, Enfield, NH.

Dashek, W.V. and McMilin, D.E. (2009) *Biological Environmental Science*, Science Publishers, Enfield, NH.

Dashek, W.V., Thomas, H.R. and Rosen, W.G. (1971) Secretory cells of lily pistils II: electron microscope cytochemistry of canal cells. *American Journal of Botany*, **58**, 909–920.

Efroni, I., Eshed, Y. and Lifschitz, E. (2010) Morphogenesis of simple and compound leaves: review. *The Plant Cell*, **22**, 1019–1032.

Elmore, J.M. and Coaker, G. (2011) The role of the plasma membrane H^+-ATPase in plant-microbe interaction. *Molecular Plants*, **4**, 416–427.

Evert, R. (2006) *Esau's Plant Anatomy Meristems, Cells and Tissues: Their Structure, Function and Development*, John Wiley & Sons, Inc., Chichester, UK.

Fosket, D.E. (1999) *Plant Growth and Development: A Molecular Approach*, Academic Press, New York, NY.

Franklin, K. and Wigge, P. (2014) *Temperature and Plant Development*, Wiley-Blackwell, Chichester, UK.

Fukuda H. 2004. Signals that control plant vascular cell differentiation. *Nature Reviews Molecular Cell Biology*, **5**, 379–391.

Furt, F., Lefebvre, B., Gullimore, J. *et al.* (2007) Plant lipid rafts. *Plant Signaling and Behavior*, **2**, 508–511.

Gillham, N.W. (1994) *Organelle Genes and Genomes*, Oxford University Press, Oxford, UK.

Glover, B.J. (2007) *Understanding Flowers and Flowering: An Integrated Approach*, Oxford University Press, Oxford, UK.

Godbold, D.L. and Hüttermann, A. (1994) *Effects of Acid Rain and Forest Processes*, Wiley-Liss, New York, NY.

Gray, J. (2004) *Programmed Cell Death in Plants*, CRC Press, Boca Raton, FL.

Greenland, A.J. (2003) *Control of Plant Development*, Garland Science, London, UK.

Grennan, A.K. (2007) Lipid rafts in plants. *Plant Physiology*, **143**, 1083–1085.

Griffin, K.L., Anderson, O.R., Gastrich, M.D., Lewis, J.D., Lin, G. *et al.* (2001) Plant growth in elevated CO_2 alters mitochondrial number and chloroplast fine structure. *Proceedings of the National Academy of Sciences of the United States of America*, **98** (5), 2473–2478.

Gunning, B.E.S. (2009) *Plant Cell Biology on DVD*, Springer, New York, NY.

Gupta, G.P. (2004) *Plant Cell Biology*, Discovery, New Delhi, India.

Hake, S., Smith, H. and Magnani, E. *et al.* (2004) The role of genes in plant development. *Annual Review of Cell and Developmental Biology*, **20**, 125–151.

Hepler, P.K., Vidali, L. and Cheung, A.Y. (2001) Polarized cell growth in higher plants. *Annual Review of Cell and Developmental Biology*, **117**, 159–187.

Herrmann, R.G. (1994) *Cell Organelles (Plant Gene Research)*, Springer, New York, NY.

Holstein, S.H.E. (2003) Clathrin and plant endocytosis. *Traffic*, **3**, 614–620.

Hongladarom, T., Shigeru, I., Honda, S.I. and Wildman, S.G. (1964) *Organelles in Living Plant Cells*. Videotape available from University of California Extension Center for Media and Independent Learning, Berkeley, CA.

Hosseini, R.-H., Khanlarian, M. and Ghorbanti, M. (2007) Effect of lead on germination, growth and activity of catalase and peroxidase enzyme in root and shoot of two cultivars of *Brassica napus* L. *Journal of Biological Sciences*, **7**, 592–598.

Howell, S.H. (1998) *Molecular Genetics of Plant Development*, Cambridge University Press, Cambridge, UK.

Huang, T., Bohlenius, H., Erikssans, S. *et al.* (2005) The mRNA of the *Arabidopsis* gene *FT* moves from leaf to shoot apex and induces flowering. *Science*, **309**, 1694–1696.

Johansson, I., Karlsson, M., Shukla, V.K., Chrispeels, M.J., Larsson, C. and Kjellbom, P. (1998) Water transport activity of the plasma membrane aquaporin PM28A is regulated by phosphorylation. *The Plant Cell*, **10**, 451–459.

Karuppanapandian, T., Moon, J.-C., Kim, C. *et al.* (2011) Reactive oxygen species in plants. Their generation, signal transduction and scavenging mechanisms. *Australian Journal of Crop Science*, **5**, 709–725.

Kemi, U., Nittyvuopo, A., Tolvaines, T. *et al.* (2013) Role of vernalization and of duplicated FLOWERING LOCUS C in the perennial *Arabidopsis lyrata*. *The New Phytologist*, **197**, 323–335.

Khateeb, W.A. (2014) Cadmium-induced changes in germination, seedlings growth, and DNA fingerprinting of '*in vitro*' grown *Cichorium pumilum* Jacq. *International Journal of Biology*, **6**, 65 pp.

Kinmonth-Schultz, H.A., Golembeski, G.S. and Imaizumi, T. (2013) Circadian clock-regulated physiological outputs: dynamic responses in nature. *Seminars in Cell and Developmental Biology*, **24** (5), 407–413.

Krol, E. and Trebacz, K. (2000) Ways of ion channel gating in plant cells. *Annals of Botany*, **86** (3), 449–469.

Kumar, S., Gupta, R., Kumar, G., Sahoo, D. and R.C. Kuha (2013) Bioethanol production from *Gracilaria verrucosa*, a red alga, in a biorefinery approach. *Bioresource Technology*, **135**, 150–156.

Kusumi, A., Fujiwara, T., Morone, N., *et al.* (2012). Membrane mechanisms for signal transduction: the coupling of the meso-scale raft domains to membrane-skeleton-induced compartments and dynamic protein complexes. *Seminars in Cell and Developmental Biology*, **23**, 126–144.

Lam, E. (2006) Plant programmed cell death. *eLS*, doi:10.1002/9780470015902.a0001689.pub2.

Lam, E. (2008) Programmed cell death in plants: orchestrating an intrinsic suicide program within walls. *Critical Reviews in Plant Sciences*, **27**, 413–423.

Lamattina, L. and Polacco, J. (2007) *Nitric Oxide in Plant Growth, Development and Stress Physiology*, Plant Cell Monographs, Springer, New York, NY.

Lammertsmaa, E.I., de Boerb, H.J., Dekkerb, S.C., Dilcherc, D.L., Lottera, A.F. and Wagner-Cremera, F. (2011) Global CO_2 rise leads to reduced maximum stomatal conductance in Florida vegetation. *Proceedings of the National Academy of Sciences of the United States of America*, **108** (10), 4035–4040.

Larsson, C. and Miller, I.M. (1990) *The Plant Plasma Membrane. Structure, Function and Molecular Biology*, Springer, New York, NY.

Lepp, N.W. (1981) *Effects of Heavy Metal Pollution on Plants*, Applied Science Publishers, London, UK.

Lersten, N.R. and Curtis, J.D. (1996) Survey of leaf anatomy, especially secretory structures, of tribe Caesalpinieae (Leguminosae, Caesalpinioideae). *Plant Systematics and Evolution*, **200**, 21–39.

Lersten, N.R. and Horner, H.T. (2005) Macropattern of styloid and druse crystals in Quillaja (Quillajaceae) bark and leaves. *International Journal of Plant Science* **166**, 705–711.

Lersten, N.R., Czlapinski, A.R., Curtis, J.D., Freckmann, R. and Horner, H.T. 2006. Oil bodies in leaf mesophyll cells of angiosperms: Overview and a selected survey. *American Journal of Botany*, **93** (12), 1731–1739.

Leshem, Y.Y., Shewfelt, R.L., Willner, C.M. and Pantoja, O. (1991) *Plant Membranes: A Science*, Elsevier, New York, NY.

Leyser, O. and Day, S. (2003) *Mechanisms in Plant Development*, Blackwell, Malden, MA.

Low, P.S. and Chandra, S. (1994) Endocytosis in plants. *Annual Review of Plant Physiology and Plant Molecular Biology*, **45**, 609–631.

Lyndon, R.F. (1998) *The Shoot Apical Meristem: Its Growth and Development*, Cambridge University Press, Cambridge, UK.

MacAdam, J.W. (2009) *Structure and Function of Plants*, Wiley-Blackwell, New York, NY.

Maeda, E. and Maeda, K. (1988) Ultrastructural studies of leaf hydathodes. *Japanese Journal of Crop Science*, **57** (4), 733–742.

Malho, R. (2006a) *The Pollen Tube: A Cellular and Molecular Perspective*, Springer-Verlag, Berlin, Germany.

Malho, R. (2006b) The pollen tube: a model system for cell and molecular biology studies. *Plant Cell Monographs*, **3**, 10.1007/7089_041.

Malmstrom, S. (2006) Movement of molecules across membranes. In: *Plant Cell Biology* (Dashek, W.V. and Harrison, M., eds), Science Publishers, Enfield, NH, pp. 131–196.

Mano, S., Miwa, T., Nishikawa, S. *et al.* (2008) The plant organelles database (PODB): a collection of visualized plant organelles and protocols for plant organelle research. *Nucleic Acids Research*, **36** Database issue, D929–D937.

Maurel, C., Verdoucq, L., Luu, D.T. *et al.* (2008) Plant aquaporins: membrane channels with multiple integrated functions. *Annual Review of Plant Biology*, **59**, 595–624.

Mazumdar, R.C. (2013) *Photoperiodism and Vernalization in Plants*, Daya Publishing House, New Delhi, India.

McManus, H. and Veit, B.E. (2001a) Meristematic tissues. In: *Plant Growth and Development*, Sheffield Academic Press, New York, NY.

Medioini, C., Holune, G., Chaboute, M.E. *et al.* (2008) *Cadmium and Copper Genotoxicity in Plants*, Springer, New York, NY.

Micol, J.L. and Hake, S. (2003) The development of plant leaves. *Plant Physiology*, **131**, 389–394.

Mokobia, C.E. and Anomohanran, O. (2005) The effect of gamma irradiation on the germination and growth of certain Nigerian agricultural crops. *Journal of Radiological Protection*, **25**, 181.

Mongrand, S., Morel, J., Laroche, J., Claverol, S., Carde, J.P., Hartmann, M.A., *et al.* (2004). Lipid rafts in higher plant cells purification and characterization of triton X-100-insoluble microdomains from tobacco plasma membrane. *The Journal of Biological Chemistry*, **279**, 36277–36286.

Moore, A.L. and Clark, R.B. (1995) *Botany, Plant, Form and Function*, Win C. Brown Publ., Dubuque, IA.

Mossor-Pietraszewska, T. (2001) Effect of aluminum on plant growth and metabolism. *Acta Biochimica Polonica*, **48** (3), 673–686.

Muller, J., Mettback, U., Menzel, D. *et al.* (2007) Molecular dissection of endosomal compartments in plants. *Plant Physiology*, **145**, 293–304.

Murray, J.A.H. (2012) Systems analysis of shoot apical meristem growth and development. *The Plant Cell*, **24**, 3907–3919.

Naidoo, Y. and Naidoo, G. (1998) *Sporobolus virginicus* leaf salt glands: morphology and ultrastructure. *South African Journal of Botany*, **64**, 198–204.

Nicolson, S.W. and Nepi, M. (2005) Dilute nectar in dry atmosphere: nectar secretion patterns in *Aloe castanea* (Asphodelaceae). *International Journal of Plant Sciences*, **166**, 227–233.

Noguchi, T., Kawano, S., Tsukaya, H. *et al.* (2014) *Atlas of Plant Cell Structure*, Springer, New York, NY.

Offler, C.E., McCurdy, D.W., Patrick, J.W. *et al.* (2003) Transfer cells: cells specialized for a special purpose. *Annual Review of Plant Biology*, **54**, 431–454.

Oparka, K.J. and Turgeon, R. (1999) Sieve elements and companion cells—traffic control centers of the phloem. *The Plant Cell*, **11**, 739–750.

Oross, J.W., Leonard, R.T. and Thomson, W.W. (1985) Flux rate and a secretion model for salt gland of grasses. *Israel Journal of Botany*, **34**, 69–77.

Paiva, E.A.S. (2009) Ultrastructure and post-floral secretion of the pericarpial nectaries of *Erythrina speciosa* (Fabaceae). *Annals of Botany*, **104**, 937–944.

Pandian, I.D. (2008) *Botanical Analysis of Plant Cell*, A.K. Pub., New Delhi, India.

Parida, A., Das, A.B. and Das, P. (2002) NaCl stress causes changes in photosynthetic pigments, proteins and other metabolic components in the leaves of a true mangrove, *Bruguiera parviflora*, in hydroponic cultures. *Journal of Plant Biology*, **45**, 28–36.

Parton, R.M., Fischer-Parton, S., Watahiki, M.K. *et al.* (2001) Dynamics of the apical vesicle accumulation and the rate of growth are related in individual pollen tubes. *Journal of Cell Science*, **114**, 2685–2695.

Patel, H.H. and Insel, P.A. (2009) Lipid rafts and caveolae and their role in compartmentation of index signaling. *Antioxidants and Index Signaling*, **11**, 1357–1372.

Pearse, B.M.F. and Robinson, M.S. (1990) Clathrin, adaptors, and sorting. *Annual Review of Cell Biology*, **6**, 151–171.

Pickard, W.F. (2008) Laticifers and secretory ducts: two other tube systems in plants. *The New Phytologist*, **177**, 877–888.

Pickett-Heaps, J.D. and Pickett-Heaps, J. (1994) *VHS: Living Cells: Structure and Diversity*, Sinauer Associates Inc., Sunderland, MA.

Pridham, J.B. (1968) *Plant Cell Organelles*, Academic Press Inc., New York, NY.

Pritchard, S.F., Rogers, H.H., Prior, S.A. *et al.* (1999) Elevated CO_2 and plant structure: a review. *Global Change Biology*, **5**, 807–837.

Quinn, P.J. (1988) Effects of temperature on cell membranes. *Symposia of the Society for Experimental Biology*, **42**, 237–258.

Reid, R.A. and Leech, R.M. (1980) *Biochemistry and Structure of Cell Organelles*, Blackie, Glasgow, UK.

Roberts, S.K. (2006) Plasma membrane anion channels in higher plants and their putative functions in roots. *The New Phytologist*, **169**, 647–666.

Rodelas, A.J.D., Regalodo, E.S., Bela-ong, D.B. *et al.* (2008) Isolation and characterization of the oil bodies and oleosin of coconut (*Cocos nucifera* L.). *The Philippine Agricultural Scientist*, **91**, 389–394.

Roshchina, V.V. and Roshchina, V.D. (2003) *Ozone and Plant Cell*, Springer, New York, NY.

Rout, G. and Das, P. (2003) Effect of metal toxicity on plant growth and metabolism: I. Zinc. *Agronomie*, **23** (1), 3–11.

Rout, G.R. and Das, P. (2009) Effect of metal toxicity in plant growths and metabolism. I, Zinc. *Sustainable Agriculture*, **33**, 873–884.

Saadati, M., Motesharezadeh, B. and Ardala, M. (2012) Study of concentration changes of proline and potassium for two varieties of pinto beans under cadmium stress. *International Research Journal of Applied and Basic Sciences*, **3**, 344–352.

Sachs, T. (2005) *Pattern Formation in Plant Tissues*, Cambridge University Press, Cambridge, UK.

Sadava, D. (1993) *Cell Biology: Organelles, Structure and Function*, Jones and Bartlett, Sudbury, MA.

Sajeva, M. and Mauseth, J.D. (1991). Leaflike structure in the photosynthetic, succulent stem of cacti. *Annals of Botany*, **68**, 405–411.

Samaj, J., Read, N.D., Volkmann, D. *et al.* (2005) The endocytic network in plants. *Trends in Cell Biology*, **15**, 425–433.

Šamaj, J., Müller, J., Beck, M., Böhm, N. and Menzel, D. (2006) Vesicular trafficking, cytoskeleton and signalling in root hairs and pollen tubes. *Trends in Plant Science*, **11** (12), 594–600.

Sauter, J.J., Dorr, I. and Kollmann, R. (1976) The ultrastructure of Strasburger cells (=albuminous cells) in the secondary phloem of *Pinus nigra* var. *austraiaca* (Hoess) Badoux. *Protoplasma*, **88**, 31–49.

Schroeder, J.I. and Hagiwara, S. (1989) Cytosolic calcium regulates ion channels in the plasma membrane of *Vicia faba* guard cells. *Nature*, **338**, 427–430.

Schulze, E.-D., Högberg, P., vanOene, H., Persson, T., Harrison, A.F., Read, D., Kjøller, A. and Matteucci, G. (2000) Interactions between the carbon- and nitrogen cycle and the role of biodiversity: a synopsis of a study along a north-south transect through Europe, in: *Carbon and Nitrogen Cycling in European Forest Ecosystems, Ecological Studies* (Schulze, E.-D., ed), Springer Verlag, Heidelberg, Germany.

Seregin, I. and Kozhevnikova, A. (2006) Physiological role of nickel and its toxic effects in higher plants. *Russian Journal of Plant Physiology*, **5**, 251–277.

Shah, K., Kumar, R.G., Verma, S. and Dubey, R.S. (2001). Effect of cadmium on lipid peroxidation, superoxide anion, germination and activities of antioxidant enzymes in rice seedlings. *The Plant Science*, **161**, 1135–1144.

Shresth, R., Gómez-Ariza, J., Brambilla, V. and Fornara, F. (2014) Molecular control of seasonal flowering in rice, arabidopsis and temperate cereals. *Annals of Botany*, **114** (7), 1445–1458.

Singer, S.J. and Nicolson, G.L. (1972) The fluid mosaic model of the structure of cell membranes. *Science*, **175**, 720–731.

Sinha, N. (1999) Leaf development in angiosperms. *Annual Review of Plant Physiology and Plant Molecular Biology*, **50**, 419–446.

Soltis, D., Soltis, P. and Leeben-Mack, J. (2006) *Developmental Genetics of the Flower*, Academic Press, New York, NY.

Steeves, T.A. and Sussex, I.M. (1989) *Patterns in Plant Development: A Molecular Approach*, Academic Press, New York, NY.

Tan, W., Lim, T. and Loh, C. (2010) A simple, rapid method to isolate salt glands for three-dimensional visualization, fluorescence imaging and cytological studies. *Plant Methods*, **6**, 24.

Taoka, K., Chai, I., Tsuji, H. *et al.* (2013) Structure and function of florigen and the receptor complex. *Trends in Plant Science*, **14**, 287–294.

Taylor, L.P. and Hepler, P. (1997) Pollen germination and tube growth. *Annual Review of Plant Physiology and Plant Molecular Biology*, **48**, 461–491.

Thomson, W.W., Faraday, C.D. and Oross, J.W. (1988) Salt glands. In: *Solute Transport in Plant Cells and Tissues* (Baker, D.A. and Hall, J.L., eds), Longman Scientific and Technical, Harlow, Essex, UK, pp. 498–537.

Tobin, A.K. (1992) *Plant Organelles: Compartmentation of Metabolism in Photosynthetic Tissue*, Cambridge University Press, Cambridge, UK.

Treshow, M. and Anderson, F.K. (1989) *Plant Stress from Air Pollution*, John Wiley & Sons, Inc., Chichester, UK.

Turck, F., Formara, F. and Coupland, G. (2008) Regulation and identity of florigen: FLOWERING LOCUS T moves center stage. *Annual Review of Plant Biology*, **59**, 573–594.

Tyerman, S.D. (1992) Anion channels in plants. *Annual Review of Plant Physiology and Plant Molecular Biology*, **43**, 351–373.

Tyree, M.T. and Zimmermann, M.H. (2002) *Xylem Structure and the Ascent of Sap*, Springer Science and Business Media, Berlin, Germany.

van Deurs, B.V., Roepstorff, K., Hommelgaard, A.M. *et al.* (2003) Caveolae: anchored, multi-functional platforms in the lipid ocean. *Trends in Cell Biology*, **13**, 92–100.

Van Doorn, W.G. (2011) Classes of programmed cell death in plants, compared to those in animals. *Journal of Experimental Botany*, **62**, 4749–4761.

Vinay, S., Nilima, L.K. and Blunni, N.T. (2009) V-ATPase in plants: an overview structure and role in plants. International Journal of Biotechnology and Biochemistry, **22**, High Beam Research 23, http:www.highbeam.com (accessed May 24, 2016).

Wayne, R.O. (2009) *Plant Cell Biology: From Astronomy to Zoology*, Academic Press, New York, NY.

Weis, E. and Berry, J.A. (1988) Plants and high temperature stress. *Symposia of the Society for Experimental Biology*, **42**, 329–346.

Western, T.L., Skinner, D.J., and G.W. Haughn. (2000) Differentiation of mucilage secretory cells of the Arabidopsis seed coat. *Plant Physiology*, **122**, 345–355.

White, J.C. and Terninko, J. (eds.) (2003) Acid rain: Are the Problems Solved? Conference proceedings. Sponsored and organized by the Center for Environmental Information, Washington, DC, May 2–3, 2001. American Fisheries Society, Bethesda, MD.

Wille, A.C. and Lucas, W.J. (1984) Ultrastructural and histochemical studies on guard cells. *Planta*, **160** (2), 129–142.

Woodward, F.I., Thompson, G.B. and McKee, I.F. (1991) The effects of elevated concentrations of carbon dioxide on individual plants, populations, communities and ecosystems. *Annals of Botany*, **67**, 23–38.

Yang, Z. and Liu, B. (2007) Celebrating plant cells: a special issue on plant cell biology. *Journal of Integrative Plant Biology*, **49**, 1089–1278.

Yang, Y., Klejinot, Y., Yu, X. *et al.* (2007) Florigen (II) mobile protein. *Journal of Integrative Plant Biology*, **49**, 1665–1669.

Yusuf, M., Faridudchin, Q., Hugart, S. and Ahmod, D. (2011) Nickel an overview of and toxicity in plants. *Environmental Toxicology*, **86**, 11–17.

Zeevaart, J.A. (2006) Florigen coming of age after 70 years. *The Plant Cell*, **18**, 1783–1789.

Zhang, X., Zhang, L., Dong, F., Gao, J., Galbraith, D.W. and Song, C.P. (2001). Hydrogen peroxide is involved in abscisic acid-induced stomatal closure in *Vicia faba*. *Plant Physiology*, **126**, 1438–1448.

Further reading

Alberts, B., Johnson, A., Lewis, J., *et al.* (2014) *Molecular Biology of the Cell*, 6th edition, Garland Science, New York.

Buchanan, B. and Gruissem, W. (2015) *Biochemistry and Molecular Biology of Plants*, Wiley, Chichester, UK.

Mano, S., Miwa, T., Nishikawa, S., Mimura, T. and Nishimura, M. (2011) The plant organelle databases 2 (PODB2): An updated resource containing movie data of plant organelle dynamics, *Plant Cell Physiology*, **52** (2), 244–253.

Mano, S., Nakamura, T., Kondo, M. *et al.* (2014) The plant organelles database 3 (PODB3) Update 2014: Integrating electron micrographs and new options for plant organelle research, *Plant Cell Physiology*, **55** (1), e1, doi: 10.1093/pcp/pct140.

Nick, P. and Opatrny, Z. (2014) *Applied Plant Cell Biology: Cellular Tools and Approaches for Plant Biotechnology*, Springer, New York, NY.

Plopper, G., Sharp, D. and Sikorski, E. (2015) *Lewin's Cells*, 3rd edition, Jones & Bartlett, New York.

CHAPTER 2

Isolation and characterization of subcellular organelles from plant cells

Milee Agarwal[1], P. Desai[2], and Harish Padh[3]

[1] Pharmacology and Toxicology Department, B. V. Patel PERD Centre, Ahmedabad, Gujarat, India
[2] Cellular and Molecular Biology Department, B. V. Patel PERD Centre, Ahmedabad, Gujarat, India
[3] Sardar Patel University, Vallabh Vidyanagar, Anand, Gujarat, India

A eukaryotic cell is a supermolecular structure comprising large macromolecular assemblies of cell organelles. The presence of organelles is one of the main features of eukaryotic cells. During evolution, to increase cellular efficiency, eukaryotic cells compartmentalized their functions by acquiring and assembling different organelles. The organellar content of difference cell types ranges from no organelle (e.g., in mature mammalian red blood cells) to many and diverse organelles (c.g., in cells active in the synthesis of proteins for secretion). Each organelle has a specific function. For example, the mitochondrion is involved in the process of energy metabolism and generation of adenosine triphosphate (ATP); chloroplasts convert light energy into chemical energy, while endoplasmic reticulum is involved in protein synthesis. In addition, plant cells also contain plastids (in the form of chloroplasts) which are engaged in the generation of ATP and reducing power (nicotinamide adenine dinucleotide phosphate (NADPH or NADH) depending on light, chemical energy, and synthesis of various metabolites, such as amino acids and fatty acids, to support the biosynthesis of macromolecules required for the construction and maintenance of a plant cell.

The recent advent of in-depth research involving the signaling cascades, metabolic gradients, and proteomics of cellular organelles has led to the development of new techniques for organelle isolation and characterization. This chapter will provide an overview of the techniques that are routinely being employed for the isolation of plant organelles as well as their characterization.

M. Agarwal and P. Desai have equally contributed to the chapter.

Isolation of subcellular organelles

The word "organelle" literally means "little organs." The term is also used to define any functionally or visibly defined subcellular compartment of a cell. Plant cell organelles differ from those of animal cell organelles: notably, cell wall, plastid, and a large vacuole; animal cells have, in general, numerous small vacuoles and lack some structures such as the chloroplast. Plant organelles are of interest, therefore, due to their diversity and difference from those of other kingdoms. The subcellular organelles of a plant cell are described diagrammatically in Figure 2.1.

The investigation of any functional aspect of cell or cellular response to any stimuli requires the study of the function of each organelle. In order to study the subcellular organelles, they need to be isolated and characterized. For isolation and characterization of mammalian cell organelles, numerous scientific reports have been published. However, in case of plants, the methods to isolate subcellular organelles have to be optimized based on the part of the plant used for isolation, content of secondary metabolites, and cell wall composition. All these factors play an important role in the composition of the buffer used for isolation, disruption methodology, and centrifugation protocol.

How you disrupt the cell depends on the cell type. On the one hand, animal cells are very easy to break since they have no cell wall protecting the membrane. So very mild shearing is sufficient to release their organelles. Plants, on the other hand, have primary and sometimes secondary cell walls, so greater shear forces or grindings are needed to break these walls. Once the organelles

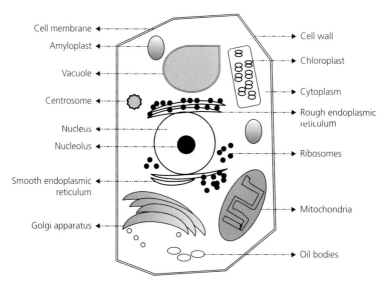

Figure 2.1 A diagrammatic representation of the higher plant cell and its organelles.

are released, they can be separated routinely by either of the two methods of centrifugation: differential centrifugation, which fractionates organelles according to their size, shape, or density; or equilibrium density centrifugation, which depends only on the different densities of organelles.

Cell disruption

The disruption of cell is one of the first and crucial steps for isolation and characterization of cell organelles. One of the earliest lysis methods used is manual disruption employing a mortar and pestle. The method uses grinding of plant cells/tissues into a buffer within a mortar and pestle over ice, sometimes cells/tissues frozen in liquid nitrogen. However, this methodology yields damaged organelles. The disruption of the cells should be controlled in order to isolate intact subcellular organelles. The removal of cell wall by enzymatic digestion and protoplast preparation is the gentlest method for the purification of intact organelles. Other common homogenization methods used include (i) Dounce homogenization, where the cells are crushed between two revolving solid surfaces; (ii) filtration, where cells are forced through smaller pores in a filter; (iii) grinding, where cells are ground by swirling with glass beads; (iv) sonication, where cells are bombarded with ultrasonic vibrations; and (v) solubilization, in which cell membranes are dissolved in detergents such as Triton X-100 (Padh, 1992). A thorough discussion of many of these techniques can be found in Goldberg (2008).

The disruption medium for homogenization also plays an important role in the maintenance of the integrity of subcellular organelles during the homogenization. The disruption medium is often modified for the osmoticum, buffering capacity, pH, ionic strength, reductant, and presence of agents that protect protein structure (bovine serum albumin or BSA, polyvinylpyrrolidone or PVP, and protease inhibitors). The osmoticum (e.g., sucrose and mannitol) maintains organelle structure by limiting physical swelling and rupture of membranes. A neutral or alkaline pH buffer, usually between 7.2 and 7.8, is used to quench the acidity from the rupture of the vacuoles, which can occur during the homogenization process (Agrawal *et al.*, 2010). Ethylene diamine tetraacetic acid inhibits functions of phospholipases and various proteases, which are Ca^{2+}- and Mg^{2+}-dependent. A reductant, such as dithiothreitol (DTT) minimizes damage from oxidants. BSA binds fatty acids and phenolic compounds and removes them. PVP is widely used to effectively sequester phenolics.

Isolation of subcellular organelles

The isolation of pure fractions of intact organelles from the cellular homogenate is very difficult. Various methods have been employed, including differential centrifugation, buoyant density-based centrifugation, affinity purification, free-flow electrophoresis (FFE), magnetic immunoabsorption, and fluorescence-activated organelle sorting (Table 2.1).

Table 2.1 Methods for isolating organelles from homogenates.

Methods	References
Centrifugation	Pasquali *et al.* (1999)
Density gradient	Nishimura *et al.* (1976)
	Timonen and Saksela (1980)
Rate-zonal (separation based on size and mass)	Cole Technical Library (2009)
Isopycnic (separation based on buoyant density)	Neuburger *et al.* (1982)
Centrifugal elutriation (separates on basis of size and mass)	Wahl and Donaldson (2001)

Differential centrifugation

Centrifugation techniques have been in use for decades for the isolation of sub-cellular organelles from cellular homogenates. Differential centrifugation is one of the widely used techniques to separate cellular organelles. The size, density, and shape influence the movement of a subcellular particle in a centrifugal field. This movement (sedimentation) results from the interaction between a particle's weight and the resistance it encounters in moving through a suspension medium and the relative centrifugal force exerted on the particle. The density of each organelle is determined by the ratio of their lipid to protein content; for example, mitochondrial inner membranes are protein-rich and are of high density, whereas endosomal membranes are lipid-rich and are of low density (Agrawal *et al.*, 2010).

The speed of centrifugation depends on the size and density of the organelle to be purified. Under a given centrifugal force, particles that are relatively large or dense will sediment more rapidly than particles that are smaller and lighter. A series of differential centrifugations are used to enrich the target organelle and selectively eliminate other compartments and contaminants—for example, non-broken cells, nuclei or chloroplasts pellet at low speed and short time of centrifugation: 500–1800 g for 10 min in the case of nuclei and 280–1500 g for 90 s to 10 min in the case of chloroplasts. In contrast, mitochondria require 12 000 g for 20 min. By applying different centrifugation speeds to the cell homogenate, enriched fractions of the organelle of interest can be obtained. For example, in the case of mitochondria, a preliminary centrifugation of the cell lysate at low speed allows the elimination of a high percentage of cell fragments, nuclei, and plastids (Table 2.2). The main advantage of differential centrifugation is that it is easily set up and can be ideally combined with analytical proteomics techniques. However, differences in the rate of sedimentation are sometimes not large enough to provide separation of one organelle from another.

Density gradient centrifugation

A second widely used procedure for separating organelles is known as density gradient centrifugation (Figure 2.2). In this procedure, subcellular particles are layered on a density gradient and subjected to a very high centrifugal force.

Table 2.2 Density ranges for some subcellular organelles.

Organelle	CsCl$_2$	Sucrose	Percoll	Opti-Pre	Metrizamid	Ficoll
Chromatin	1.7–1.95	1.357(SG) 1.6–1.75(S)			1.18–1.79	
Chloroplasts		1.21(S)				
Mitochondria		1.13–1.19(S)	1.07–1.11			1.136
Peroxisomes		1.23(S)	1.03–1.08	1.2	1.24–1.27	

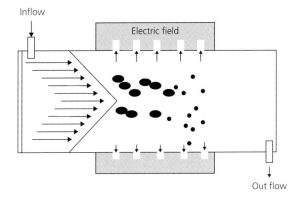

Figure 2.2 Basic principle of free-flow electrophoresis.

Usually, the density gradient is formed by layering increasing concentrations of solution (e.g., sucrose and Ficoll) in a centrifuge tube.

During centrifugation, organelles initially layered on the density gradient will sediment until they arrive at the region of the gradient where the density of the suspension is equal to their own. At this point, an equilibrium condition is reached between the downward centrifugal force and the particle's tendency to float due to buoyancy, and sedimentation halts. Hence, this procedure is also known as density equilibrium centrifugation. Although differences in composition of subcellular components affect relative densities of fractions, the degree of separation obtained also depends on the nature of the gradient medium used. Although sucrose is the most commonly used gradient medium, there are many other alternatives, for example, Ficoll, Percoll, Nycodenz, or Metrizamide. Discontinuous gradients and step gradients have been applied successfully for the separation. When used alone, neither differential nor density gradient centrifugation normally provides preparations containing organelles of sufficient purity. A common practice is to use both types of centrifugation procedures in sequence. However, even with this approach, obtaining desired organelles free from contaminants can require additional steps. With regard to retrieving the organelles from gradients, centrifuge tubes can be punctured at the base and functions of equivalent volumes collected on the bands can be aspirated.

The density of the gradient can be measured by a refractometer, which relies upon refractive indices. These can be plotted against fraction numbers to yield the shape of the gradient.

Centrifugal elutriation is a noninvasive technique used for separating large numbers of cells based on their size and mass (Wahl and Donaldson, 2001).

Affinity chromatography

Affinity chromatography utilizes two-phase systems for phase separation. These systems occur when solutions of two different water-soluble polymers are mixed above critical concentrations (described by D. Southern and D. Hollis in 1986), then the two immiscible liquid phases spontaneously separate. The attributes of these phases are low osmotic pressure, high water content, and a liquid–liquid interface between the phases. The most common phases are polyethylene glycol and dextran. Salts can be added for buffering and tonicity. Two-phase aqueous systems (Van Alstine, 2000) are very useful for organelle preparation because of the mild conditions of the phases.

Free-flow electrophoresis

FFE is a matrix-free electrophoresis lacking a solid matrix, for example, acrylamide. This is an old technique that in recent times is being adopted to fill a particular niche in subcellular organelle separation. It is a continuous separation technique that delivers organelles with different electrophoretic mobilities to different collection points. The flow is collected in 96 capillaries at the termination of the separation chamber (Nissum and Foucher, 2008). FFE can be of two types: zonal electrophoresis (ZE), or isoelectric focusing (IEF) mode. ZE-FFE mode is generally used for organelle separation. An electric field, applied perpendicularly to the direction of a laminar flow profile, causes displacement of the organelles along the electric field. This displacement is related to the organelles' surface properties that are associated with their electrophoretic mobilities as shown in Figure 2.2. Thus, FFE can be used to separate and identify subtypes of different organelles possessing different electrophoretic mobilities. Weber and Wildgruber (2008) and Zischka *et al.* (2003) have reported that FFE can be employed for organelle separation. FFE in ZE mode does not rely on size or density, but it provides separation electrophoretically by using differences in the surface charge of the organelles. To our knowledge, ZE-FFE has been reported once on plant material for the production of highly purified vesicles of the *Arabidopsis* plasma membrane and the tonoplast by N. Bardy and coworkers in 1998. The presence of plastids in plants adds an additional challenge for any mitochondrial purification technique, as plastids are known to be the major source of contamination for mitochondria isolated from photosynthetic material. Reports employing ZE-FFE for the purification of plant mitochondria from green and nongreen material demonstrate the value of this approach by identifying impurities in mitochondrial preparations (Eubel *et al.*, 2007).

Flow field-flow fractionation

Flow field-flow fractionation (FFIFF) is one member of the FFE family, which is capable of fractionating particles, cells, proteins, or DNA according to their size and shape (Kang *et al.*, 2008). In FFIFF, separation is carried out in an unobstructed channel (i.e., without packing material) with an external field (i.e., with a crossflow moving across channel) driven in the direction perpendicular to the migration flow (i.e., along the channel axis). When the crossflow is applied to sample components injected into the FFIFF channel, sample materials are hydrodynamically driven toward one wall (the accumulation wall) of the channel, while simultaneously they diffuse away from the wall due to diffusion. Owing to the balance of the two counter-directed transport mechanisms, sample components find their equilibrium states at a given distance from the wall. The different equilibrium heights reached by the sample components are vertically distributed from the channel wall, according to differences in the diffusion coefficients (or in the hydrodynamic diameters) of sample species. Components with lower molecular weight (MW) (or smaller hydrodynamic diameter) have larger diffusion coefficient values, and they migrate faster at equilibrium positions that are higher than those reached by components of higher MW values. Thus, separation in FFIFF is achieved in the order of increasing MW or hydrodynamic diameter of sample components. Since FFIFF takes place in an open channel, it is generally suited to the gentle separation of organelles. Sedimentation field-flow fractionation (SdFIFF) is described by S.M. Mozersky and coworkers in 1988 as the FFIFF sub-technique utilizing centrifugal acceleration as the driving force for sample retention. SdFIFF requires a highly concentrated sucrose solution to induce the necessary density difference between the organelles and the carrier solution. Attempts to use SdFIFF for micro-preparative isolation of subcellular fractions containing mitochondria, microsomes, Golgi membranes, and plasma membranes of corn roots have been reported.

Affinity purification

Affinity purification of organelles relies on the availability of antibodies against proteins found on the surface of organelles of interest. The availability of antibodies has made immuno-purified organelles a desirable method for isolating organelles. Recent methods have reported the use of a magnetic field to retain organelles that bind to the antibodies attached to magnetic beads. After washes and removal of unwanted contaminants, the magnetic field is removed to recover a fraction containing the organelle of interest (Truernit and Hibberd, 2007). In the absence of an antibody against the native organelle-specific membrane protein, a recombinant protein can be expressed on a specific organelle membrane, and antibody against this recombinant protein can be used for organelle isolation and purification. The first affinity purification of chloroplasts expressing yellow fluorescent protein (YFP) OEP-14 on the surface of *Arabidopsis thaliana* and *Nicotiana tabacum* leaves has been reported by Truernit and Hibberd

(2007). For immuno-purification, magnetic beads coated with anti-YFP antibodies were used. The YFP was expressed in only certain types of cells, thereby selectively isolating chloroplasts from only those cell types. The overall yield of chloroplasts by affinity purification was high, and close to a half of the organelles had their double membranes intact. The study showed that a genetically engineered protein designed to target the surface of a specific organelle type can be used for affinity purification of organelles.

Fluorescence-activated organelle sorting

Fluorescence-activated organelle sorting (FAOS) uses a flow cytometer to detect and sort organelles with specific fluorescence and scattering characteristics. The technique requires labeling of organelle-specific proteins with fluorescently tagged antibodies or expression of organelle-specific fluorescent proteins. Grebenok *et al.* (1997) showed that green fluorescent protein (GFP) can be effectively targeted to nuclei of various higher and lower plant species by using a nuclear localization signal from an orphan tobacco transcription factor. In order to increase the size of the chimeric GFP containing the nuclear localization signal beyond that of the passive exclusion limit of nuclear pores, β-glucuronidase is included as a passive stuffer. The fluorescent nuclei expressing GFP were readily detected and sorted in homogenates of transgenic tobacco plants. FAOS technique has also been reported to sort *Arabidopsis thaliana* nuclei that express GFP-tagged histone protein (Zhang *et al.*, 2008).

Magnetic immunoabsorption

One of the most promising organelle isolation methods described by S.M. Jones and coworkers in 1994 is the use of magnetic beads coated with antibodies. Dynacal manufactures antibody-coated beads (Neurauter *et al.*, 2007) which have proved useful for the isolation of peroxisomes. Because there appear to be proteins unique to certain organelles, one could raise an antibody to an organelle protein and then coat the beads with the antibody. Then, the antibody-coated beads could be added to cellular homogenates and the organelle can be purified under the applied magnetic field.

Isoelectric focusing

IEF is a procedure that has been mainly applied to protein isolation (Garfin and Ahuja, 2005). This method separates proteins based on their isoelectric points, that is, the pH at which a protein lacks a net negative charge. L.R. Griffing and R.S. Quatrano described in 1984 an IEF procedure for purifying certain plant cell membranes. The Golgi apparatus, mitochondria, and putative plasmalemma possess unique isoelectric points. Instead of a horizontal IEF apparatus, they employed an IEF column which discriminated between the surface charges of plant cell organelles.

Nanotrap technology

Nanotrap technology offers the possibility of yielding isolated organelles should the technology be developed for bulk organelle preparation (Rothbauer *et al.*, 2008). Carbon nanotubes are capable of penetrating plant cell walls. However, it is difficult to discern how many cells would have to be penetrated to yield organelles in bulk let alone a specific organelle. A more promising approach involves the Nanotrap sample processing kit developed by CERES nano. This kit offers the advantage of rapid isolation. The possible biological applications of nanotubes have been discussed by Yang *et al.* (2007)

Identification and characterization of isolated organelles

For modern research involving subcellular organelle proteome analysis, the level of confidence largely depends on the extent of purification that has been achieved. The sensitivity of detection of contamination can determine the success of an experiment. A large number of diverse methods have been developed to assess enrichment of the target organelle as well as the presence of contaminating organelles (Figure 2.3).

Microscopic techniques

Microscopic observation provides a quick and informative method for evaluating the purity and integrity of subcellular organelle fractions. Microscopic and biological techniques are used to identify the subcellular organelles based on their morphological characteristics. Various microscopic techniques including electron tomography, fluorescence microscopy, and transmission electron microscopy (TEM) have been used to identify the preparation of subcellular organelles as well as assess the purity of the preparations.

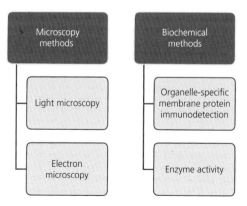

Figure 2.3 Different methods to determine the purity of the subcellular organelle fractions.

Bright-field microscopy

For bright-field microscopy, parallel visible light derived from a tungsten lamp impinges upon an object. Then, some of the light continues as undiffracted light while other light is scattered. The latter diverges at various angles as diffracted light. This is accomplished by employing two types of magnifying lenses: ocular and objective. The maximum magnification of bright-field microscopy is 1000×. I.K. Lichtscheidl and I. Foissner found in 1996 that this type of microscopy is of limited use in identifying isolated subcellular organelles. However, when fluorescent probes are employed, some degree of specificity for some organellar preparations can be obtained.

Fluorescence microscopy

Fluorescence microscopy is based on the principle of fluorescence. It employs various useful fluorescent dyes (Table 2.3) which can be conjugated to antibodies against organelle-specific enzymatic and nonenzymatic proteins (Lippincott-Schwartz and Patterson, 2003). Due to their subjective manner, microscopy techniques can be used to assess the morphology and integrity of subcellular organelle fractions, but it is sometimes difficult to quantify the purity of fractions (Stephens and Allan, 2003).

Dark-field microscopy

Dark-field microscopy is a specialized optical illumination which employs oblique illumination. This permits specimen-enhanced contrast that cannot be obtained with bright-field illumination. Direct light is blocked by an opaque step position within the substage condenser. However, limited light can pass through the specimen from oblique angles. This light can be deflected, refracted, and reflected into microscope's objective that yields a bright image superimposed on a dark background, as reported by I.K. Lichtscheidl and I. Foissner in 1996. Thus the structural features of some of the isolated subcellular organelles can be elucidated by dark-field microscopy.

Table 2.3 Fluorescent dyes used for the characterization of different subcellular organelles.

Fluorescent dye	Target organelle/structure
MitoTracker (molecular probes)	Mitochondria
Acridine Orange (molecular probes)	Nuclei
ER-Tracker (molecular probes)	Endoplasmic reticulum
BODIPY FL-C5 (molecular probes)	Golgi apparatus
CellMask Deep Red (molecular probes)	Plasma membrane
TubulinTracker Green (molecular probes)	Tubulin
Calcofluor White Stain (sigma)	Cell wall
Nile red	Oil bodies

Nomarski microscopy

Nomarski microscopy yields a two-dimensional image that is generated by the interference of mutually coherent waves that possess a lateral displacement of a few lengths of a nanometre and one-phase shuffled relative to each other (Yochem, 2006). A specimen is illuminated with plane polarized light that is divided into two orthogonally polarized mutually coherent components by a Wollaston prism. The two components are recombined into a single beam via a second prism and analyzer. The value of Nomarski microscopy is the ability to view transparent specimens not visible by bright field.

Electron tomography

Electron tomography (ET) has contributed significantly to advancing the understanding of plant cell organelle geometry and size (Donohoe *et al.*, 2006). When combined with cryo-fixation/freeze substitution, it can yield a three-dimensional structure and membrane trafficking (Sosinsky *et al.*, 2008). Other application of ET has been the formation of the plant endosome/prevascular components and the organization of photosynthetic membranes (Haas and Otegui, 2007). The methods for ET which can enhance resolution 30–80 fold over light microscopy have been discussed by Frank (2006). ET data can be analyzed by an IMOD software.

Transmission electron microscopy

TEM permits the direct visualization of isolated organelles, but this technique requires considerable sample preparation as depicted in Figure 2.4. TEM enables to characterize the purity and quality of the organelle preparation by comparing isolated organelles to those within the cell (Kuo, 2007). By combining TEM with biological techniques, such as marker enzyme analysis, the identity of the isolated organelle can be determined with considerable confidence. In addition to traditional TEM, negative staining of certain organelles can be performed as described in Figure 2.5.

Scanning electron microscopy

In scanning electron microscopy (SEM), electrons do not pass through the sample that is imaged by a reflected beam. In TEM, the electron beam traverses the spectrum of a sample. The electrons are derived from a filament. Thorough discussions of TEM and SEM occur in Dashek (2000). The observation of plant or animal cell organelles by SEM based on surface irregularities has been demonstrated by others using various methods, and the same has been employed for the plant cell organelles too. This sample preparation is as depicted in Figure 2.6. Due to more chances of artifact formation, the identification of organelles using SEM requires careful processing of the samples and extensive technical expertise. A variant of SEM is field emission SEM (FE-SEM). A cold field emission electron

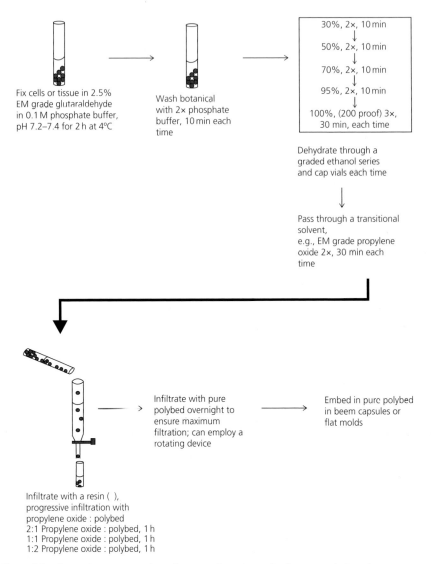

Figure 2.4 Schematic representation of preparation of samples for transmission electron microscopy.

source is employed (Bonard *et al.*, 2002). This allows for higher image resolution, measured signal-and-noise ratio, and enhanced depth of field.

Soft X-ray microscopy

C. Jacobsen reported in 1999 that soft X-ray microscopy accomplishes similar to that with light or electrons; however, the integration of X-rays with matter and X-ray wavelengths permit a resolution less than 25 mm for objects as thick as

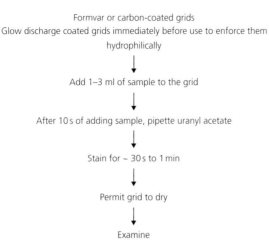

Formvar or carbon-coated grids
Glow discharge coated grids immediately before use to enforce them
hydrophilically

↓

Add 1–3 ml of sample to the grid

↓

After 10 s of adding sample, pipette uranyl acetate

↓

Stain for ~ 30 s to 1 min

↓

Permit grid to dry

↓

Examine

Figure 2.5 Schematic representation of preparation of samples for negative staining transmission electron microscopy. Source: Adapted from http://www.protocol-online.org/cgi-bin/prot/view_cache.cgi?ID=212.

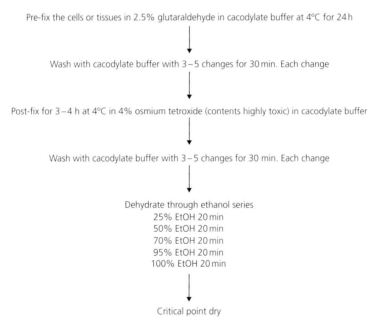

Pre-fix the cells or tissues in 2.5% glutaraldehyde in cacodylate buffer at 4°C for 24 h

↓

Wash with cacodylate buffer with 3–5 changes for 30 min. Each change

↓

Post-fix for 3–4 h at 4°C in 4% osmium tetroxide (contents highly toxic) in cacodylate buffer

↓

Wash with cacodylate buffer with 3–5 changes for 30 min. Each change

↓

Dehydrate through ethanol series
25% EtOH 20 min
50% EtOH 20 min
70% EtOH 20 min
95% EtOH 20 min
100% EtOH 20 min

↓

Critical point dry

Figure 2.6 Schematic representation of preparation of samples for scanning electron microscopy.

10 cm. Also, specimens in an aqueous medium can be examined. With regard to the procedure, X-rays from a bending magnet of the synchrotron are focused via a condenser lens into the object. Amperometric electrochemical detection (AECD) system accepts a magnified image produced by an objective lens (Aarhus University 2006, The X-ray Microscope at ASTRID).

Wide-field microscopy

Wide-field microscopy (WFM) is the illumination and imaging of the entire microscopic field of view permitting minimal photodamage during live cell imaging (Swedlow and Platani, 2002). Other advantages of WFM are optimizing high isolation warping under less-than-ideal condition, enhancing three-dimensional sharpness, optimizing contrast for thick specimens. WFM utilizes special optics.

Magnetic resonance force microscopy

J.A. Sidles and his associates found in 1995 that magnetic resonance force microscopy (MRFM) is an imaging technique that obtains magnetic resonance images at manometer scales and perhaps atomic scale. The value of this technique is the ability to visualize protein structures that cannot be observed by X-ray crystallography or protein nuclear magnetic resonance spectroscopy (Vosegaard and Nielsen, 2004).

Multiphoton microscopy

Multiphoton microscopy (MM) is based on two-photon excitation of fluorescence (Williams *et al.*, 2001; Zipfel *et al.*, 2003). This type of microscopy allows for collection of images hundreds of millimeters into tissues. Its advantages are low photon toxicity in greater spatial and temporal resolution in contrast to *in vivo* imaging.

Marker enzymes

Isolation of any organelle requires a reliable test for the presence of the organelle. Typically, this is done by following the activity of an enzyme that is known to be localized exclusively in the target organelle; such enzymes are known as marker enzymes (Table 2.4). For example, the enzyme acid phosphatase (that

Table 2.4 Organelle-specific enzymes that can be used for characterization of subcellular organelle fractions.

Organelle	Enzyme
Mitochondria	Succinate dehydrogenase
Lysosomes	Acid phosphatase
Chloroplast	Sialyl transferase
Endoplasmic reticulum	Glucose-6-phosphatase
Nucleus	RNA polymerase II

cleaves terminal phosphate group from substrates and has a pH optimum in the acidic range) is localized in lysosomes, while the enzyme succinate dehydrogenase is localized in mitochondria. By monitoring the site where each enzyme activity is found during a cell fractionation protocol, one can monitor the fractionation of lysosomes and mitochondria, respectively. Marker enzymes also provide information on the biochemical purity of the fractionated organelles. The presence of unwanted marker enzyme activity in the preparation indicates the level of contamination by other organelles, while the degree of enrichment for the desired organelle is determined by the specific activity of the target marker enzyme. Although marker enzymes reveal much concerning the purity of the organelle preparation, electron microscopy is generally used as a final step to assess the preparation's purity and the morphology of the isolated organelle.

Summary

Plant cell biology includes the study of the different components of cell, that is, subcellular organelles. Pure preparations of subcellular organelles are a prerequisite for studying the various aspects including their functional and structural genomics as well as proteomic profiling. There are different methods that are currently being used for isolation of pure fractions of subcellular organelles based on the physical, chemical, and biological properties of the organelles. No method is universal, and usually using a combination of method such as density gradient centrifugation with affinity purification will yield an almost pure preparation. Before moving forward with experiments, the isolated fractions need to be characterized. Microscope-based techniques based on the size, shape, conformation, as well as organelle-specific protein expression (marker enzymes and nonenzymatic proteins) are used to characterize organelle preparations. However, most of the methods of characterization are qualitative and need expertise. Thus, there is still need for further improvement in the methodologies for both isolation and characterization of plant organelles.

References

Agrawal, G.K., Bourguignon, J., Rolland, N. *et al.* (2010) Plant organelle proteomics: Collaborating for optimal cell function. *Mass Spectrometry Reviews*, 10.1002/mas.20301.

Bonard, J.-M., Dean, K.A., Coll, B.F. and Klinke, C. (2002) Field emission of individual carbon nanotubes in the scanning electron microscope. *Physical Review Letters*, **89**, 197602-1–197602-4.

Cole Technical Library. (2009) Basics of centrifugation. http://www.coleparmer.in/TechLibraryArticle/30. Accessed March 2, 2016.

Dashek, W.V. (2000) *Methods in Plant Electron Microscopy and Cytochemistry*, Humana Press Inc., Totowa, NJ.

Donohoe, B.S., Mogelsvang, S. and Staehelin, L.A. (2006) Electron tomography of ER, Golgi and related membrane systems. *Methods*, **39**, 154–162.

Eubel, H., Lee, C.P., Kuo, J. *et al.* (2007) Free-flow electrophoresis for purification of plant mitochondria by surface charge. *The Plant Journal*, **52**, 583–594.

Frank, J. (2006) Introduction: Principles of electron tomography. In: *Electron Tomography*, Springer, New York, pp. 1–15.

Garfin, D. and Ahuja, S. (2005) *Handbook of Isoelectric Focusing and Proteomics*, Academic Press, New York, NY.

Grebenok, R.J., Pierson, E.A., Lambert, G.M. *et al.* (1997) Green-fluorescent protein fusions for efficient characterization of nuclear localization signals. *The Plant Journal*, **11**, 573–586.

Goldberg, S. (2008) Mechanical/physical methods of cell disruption and tissue homogenization. *Methods in Molecular Biology*, **424**, 3–22.

Haas, T.J. and Otegui, M.S. (2007) Electron tomography in plant cell biology. *Journal of Integrative Plant Biology*, **49**, 1091–1099.

Kang, D., Oh, S., Reschiglian, P. and Moon, M.H. (2008) Separation of mitochondria by flow field-flow fractionation for proteomic analysis. *Analyst*, **133**, 505–515.

Kuo, J. (2007) *Electron Microscopy: Methods and Protocols*, Second edition, Humana Press Inc., Totowa, NJ.

Lippincott-Schwartz, J. and Patterson, G.H. (2003) Development and use of fluorescent protein markers in living cells. *Science*, **300**, 87–91.

Neuburger, M., Journet, E.-P., Bligny, R., Carde, J.-P. and Douce, R. (1982) Purification of plant mitochondria by isopycnic centrifugation in density gradients of Percoll. *Archives of Biochemistry and Biophysics*, **217**, 312–323.

Neurauter, A.A., Bonyhadi, M., Lien, E. *et al.* (2007) Cell isolation and expansion with Dynabeads. In: *Cell Separation: Fundamentals, Analytical and Preparative Methods* (Kumar, A., Galaev, I.Y. and Mattiasson, B., eds), Springer, New York, NY.

Nishimura, M., Graham, D. and Akazawa, T. (1976) Isolation of intact chloroplasts and other cell organelles from spinach leaf protoplasts. *Plant Physiology*, **58**, 309–314.

Nissum, M. and Foucher, A.L. (2008) Analysis of human plasma proteins: A focus on sample collection and separation using free-flow electrophoresis. *Expert Review of Proteomics*, **5**, 571–587.

Padh, H. (1992) Organelle isolation and marker enzyme assay. In: *Tested Studies for Laboratory Teaching*, Volume **13** (Goldman, C.A., ed.). Association for Biology Laboratory Education (ABLE), Laramie, pp. 129–146.

Pasquali, C., Fialka, I. and Huber, L.A. (1999) Subcellular fractionation, electromigration analysis and mapping of organelles. *Journal of Chromatography*, **722**, 89–102.

Rothbauer, U., Zolghadr, K., Muyldermans, S. *et al.* (2008) A versatile nanotrap for biochemical and functional studies with fluorescent fusion proteins. *Molecular and Cellular Proteomics*, **7**, 282–289.

Sosinsky, G.E., Crum, J., Jones, Y.Z. *et al.* (2008) The combination of chemical fixation procedures with high pressure freezing and freeze substitution preserves highly labile tissue ultrastructure for electron tomography applications. *Journal of Structural Biology*, **161**, 359–371.

Stephens, D.J. and Allan, V.J. (2003) Light microscopy techniques for live cell imaging. *Science*, **300**, 82–86.

Swedlow, J.R. and Platani, M. (2002) Live cell imaging using wide-field microscopy and deconvolution. *Cell Structure and Function*, **27**, 335–341.

Timonen, T. and Saksela, E. (1980) Isolation of human NK cells by density gradient centrifugation. *Journal of Immunological Methods*, **36**, 285–291.

Truernit, E. and Hibberd, J.M. (2007) Immunogenic tagging of chloroplasts allows their isolation from defined cell types. *The Plant Journal*, **50**, 926–932.

Van Alstine, J.M. (2000) Eukaryotic cell partition: Experimental considerations. In: *Aqueous Two-Phase Systems: Methods and Protocols*, Volume **11**, Methods in Biotechnology (Hatti-Kaul, R. and Totowa, N.J., Eds.). New York: Humana Press Inc., pp. 119–142.

Vosegaard, T. and Nielsen, N.C. (2004) Improved pulse sequences for pure exchange solid-state NMR spectroscopy. *Magnetic Resonance in Chemistry*, **42**, 285–290.

Wahl, A.F. and Donaldson, K.L. (2001) Centrifugal elutriation to obtain synchronous populations of cells. *Current Protocols in Cell Biology*, Chapter 8, Unit 8.5., 10.1002/0471143030. cb0805s02.

Weber, G. and Wildgruber, R. (2008) Free-flow electrophoresis system for proteomics applications. *Methods in Molecular Biology*, **384**, 703–716.

Williams, R.M., Zipfel, W.R. and Webb, W.W. (2001) Multiphoton microscopy in biological research. *Current Opinion in Chemical Biology*, **5**, 603–608.

Yang, W., Thordarson, P., Gooding, J.J., Ringer, S.P. and Braet, F. (2007) Carbon nanotubes for biological and biomedical applications. *Nanotechnology*, **18**, 412001. 10.1088/0957-4484/18/41/412001

Yochem, J. (2006) Nomarski images for learning the anatomy, with tips for mosaic analysis. The *C. elegans* Research Community, *WormBook*, doi/10.1895/wormbook.1.100.1

Zhang, C., Barthelson, R.A., Lambert, G.M. and Galbraith, D.W. (2008) Global characterization of cell-specific gene expression through fluorescence-activated sorting of nuclei. *Plant Physiology*, **147**, 30–40.

Zipfel, W.R., Williams, R.M. and Webb, W.W. (2003) Nonlinear magic: Multiphoton microscopy in the biosciences. *Nature Biotechnology*, **21**, 1369–1377.

Zischka, H., Weber, G., Weber, P.J.A. *et al.* (2003) Improved proteome analysis of *Saccharomyces cerevisiae* mitochondria by free-flow electrophoresis. *Proteomics*, **3**, 906–916.

Further reading

Baltes, N.J. and Voytas, D.F. (2015) Enabling plant synthetic biology through genome engineering. *Trends in Biotechnology*, **33** (2), 120–131.

de Oliveira Dal'Molin, G.C., Quek, L.E., Saa, P.A. and Nielsen, L.K. (2015) A multi-tissue genome-scale metabolic modeling framework for the analysis of whole plant systems. *Frontiers in Plant Science*, **6**, 4, http://dx.doi.org/10.3389/fpls.2015.00004.

Harford, J.B. and Bonifacino, J.S. (2011) Subcellular fractionation and isolation of organelles. In: *Current Protocols in Cell Biology*, John Wiley & Sons, Inc., New York.

McCormack, M.E., Lopez, J.A., Crocker, T.H. and Mukhtar, M.S. (2016) Making the right connections: Network biology and plant immune system dynamics. *Current Plant Biology*, **5**, 2–12.

Righetti, K., Ly Vu, J., Pelletier, S. *et al.* (2015) Inference of longevity-related genes from a robust coexpression network of seed maturation identifies regulators linking seed storability to biotic defense-related pathways. *The Plant Cell*, **27**, 2692–2708.

CHAPTER 3

Endoplasmic reticulum

William V. Dashek

Retired Faculty, Adult Degree Program, Mary Baldwin College, Staunton, VA, USA

Structure

The endoplasmic reticulum (ER) consists of two parallel membranes with an intervening lumen (Figure 3.1). The ER is of two types: rough (RER, Figure 3.2) and smooth (SER, Figure 3.3). While the RER is "studded" with ribosomes, the SER is not (Staehelin, 1997). The RER is connected to the nuclear envelope at one end and is continuous with the SER at the other end. An important current review of the plant ER is that of Chen *et al.* (2012).

The 80S eukaryotic ribosome (Figure 3.4) consists of two subunits: 60S and 40S (Ramakrishnan, 2002; Blaha, 2004; Lafontaine and Tollervey, 2001; Nierhaus and Wilson, 2004). The large subunit contains 5S, 5.85S, and 28S rRNA, whereas the small subunit houses 18S rRNA (Table 3.1). There appear to be 49 and 33 proteins for the large and small subunits, respectively. Ribosomes can exist as polysomes free of the RER (Figure 3.5). Proteome analysis of *Arabidopsis* cytosolic ribosomes revealed 87 different proteins (Carroll *et al.*, 2008). Ribosomes can also exist within chloroplasts and mitochondria.

Chemical composition

The RER is complex array of ribosomal, membrane, and lumenal proteins as revealed by proteome analysis. Many of the lumenal proteins are subjected to posttranslational modification (Battey *et al.*, 1993; Tekoah, 2005; Kwon *et al.*, 2006) and are then transported to other cellular destinations. Are there any resident proteins in the ER? The chaperones are proteins that bind *in vivo* to either misfolded or unfolded proteins (Gupta and Tuteja, 2011). Calnexin is an integral membrane protein that binds Ca^{++} and may function as a chaperone for the transport of ER proteins to the plasmodesmata (Johnson and Hakansson, 2003). Calreticulin is a ubiquitous, lumenal-binding protein (Mariani *et al.*, 2009) which

Plant Cells and their Organelles, First Edition. Edited by William V. Dashek and Gurbachan S. Miglani.

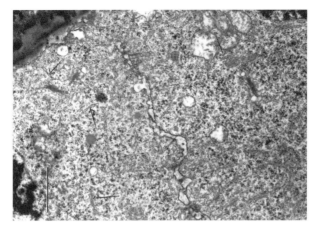

Figure 3.1 Electron micrograph of a plant cell illustrating the distribution of endoplasmic reticulum. Note arrows. Source: Reproduced with permission of J. Mayfield.

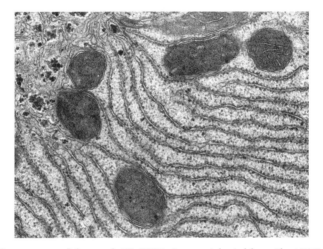

Figure 3.2 Ultrastructure of the rough ER (RER). Source: Adapted from Plant RER micrographs occur at Evert (2006).

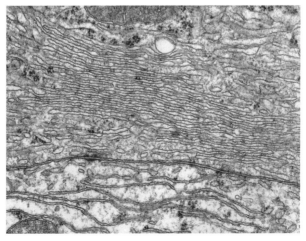

Figure 3.3 Micrograph of the smooth ER (SER). Source: From medcell-med.edu/histology/ cell_lab/smooth.

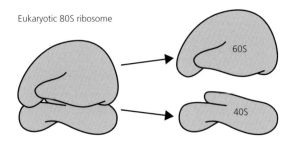

Eukaryotic 80S ribosome

60S

40S

18S rRNA
33 proteins

5S, 5.8S, 28S
rRNA
49 proteins

1. The small subunit has a sedimentation coefficient of
    40S and consists of 18S rRNA together with 33
    different proteins.

2. The large subunit has a sedimentation coefficient of
    60S consists of 5S, 5.8S, and 28S rRNA together,
    with about 49 different proteins.

Figure 3.4 Structure of eukaryotic ribosome. Source: From http://www.pc.maricopa.edu.

Table 3.1 Characteristics of plant cytosolic ribosomes.

Subunit	Sediment coefficient	Number of nucleotides	Subunit proteins
Small 40S	18S	1926	33
Large 60S	5S	120	49
	5.8S	157	
	25S	3580	

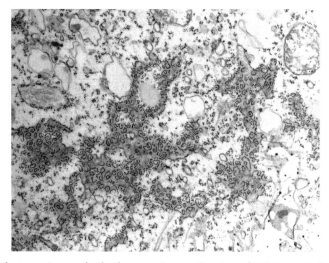

Figure 3.5 Electron micrograph of polysomes. Source: Courtesy of H.T. Bonnett, Jr. and E.H. Newcomb.

appears to function in glycoprotein folding (Crofts *et al.*, 1999). It is the most abundant resident protein. Other abundant resident ER proteins are reticuloplasmins or folding enzymes (Crofts *et al.*, 1999) and translocon, a conserved protein-conducting channel (Osborne *et al.*, 2005). Translocons consist of several ER membrane proteins (Johnson, 2009). These proteins form an aqueous pore that is both structurally and functionally dynamic (Johnson and van Waes, 1999). Johnson and van Waes (1999) stated that translocons can cycle between bound and free ribosomes. Examples of reticuloplasmins are binding immunoglobulin protein (BIP) and endoplasmin (Denecke *et al.*, 1993, 1995). Béthune *et al.* (2006) presented an updated view of the composition of the ER noting that its lumenal environment is very different from that of the cytosol.

Biogenesis

The mechanisms responsible for ER formation and morphology are not clear (Voeltz *et al.*, 2002). In yeasts, ER tubules, microtubules, and molecular motors are important (Hensel, 1987; Du *et al.*, 2004; Gupton *et al.*, 2006; Vedrenne and Hauri, 2006). The motors appear to be F-actin and myosins (Ueda *et al.*, 2010). ER exit sites (ERES) are important in protein export (Hanton *et al.*, 2007). The molecular mechanisms for their biogenesis and maintenance are not fully understood (Farhan *et al.*, 2008). Considerable research is required to establish the precise events underlying ER biogenesis. Ribosome synthesis involves processing of a polycistronic pre-ribosomal RNA to the mature rRNA of the 40S subunit and the 5.8S/28S rRNA of the 60S subunit (Fatica and Tollervey, 2002; Dlakić, 2005). RNA polymerase I drives the synthesis, cleavage, and modification of pre-RNA in eukaryotes; ribosome assembly requires about 200 essential factors (Chen *et al.*, 2011). Chen *et al.* (2011) point out that the ER network can rearrange in response to developmental cues. This rearrangement involves continuous fusion and fission reaction. The shaping of the cell tubular end appears to be mediated by the membrane proteins, reticulons, and atlastin (Tollervey *et al.*, 2011; Zheng and Chen, 2011; Lee *et al.*, 2013).

Functions

Protein synthesis

The transcriptional and translational steps resulting in protein synthesis (Figure 3.6) have been discussed many times (see Vitale *et al.*, 1993). Therefore, the present discussion centers on initiating, elongating, and terminating factors during translation (Tables 3.2 and 3.3). Also presented are posttranslational events of newly synthesized proteins (Table 3.4), novel protein synthesis inhibitors (Table 3.5) and *in vitro* protein synthesis (Figure 3.7).

Transcription

Initiation—RNA polymerase binds DNA; attachment of nucleotides

Elongation—Additional nucleotides added; RNA polymerase migrates down DNA

Termination—RNA polymerase reaches terminator; newly synthesized RNA released; release of

polymerase from DNA

Translation

Initiation—Small ribosomal subunit and initiator codon attach to start codon; large subunit binds to RNA

Elongation—Delivery of charged tRNA; transpeptidase activity (peptide bond formation); translocation

Termination—Release factor senses the triplet at a stop codon and peptide synthesis stops; ribosomal

subunit separates from the mRNA

Figure 3.6 Summary of transcription and translation of protein synthesis.

Table 3.2 Translation initiation factors and their functions.

Factor	Function	Reference
e1F4F and e1Fiso4F	Cap-binding complexes, initiation factor complex	Rrowning (2004)
e1F3	Complex initiation factor	Browning *et al.* (2001)
e1F4B	Least conserved factor of regulation of protein synthesis	Metz *et al.* (1999)
e1F2	Most recalcitrant factor, mechanism of regulation of protein synthesis unclear	Langland *et al.* (1999)

Table 3.3 Elongation and termination protein synthesis factors.

Factor	Function	Reference
Elongation		
eEF1α	Bring the aminoacylated t-RNAs to the ribosome	Aguilar *et al.* (1991) and Monnier *et al.* (2001)
eEF1Bδ	Exchanges GDP for GTP; nucleotide protein required to recycle above nucleotide phosphates	Manevski *et al.* (2000)
eEF2	Required for translocation	Yin *et al.* (2003)
Termination		
eRF eRF1 eRF3	Release of completed polypeptide chain	Drugeon *et al.* (1997)

Table 3.4 Protein synthesis inhibitors.

Inhibitor	Action	Reference
Anisomycin	Blocks transcription factors	Zinck *et al.* (1995)
Chloramphenicol	Blocks translation, binds to 70S ribosomes	www.emdbiosciences.com
Chlorolissoclimides	Blocks translation	www.cs.stedwards.edu/chem/ Chemistry/CHEM43/CHEM43/Protinhib/ FUNCTION.HTML Robert *et al.* (2006)
Cycloheximide	Blocks translational elongation	www.cs.stedwards.edu/chem/ Chemistry/CHEM43/CHEM43/Protinhib/ FUNCTION.HTML
Erythromycin	Blocks translation	Bmb.leeds.ac.uk
Fusidic acid	Inhibits elongation of factor G at the ribosome	Borg *et al.* (2015)
Hygromycin B	Binds to ribosome preventing initiation	Hygromycin.com
Linezolid	Interferes with the interaction of mRNA and ribosomal subunits	Swaney *et al.* (1998)
Sparsomycin	Inhibits peptidyltransferase	
Streptomycin	Inhibits translation	Biology-Online.org
Tobramycin	Prevents initiation	Ryu *et al.* (2002)
	Binds to ribosome	Carrasco *et al.* (1986)

Table 3.5 Types of ribosome-inactivating proteins (RIPs) based on physical characteristics.

Type	Characteristic	Reference
Type I Majority of RIPs (Pokeweed antiviral factor, saporin, barley translation inhibitor, gelonin dianthin)	Monomeric enzymes, either single proteins or peptides	Stirpe *et al.* (1992) and Hartley and Lord (2004)
Type II (abrin, ricin)	Highly toxic heterodimeric proteins with enzymatic and galactose-lectin characteristics, bind to cell surfaces	Stripe *et al.* (1992), Nielsen and Boston (2001) and Lord and Hartley (2010)
Type III (barley 60 kDa jasmonate-induced protein)	Molecules that act as inactive precursors less prevalent than I and II	Nielsen and Boston (2001)

Initiation factors are proteins required for accurate translation (Figure 3.8—Preiss and Hentze, 2003). The specific functions of these factors are depicted in Table 3.2. With regard to elongation, a relatively small number of factors are required (Table 3.3). The main factors are eEF1 and eEF2 (Monnier *et al.*, 2001). The elongation factor of eEF1 is composed of eF1α and eEF1β (Monnier *et al.*, 2001). The value of the release factors has been addressed by Moreira *et al.* (2002). Termination occurs via stop codons, for example, UAA, UAG, and UGA (Tuite and Stansfield, 1994).

- Harvest seedlings of rice, maize, or mung bean.

- Cut shoots into ~0.5 cm sections at 4 °C.

- Homogenize tissues into a buffer in a Waring blender.

- Filter homogenate through cheesecloth and Miracloth.

- Centrifuge filtrate at 14,000×g for 15 min.

- Add Triton X and N-adeoxycholate to the supernatant and shake for 3–5 min.

- Centrifuge at 25,000×g for 15 min.

- Overlay the supernatant onto a sucrose cushion, centrifuge, 90,000×g, 6.5 hr, 4 °C.

- Recover proteins and store at −80 °C.

- Translate isolated polysomes with a Promega's wheat germ extract.

- Analyze proteins with SDS-PAGE.

Figure 3.7 *In vitro* protein synthesis. Source: Adapted from Dai *et al.* (1996).

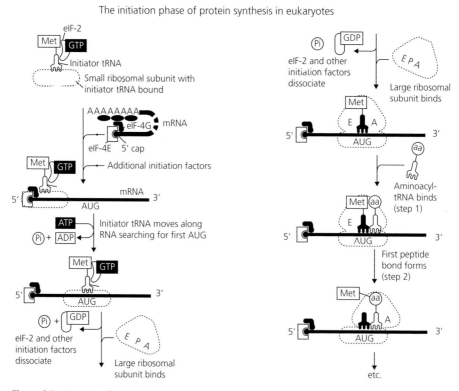

Figure 3.8 Diagram of initiation factors. Source: From http:www.edu.medmaccenterColler. htm. Other diagrams of translation in eukaryotes occur at: http:biologyonline.org and Preiss, T. and Unbehaum *et al.* (2004) compare a ribosomal subunit joining model for prokaryotes and eukaryotes.

Posttranslational events

Protein folding

Newly synthesized proteins are folded to yield their proper configurations (Obalinsky, 2006). Nölting and Andert (2000) have proposed three models of protein folding (Figure 3.9). These are the framework, hydrophobic collapse, and nucleation–condensation models. Chaperones (Schroder, 2010) aid in the folding of select proteins, thereby preventing misfolding. When just synthesized, a specific *N*-linked glycan structure protects proteins, but there can be folding defects. These defects can result from ultraviolet radiation, chemical assaults, or metabolic by-products (Hirsch *et al.*, 2009). Misfolded proteins are transported across the ER membrane, subjected to ubiquitylation (the covalent attachment of ubiquitin to a substrate protein) and degraded by the 26S proteasome, a complex of a proteolytically active 20S core and 19S regulatory particles (Schmidt *et al.*, 2005; Isono *et al.*, 2007). Ubiquitin is a 76 amino acid protein that attaches

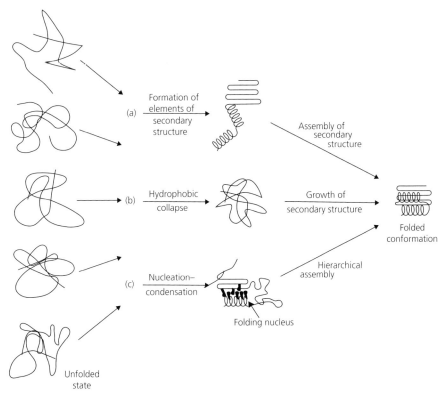

Figure 3.9 Models from protein folding. (a) Framework model, (b) hydrophobic collapse model, and (c) nucleation–condensation mechanism. Source: Adapted from Nölting and Andert (2000).

Table 3.6 Summary of posttranslational events.

Event	Reference
Glycosylation	Bill *et al.* (1998), Tekoah (2005), Henquet *et al.* (2008)
Methylation	Paik *et al.* (2007) and Niemi *et al.* (1990)
Palmitoylation	Hemsley and Grierson (2008)
Phosphorylation	Zhou *et al.* (2007)
SUMOylation	Zhou *et al.* (2007)
Ubiquitylation	Hirsch *et al.* (2009)

Krishna and Wold (1993) review posttranslational modifications of proteins.

to a substrate via E1 (ubiquitin-activating enzyme), E2 (ubiquitin-conjugating enzyme), and E3 (ubiquitin–protein ligase). The ubiquitin–proteasome complex is usually referred to as the ER-associated degradation (ERAD). This complex delivers misfolded proteins to the cytosol (Smith *et al.*, 2011). A summary of posttranslational events is given in Table 3.6.

Glycosylation
Plants can attach both *N*- and *O*-linked glycans (Figure 3.10, Fitchette-Laine *et al.*, 1997; Tekoah, 2006) to proteins. The former "contain 2-(1,3)-linked fucose (Fuc) attached to the proximal Glc Nac residue and/or a β(1,2)-xylose (Xyl) attached to the D linked mannose (Man) of the glycan core." As mentioned earlier, glycosylation of newly synthesized proteins protects them from misfoldings.

Methylation
A well-known posttranslational modification is the methylation of arginine, lysine, and histidine (Figure 3.10) as well as the carboxyls of glutamic and aspartic acids (Verdoucq *et al.*, 2006). Methylation can increase certain activities of proteins, for example, epigenetic regulation. Methylation is mediated by methyltransferases with the main methyl donor being *S*-adenosylmethionine.

Nitrosylation
Protein nitrosylation (Figure 3.10) permits nitric oxide (NO) to function through posttranslational modification of proteins (Gupta, 2011; Brown *et al.*, 2014). Nitrosylated caspases can be stored in the mitochondrial intermembrane space. These enzymes are involved in apoptosis (Belenghi *et al.*, 2007).

Palmitoylation (thioacylation or S-acylation)
This reversible process (Figure 3.10) attaches 16C saturated fatty acids to specific cysteine residues through a thioester bond (Hemsley *et al.*, 2008). The enzyme

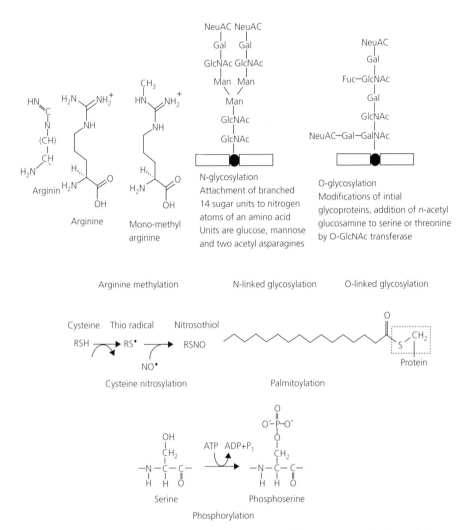

Arginine methylation   N-linked glycosylation   O-linked glycosylation

Cysteine nitrosylation   Palmitoylation

Phosphorylation

Figure 3.10 Representation of post-translational protein modifications, namely, glycosylation, methylation, nitrosylation, palmitoylation, and phosphorylation.

responsible for palmitoylation is S-acyltransferase. Palmitoylation may be important in protein trafficking.

Phosphorylation

This process (Figure 3.10), a common posttranslational event, is mediated by protein kinases (Lee *et al.*, 2011). Phosphorylation and dephosphorylation by protein phosphatases can regulate cell division, signal transduction, and cell growth and development. A plant protein phosphorylation database, P3DB, exists (Gao *et al.*, 2009). Serine and tyrosine are the most common amino acids that are phosphorylated.

Prenylation

Protein prenylation (Figure 3.11) centers about the covalent attachment of C-15 isoprene farnesyl or C-20 isoprene geranylgeranyl groups to certain protein C-termini (Yang *et al.*, 1993). The adherence of certain proteins to membranes is promoted by prenylation (Caldeira *et al.*, 2001). Certain nuclear lamins are prenylated.

Sulfation

Sulfation of plant proteins occurs at tyrosine residues (Figure 3.11—Monigatti *et al.*, 2006). Sulfation involves the enzyme tyrosylprotein sulfotransferase (Hanai *et al.*, 2000). Sulfation (addition of sulfur) occurs in secreted and trans-membrane proteins.

SUMOylation

SUMOylation (Figure 3.11) is an ubiquitin-like protein conjugation process (Miura *et al.*, 2007; Park *et al.*, 2011). SUMO is an ubiquitin-like modifier that plays a role in SUMO conjugation and deconjugation of protein substrates

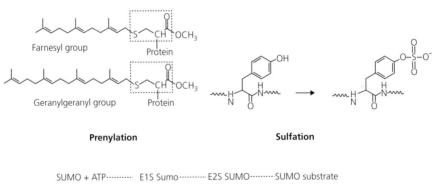

Figure 3.11 Representation of post-translational protein modifications, namely, prenylation, sulfation, SUMOylation, and ubiquitylation.

Table 3.7 Other types of posttranslational modifications.

Type	Process	Reference
Acetylation	Addition of an acetyl group	Sylvestersen *et al.* (2014)
Alkylation	Addition of alkyl group	Sylvestersen *et al.* (2014)
Biotinylation	Acylation of conserved lysine residues	Sylvestersen *et al.* (2014)
Glutamylation	Covalent linkage of glutamic acid to tubulin	Edde *et al.* (1990)
Isoprenylation	Isoprenoid group addition	Hemmerlin (2013)
Phosphopantetheinylation	Addition of a 4′-phosphopantetheinyl group from COA	Lambalot *et al.* (1996)

(Miura *et al.*, 2007). Furthermore, SUMO is a reversible posttranslational protein modifier (Geiss-Friedlander and Melchior, 2007). SUMOylation appears to be involved in cell cycle activity, DNA repair, subnuclear localization, enzymatic activity (Miura *et al.*, 2007) and responses to abiotic environmental stress (Park *et al.*, 2011). Finally, Table 3.7 presents after posttranslational processes. These are involved in a variety of plant developmental events.

Ubiquitylation

This process (Figure 3.11) is significant in the degradation of misfolded proteins (Hirsch *et al.*, 2009). Ubiquitin is anchored in the ER membrane (see earlier discussion). Ubiquitylation plays a role in protein degradation via the proteasome as discussed earlier.

Inhibitors

The classic inhibitors of protein synthesis are puromycin, chloramphenicol, cycloheximide, and streptomycin. Chloramphenicol inhibits chloroplast protein synthesis but not cytoplasmic protein synthesis. Over the years, a number of other compounds with diverse mechanisms of action have been found to inhibit protein synthesis (Table 3.4). One recent class is the ribosome-inactivating proteins or RIPs (Table 3.5, Bass *et al.*, 1992; Stirpe *et al.*, 1992). The RIPs act as N-glycosidases and delete a specific adenine in a conserved loop of rRNA. The result is that the ribosome is unable to bind elongation factor 1α, and therefore translation cannot occur. A very potent example of an RIP is ricin derived from castor bean (Harley and Beevers, 1982; Frigerio and Roberts, 1998). Ricin can inhibit synthesis by plant ribosomes by 50% (Harley and Beevers, 1982).

Another recently discovered class of protein synthesis inhibitors is micro-RNA or miRNA (Carrington and Ambros, 2003; Bartel, 2004; Mallory and Vaucheret, 2006; Sunkar and Jagadeeswaran, 2008). The miRNAs are endogenous RNAs of 22 nucleotides that target mRNAs for cleavage or translational repression (Bartel, 2004). There are two types of small RNAs: miRNAs and siRNAs (Voinnet, 2009). The latter can direct posttranslational gene silencing via mRNA degradation or gene silencing through DNA methylation or histone modifications (Sunkar and Zhu, 2007).

In vitro protein synthesis

Protein synthesis can occur *in vitro* by supplying reagents requisite for such synthesis (Dai *et al.*, 1996). An example of a cell-free plant protein synthesis system is depicted in Figure 3.7. Such systems provide a tool for investigating the components for protein synthesis.

Other functions

Other functions include SER-mediated lipid synthesis (Robinson and Pimpl, 2014), generation of protein bodies (Herman and Larkins, 1999) and oil bodies (Galili *et al.*, 1998), and cell-to-cell communication via plasmodesmata (Galili *et al.*, 1998). Also, the plant ER can accumulate hydrolytic enzymes within ER bodies (Yamada *et al.*, 2011) in response to pathogens (Galili, 2004). In addition, the ER can function as a storage site for Ca^{++} (Denecke, 2001; Wyatt *et al.*, 2002). Finally, the ER can house monooxygenases that are able to convert ent-kaurene to GA12, a precursor of the family of gibberellins (Rademacher, 2000; Lange and Lange, 2006).

Another function of the ER is auxin homeostasis (Ding *et al.*, 2012). This is accomplished by PIN-Formed (PIN) proteins (Huang *et al.*, 2010). A group of PIN proteins (PIN 6 and 8) localize to the ER (Křeček *et al.*, 2009; Viaene *et al.*, 2013).

References

Aguilar, F., Montardon, P.E. and Stutz, E. (1991) Two genes encoding the soybean translation elongation factor eEF1α are transcribed in seedling leaves. *Plant Molecular Biology*, **17**, 351–360.

Bartel, D.P. (2004) Micro RNAs: genomics, biogenesis, mechanism and function. *Cell*, **116**, 287–297.

Bass, H., Webster, C., OBrian, G.R. *et al.* (1992) A maize ribosome-inactivating protein is controlled by the transcriptional activator *Opaque-2*. *The Plant Cell*, **4**, 225–234.

Battey, N.G., Dickinson, H.G. and Hetherington, A.M. (1993) *Post-translational Modifications in Plants*, Cambridge University Press, Cambridge, UK.

Belenghi, B., Romero-Puertas, M.C., Vercammen, D., *et al.* (2007) Metacaspase activity of *Arabidopsis thaliana* is regulated by S-nitrosylation of a critical cysteine residue. *The Journal of Biological Chemistry*, **282**, 1352–1358.

Béthune, J., Kol, M., Hoffmann, J., *et al.* (2006) Coatomer, the coat protein of COPI transport vesicles, discriminates endoplasmic reticulum residents from p24 proteins. *Molecular and Cellular Biology*, **26**, 8011–8021.

Bill, R.M., Reves, L. and Wilson, I.B.H. (1998) *Protein Glycosylation*, Kluwer Academic Publishers, Boston, MA.

Blaha, G. (2004) Structure of the ribosome. In: *Protein Synthesis and Ribosome Structure* (eds. Nierhaus, K.H. and Wilson, D.N., eds) Wiley-VCH, Weinheim, pp. 85–106.

Borg, A., Holm, M., Shiroyama, I., *et al.* (2015) Fusidic acid targets elongation factor G in several stages of translocation on the bacterial ribosome. *The Journal of Biological Chemistry*, **290** (6), 3440–3454.

Brown, A.J.P., Budge, S., Kaloriti. D., *et al.* (2014) Stress adaptation in a pathogenic fungus. *The Journal of Experimental Biology*, **217**, 144–155.

Browning, K.S. (2004) Plant translation initiation factors: it is not easy to be green. *Biochemical Society Transactions*, **32** (4), 589–591.

Browning, K.S., Gallie, D.R., Hershey, J.W.B., *et al.* (2001) Unified nomenclature for the subunits of eukaryotic initiation factor 3. *Trends in Biochemical Sciences*, **26**, 284.

Caldeira, M.C., Ryel, R.J., Lawton, J.H. and Pericra, J.S. (2001) Mechanisms of positive biodiversity-production relationships: insights provided by d13C analysis in experimental Mediterranean grassland plots. *Ecology Letters*, **4**, 439–443.

Carrasco, L., Jimenz, A. and Vasquez, D. (1986) Formation of a functional ribosome-membrane junction during translocation requires the participation of a GTP-binding protein. *Journal of Cell Biology*, **103**, 2253–2261.

Carrington, J.C. and Ambros, V. (2003) Role of microRNAs in plant and animal development. *Science*, **301**, 336–338.

Carroll, A.J., Heazlewood, J.L., Ito, J. and Millar, A.H. (2008) Analysis of the *Arabidopsis* cytosolic ribosome proteome provides detailed insights into its components and their post-translational modification. *Molecular and Cellular Proteomics*, **7**, 347–369.

Chen, J., Stefano, G., Brandizzi, F. and Zheng, H. (2011) Arabidopsis RHD3 mediates the generation of the tubular ER network and is required for Golgi distribution and motility in plant cells. *Journal of Cell Science*, **124**, 2241–2252.

Chen, L.Q., Qu, X.Q., Hou, B.H., *et al.* (2012) Sucrose efflux mediated by SWEET proteins as a key step for phloem transport. *Science*, **335**, 207–211.

Crofts, A.J., Lebornge-Castel, N., Hillmer, S. *et al.* (1999) Saturation of the endoplasmic reticulum retention machinery reveals anterograde bulk flow. *The Plant Cell*, **11**, 2233–2247.

Dai, H., Lo, Y.-S., Lin, Y.-H. *et al.* (1996) *In vitro* polysome translation analysis of heat shock proteins in higher plants. *Botanical Bulletin of Academia Sinica*, **37**, 261–264.

Denecke, J. (2001) Plant endoplasmic reticulum. In: *Encyclopedia of Life Sciences*, John Wiley & Sons, Ltd, New York, NY.

Denecke, J., Ek, B., Caspers, M., *et al.* (1993) Analysis of sorting signals responsible for the accumulation of soluble reticuloplasmins in the plant endoplasmic reticulum. *Journal of Experimental Botany*, **44**, 213–221.

Denecke, J., Carlsson, L.E., Vidal, S., *et al.* (1995) The tobacco homolog of mammalian calreticulin is present in protein complexes *in vivo*. *The Plant Cell*, **7**, 391–406.

Ding, Z., Wang, B., Moreno, I., *et al.* (2012) ER-localized auxin transporter PIN8 regulates auxin homeostasis and male gametophyte development in Arabidopsis. *Nature Communications*, **3**, 941. doi:10.1038/ncomms1941.

Dlakić, M. (2005) The ribosomal subunit assembly line. *Genome Biology*, **6**, 234. doi:10.1186/gb-2005-6-10-234.

Drugeon, G., Jean-Jean, O., Frolova, L. *et al.* (1997) Eukaryotic release function1 (eRF1) abolishes readthrough and competes with suppressor + RNAs at all three termination codons in messenger RNA. *Nucleic Acids Research*, **25**, 2254–2258.

Du, Y., Ferro-Novick, S. and Novick, P. (2004) Dynamics and inheritance of the endoplasmic reticulum. *Journal of Cell Science*, **117**, 2871–2878.

Edde, B., Rossier, J., Gros, F., *et al.* (1990) Posttranslational glutamylation of (alpha)-tubulin. *Science*, **247**, 4938, 83–85.

Evert, R. (2006) *Esau's Plant Anatomy: Meristems, Cells and Tissues of the Plant Body: Their Structure, Function and Development*, Wiley, Chichester, UK.

Farhan, H., Weiss, M., Tani, K. *et al.* (2008) Adaptation of endoplasmic reticulum exit sites to acute and chronic increases in cargo load. *The EMBO Journal*, **27**, 2043–2054.

Fatica, A. and Tollervey, D. (2002) Making ribosomes. *Current Opinion in Cell Biology*, **14**, 313–318.

Fitchette-Laine, A., Denmat, L., Lerouge, P. and Faye, L. (1997) Analysis of N- and O-glycosylation of plant proteins. *Methods in Biotechnology*, vol. **3**, Humana Press, Totawa, NJ, pp. 271–290.

Frigerio, L. and Roberts, L.M. (1998) The enemy within: ricin and plant cells. *Journal of Experimental Botany*, **49**, 1473–1480.

Galili, G. (2004) ER-derived compartments are formed by highly regulated processes and have special functions in plants. *Plant Physiology*, **136**, 3411–3413.

Galili, G., Sengupta-Gopala, C. and Ceriotti, A. (1998) The endoplasmic reticulum of plant cells and its role in protein maturation and biogenesis of oil bodies. *Plant Molecular Biology*, **38**, 1–29.

Gao, D., Inuzuka, H., Tseng, A., *et al.* (2009) Phosphorylation by Akt1 promotes cytoplasmic localization of Skp2 and impairs APCCdh1-mediated Skp2 destruction. *Nature Cell Biology*, **11**, 4, 397–408.

Geiss-Friedlander, R. and Melchior, F. (2007) Concepts in sumoylation: a decade on. *International Review of Molecular Biology*, **12**, 947–956.

Gupta, K.J. (2011) Protein S-nitrosylation in plants: photorespiratory metabolism and NO signaling. *Science Signaling*, **4**, 154, jc1. doi:10.1126/scisignal.2001404.

Gupta, D. and Tuteja, N. (2011) Chaperones and foldases in endoplasmic reticulum stress signaling in plants. *Plant Signaling and Behavior*, **6**, 232–236.

Gupton, S.L., Collings, D.A. and Allen, N.S. (2006) Endoplasmic reticulum targeted GFP reveals ER organization in tobacco NT-1 cells during cell division. *Plant Physiology and Biochemistry*, **44**, 95–105.

Hanai, H., Nakayama, D., Yang, H., *et al.* (2000) Existence of a plant tyrosylprotein sulfotransferase: novel plant enzyme catalyzing tyrosine O-sulfation of preprophytosulfokine variants *in vitro*. *FEBS Letters*, **470**, 97–101.

Hanton, S.L., Chatre, L., Renna, L., *et al.* (2007) *De novo* formation of plant endoplasmic reticulum export sites is membrane cargo induced and signal mediated. *Plant Physiology*, **143**, 4, 1640–1650.

Harley, S.M. and Beevers, H. 1982. Ricin inhibition of *in vitro* protein synthesis by plant ribosomes. *Proceedings of the National Academy of Sciences of the United States of America*, **79**, 5935–5938.

Hartley, M.R. and Lord, J.M. (2004) Cytotoxic ribosome-inactivating lectins from plants. *Biochimica et Biophysica Acta*, **1791** (1–2), 1–14.

Hemmerlin, A. (2013) Post-translational events and modifications regulating plant enzymes involved in isoprenoid precursor biosynthesis. *Plant Science*, **204**, 41–54.

Hemsley, P.A. and Grierson, C.S. (2008) Multiple roles for protein palmitoylation in plants. *Trends in Plant Science*, **13** (6), 295–302.

Hemsley, P.A., Taylor, L. and Grierson, C.S. (2008) Assaying protein palmitoylation in plants. *Plant Methods*, **4**, 2.

Henquet, M., Lehle, L., Schreuder, M., *et al.* (2008) Identification of the gene encoding the alpha1,3-mannosyltransferase (ALG3) in Arabidopsis and characterization of downstream n-glycan processing. *The Plant Cell*, **20** (6), 1652–1664.

Hensel, W. (1987) Cytodifferentiation of polar plant cells: formation and turnover of endoplasmic reticulum in root statocytes. *Experimental Cell Research*, **172**, 377–384.

Herman, E.M. and Larkins, B.A. (1999) Protein storage bodies and vacuoles, *The Plant Cell*, **11** (4), 601–613.

Hirsch, C., Gauss, R., Horn, S.C. *et al.* (2009) The ubiquitylation machinery of the endoplasmic reticulum. *Nature*, **458**, 453–460.

Huang, F., Zago, M.K., Abas, L., *et al.* (2010) Phosphorylation of conserved PIN motifs directs Arabidopsis PIN1 polarity and auxin transport. *The Plant Cell*, **22**, 1129–1142.

Isono, E., Nishihara, K., Saeki, Y. *et al.* (2007) The assembly pathway of the 19S regulatory particles of the yeast 26S proteasome. *Molecular Biology of the Cell*, **18**, 569–580.

Johnson, A.E. (2009) The structural and functional coupling of two motor machines, the ribosome and the translocon. *Journal of Cell Biology*, **185**, 765–767.

Johnson, A.E. and van Waes, M.A. (1999) The translocon: a dynamic gateway at the ER membrane. *Annual Review of Cell and Developmental Biology*, **15**, 799–842.

Johnson, S.J. and Hakannsson, K.O. (2003) Biochemical and molecular properties of calreticulin. In: *Calreticulin*, 2nd Edition (Eggelton, P. and Michalak, M., eds), Kluwer Academic Press, New York, NY, S9–S18.

Křeček, P., Skůpa, P., Libus, J., *et al.* (2009) The PIN-FORMED (PIN) protein family of auxin transporters. *Genome Biology*, **10**, 249. doi:10.1186/gb-2009-10-12-249.

Krishna, R.G. and Wold, F. (1993) Post-translational modification of proteins. In: *Methods in Protein Sequence Analysis in Enzymology*, vol. **67** (Imuhori, K. and Saklyarun, F., eds) Springer, New York, pp. 167–172.

Kwon, S.J., Choi, E.Y., Choi, Y.J. *et al.* (2006) Proteomics studies of post-translational modifications in plants. *Journal of Experimental Botany*, **57**, 1547–1551.

Lafontaine, D.L. and Tollervey, D. (2001) The function and synthesis of ribosomes, *Nature Review Molecular and Cell Biology*, **2** (7), 514–520.

Lambalot, R.H., Gelwing, A.M., Flugel, R.S. *et al.* (1996) A new enzyme superfamily- the phospho-pantetheinyl transferases. *Chemistry and Biology*, **3**, 923–936.

Lange, M.J.P. and Lange, T. (2006) Gibberellin biosynthesis and the regulation of plant development. *Plant Biology*, **3**, 281–290.

Langland, J.O., Kao, P.N. and Jacob, B.L. (1999) Nuclear factor 90 of activated T cells; a double-stranded RNA protein and substrate fir the double-stranded RNA-dependent protein kinase, PKR. *Biochemistry*, **38**, 6361–6368.

Lee, T.-Y., Bretaña, N.A. and Lu, C.-T. (2011) PlantPhos: using maximal dependence decomposition to identify plant phosphorylation sites with substrate specificity. *BMC Informatics*, **12**, 261.

Lee, H., Sparkes, I., Gattolin, S., *et al.* (2013) An Arabidopsis reticulon and the atlastin homologue RHD3-like2 act together in shaping the tubular endoplasmic reticulum. *New Physiology*, **197** (2), 481–489.

Lord, J.M. and Hartley, M.R. (eds.) (2010) *Toxic Plant Proteins*, Volume **18**, Springer, Berlin.

Mallory, A.C. and Vaucheret, H. (2006) Functions of microRNAs and related small RNA in plants. *Nature Genetics*, **S38**, S31–S36.

Manevski, A., Bertoni, G., Bardet, C., *et al.* (2000) In synergy with various cis-acting elements, plant insterstitial telomere motifs regulate gene expression in Arabidopsis root meristems. *FEBS Letters*, **483**, 43–46.

Mariani, P., Navazio, L. and Zuppini, A. (2000–2013) *Calreticulin and the Endoplasmic Reticulum in Plant Cell Biology*, in *Calreticulin*, 2nd Ed., Landes Bioscience, Austin, TX.

Metz, A., Wang, K.C.H., Malmstrom, S. and Browning, K.S. (1999) Eukaryotic initiation factor 4B from wheat and *Arabidopsis thaliana* is a member of a multigene family. *Biochemical and Biophysical Research Communications*, **266**, 314–321.

Miura, K., Jin, J. and Hasegana, P.M. (2007) Sumoylation, a post-translational regulatory process in plants. *Current Opinion in Plant Biology*, **10**, 495–502.

Monigatti, F., Hekking, B. and Steen, H. (2006) Protein sulfation analysis. *BBA – Proteins and Proteomics*, **1764**, 1904–1913.

Monnier, A., Bellé, R., Morales, J. *et al.* (2001) Evidence for regulation of protein synthesis at the elongation step by CDK1/cyclin B phosphorylation. *Nucleic Acids Research*, **29**, 1453–1457.

Moreira, D., Kervestin, S., Jean-Jean, O. and Philippe, H. (2002) Evolution of eukaryotic translation elongation and termination factors: variations of evolutionary rate and genetic code deviations. *Molecular Biological Evolution*, **19**, 189–200.

Nielsen, K. and Boston, R.S. (2001) Ribosome-inactivating proteins: a plant perspective. *Annual Review of Plant Physiology and Plant Molecular Biology*, **52**, 785–816.

Niemi, K.J., Adler, J. and Selman, B.R. (1990) Protein methylation in pea chloroplasts. *Plant Physiology*, **93**, 1235–1240.

Nierhaus, K.H. and Wilson, D.N. (2004) *Protein Synthesis and Ribosome Structure: Translating the Genome*, Wiley-VCH, New York, NY.

Nölting, B. and Andert, K. (2000) Mechanism of protein folding, *Proteins*, **41**, 288–298.

Obalinsky, T.R. (2006) *Protein Folding: New Research*, Nova Science Publishers, Hauppauge, NY.

Osborne, A.R., Rapport, T.A. and Vanderberg, B. (2005) Protein translocation by sec 61/sec y channel. *Annual Review Cell and Developmental Biology*, **21**, 529–550.

Paik, W.K., Paik, D.C. and Kim, S. (2007) Historical review: the field of protein methylation. *Trends in Biochemical Sciences*, **32** (3), 146–152.

Park, H.J., Kim, W.Y., Park, H.C. *et al.* (2011) SUMO and SUMOylation in plants. *Molecular Cell*, **32**, 305–316.

Preiss, T. and Hentze, M.W. (2003) Sorting the protein synthesis machine: eukaryotic translation initiation. *Bioessays*, **25**, 1201–1211.

Rademacher, W. (2000) Growth retardants: effects on gibberellins biosynthesis and other metabolic pathways. *Annual Review of Plant Physiology Plant Molecular Biology*, **51**, 501–531.

Ramakrishnan, V. (2002) Ribosome structure and the mechanism of translation. *Cell*, **108** (4), 557–572.

Robert, F., Gao, H.Q., Donia, M. *et al.* (2006) Chlorolissoclimides: new inhibitors of eukaryotic protein synthesis. *RNA*, **12**, 717–724.

Robinson, D.G. and P. Pimpl (2014) Clathrin and post-Golgi trafficking: a very complicated issue. *Trends in Plant Science*, **19**, 134–139.

Ryu, D.H., Litovchick, A. and Rando, R.R. (2002) Stereospecificity of aminoglycoside-ribosomal interactions. *Biochemistry*, **41** (33), 10499–10509.

Schmidt, M., Haas, W., Crosas, B., *et al.* (2005) The HEAT repeat protein Blm10 regulates the yeast proteasome by capping the core particle. *Nature Structural and Molecular Biology*, **12**, 294–303.

Schroder, M. (2010) *Molecular Chaperones of the Endoplasmic Reticulum*, Nova Science Publishers, New York, NY.

Smith, M.H., Ploegh, H.L. and Weissmar, J.S. (2011) Road to rain: targeting proteins for degradation in the endoplasmic reticulum. *Science*, **334**, 1086–1090.

Staehelin, L.A. (1997) The plant ER: a dynamic organelle composed of a large number of discrete functional domains. *The Plant Journal*, **11**, 1151–1165.

Stripe, F., Barbieri, L., Battelli, M.G., *et al.* (1992) Ribosome inactivating proteins from plants: present status and future prospects. *Biotechnology*, **10** (4), 405–412.

Sunkar, R. and Jagadeeswaran, G. (2008) *In silico* identification of conserved microRNAs in large number of diverse plant species. *BMC Plant Biology*, **8**, 37.

Sunkar, R. and Zhu, J.-K. (2007) Micro RNAs and short-interfering RNAs in plants. *Journal of Integrative Plant Biology*, **49**, 817–826.

Swaney, S.M., Aoki, H., Ganoza, M.C. and Shinabarger, D.L. (1998) The oxazolidinone linezolid inhibits initiation of protein synthesis in bacteria. *Antimicrobial Agents and Chemotherapy*, **42** (12), 3251–3255.

Sylvestersen, K.B., Horn, H., Jungmichel, S., *et al.* (2014) Proteomic analysis of arginine methylation sites in human cells reveals dynamic regulation during transcriptional arrest. *Molecular Cell Proteomics*, **13** (8), 2072–2088.

Tekoah, Y. (2005) Post-translational modifications to plants-glycosylation. In: *Encyclopedia of Genomics, Proteomics and Bioinformatics*, John Wiley & Sons, Ltd, New York, NY.

Tekoah, Y. (2006) Posttranslational modifications to plants – glycosylation, Part 3. Proteomics, 3.5. Proteome Diversity – Short Specialist Review. *Online Encyclopedia of Genetics, Genomics, Proteomics and Bioinformatics*, doi:10.1002/047001153X.g305321.

Tollervey, J.R., Curk, T., Rogelj, B, *et al.* (2011) Characterizing the RNA targets and position-dependent splicing regulation by TDP-43. *Nature Neuroscience*, **14**, 452–458.

Tuite, M.F. and Stansfield, I. (1994) Termination of protein synthesis. *Molecular Biology Reproduction*, **19** (3), 171–181.

Ueda, H., Yokota, E., Kutsuna, N. *et al.* (2010) Myosin-dependent endoplasmic reticulum motility and F-actin organization in plant cells. *Proceedings of the National Academy of Sciences of the United States of America*, **107**, 6894–6899.

Unbehaum, A., Borukhov, S.I., Hellen, C.U. and Pestova, T.V. (2004) Release of initiation factors from 48S complexes during ribosomal subunit joining and the link between establishment of codon-anticodon base-pairing and hydrolysis of eIF-2-bound GTP. *Genes and Development*, **18**, 3078–3093.

Vedrenne, C. and Hauri, H. (2006) Morphogenesis of the endoplasmic reticulum: beyond active membrane expansion, *Traffic*, **7** (6), 639–646.

Verdoucq, S.V., Vinh, S.N., Pflieger, D. and Maurel, C. (2006) Methylation of aquaporins in plant plasma membrane. *Biochemical Journal*, **400**, 189–197.

Viaene, T., Delwiche, C.F., Rensing, S.A. and Frim, J. (2013) Origin and evolution of PIN auxin transporters in the green lineage. *Trends in Plant Science*, **18**, 5–10.

Vitale, A., Ceriottie, A. and Denecke, J. (1993) The role of the endoplasmic reticulum in protein synthesis, modification and intracellular transport. *The Journal of Experimental Biology*, **4**, 1417–1444.

Voeltz, G.K., Rolls, M.M. and Rapoport, T.A. (2002) Structural organization of the endoplasmic reticulum, *EMBO Reports*, **3** (10), 944–950.

Voinnet, O. (2009) Origin, biogenesis, and activity of plant microRNAs. *Cell*, **136** (4), 669–687.

Wyatt, S.E., Tsou, P.-L. and Robertson, D. (2002) Expression of the high capacity calcium-binding domain of calreticulin increases bioavailable calcium stores in plants. *Transgenic Research*, **11**, 1–10.

Yamada, K., Hara-Nishimura, I. and Nishimura, M. (2011) Unique defense strategy by the endoplasmic reticulum body in plants. *Plant and Cell Physiology*, **52**, 2039–2049.

Yang, Z., Cramerm C.L. and Watson, J.C. (1993) Protein farnesyltransferase in plants. Molecular cloning and expression of a homolog of the beta subunit from the garden pea. *Plant Physiology*, **101**, 667–674.

Yin, X., Fontoura, B.M.A., Morimoto, T. and Carroll, R.B. (2003) Cytoplasmic complex of eEF2 and p53. *Journal of Cellular Physiology*, **196**, 474–482.

Zheng, H. and Chen, J. (2011) Emerging aspects of ER organization in root hair tip growth: lessons from RHD3 and Atlastin. *Plant Signaling and Behavior*, **6** (11), 1710–1713.

Zhou, Y., Takahashi, E., Li, W., *et al.* (2007) Interactions between the NR2B receptor and CaMKII modulate synaptic plasticity and spatial learning. *Journal of Neuroscience*, **27** (50), 13843–13853.

Zinck, R., Cahill, M.A., Kracht, M., *et al.* (1995) Protein synthesis inhibitors reveal differential regulation of mitogen-activated protein kinase and stress-activated protein kinase pathways that converge on Elk-1. *Molecular and Cellular Biology*, **15** (9), 4930–4938.

Further reading

Carrol, A.J. (2013) The Arabidopsis cytosolic ribosomal proteome: from form to function, *Frontiers in Plant Science*, **4**, 32, http://dx.doi.org/10.3389/fpls.2013.00032.

Chaudhari, N., Talwar, P., Parimisetty, A., d'Hellencourt, C.L. and Ravanan, P. (2014) A molecular web: endoplasmic reticulum stress, inflammation, and oxidative stress. *Frontiers in Cellular Neuroscience*, http://dx.doi.org/10.3389/fncel.2014.00213.

Deng, Y. and Rath, V.I. (2013) Endoplasmic reticulum (ER) stress response and its physiological roles in plants. *International Journal of Molecular Science*, **14** (4), 8188–8212.

English, A.R. and Voeltz, G.K. (2013) Endoplasmic reticulum structure and interconnections with other organelles. *Cold Spring Harbor Perspectives in Biology*, doi:10.1101/cshperspect.a013227.

Gamerdinger, M., Hanebuth, M.A., Frickey, T. and Deuerling, E. (2015) The principle of antagonism ensures protein targeting specificity at the endoplasmic reticulum. *Science*, **348** (6231), 201–207.

Hawes, C., Kiviniceoni, P. and Kriechbaumer, V. 2014. The endoplasmic reticulum: a dynamic and well-connected organelle. *Journal of Integrated Plant Biology*, **57**, 50–62.

Howell, S.H. (2013) ER stress response in plants, *Annual Review of Plant Biology*, **64**, 477–499.

Marchi, S., Patergnani, S. and Pinton, P. (2014) The endoplasmic reticulum–mitochondria connection: one touch, multiple functions. *Biochimica et Biophysica Acta (BBA) – Bioenergetics*, **1837** (4), 461–469.

Stefano, G., Renna, L. and Brandizzi, F. (2014) The endoplasmic reticulum exerts control over organelle streaming during cell expansion. *Journal of Cell Science*, **27**, 947–953.

CHAPTER 4

The Golgi apparatus

D. Davis, T.E. Wilkop, and Georgia Drakakaki

Department of Plant Sciences, University of California Davis, Davis, CA, USA

The Golgi apparatus

History

The discovery of the Golgi apparatus, or as it was first dubbed "apparato reticolare interno," is attributed to the Italian cytologist Camillo Golgi (Figure 4.1). His staining method for light microscopy specimens pioneered for the study of nerve cells proved to be vital for the discovery of the "internal reticular apparatus" in 1898 (Morré and Mollenhauer, 2009).

A dedicated student of science, Golgi (born on July 9, 1843, in Corteno, Italy) studied medicine and took up a faculty position at the University of Pavia in 1865, which he held to his death in 1926. It was during his work as a physician that he developed his chromate of silver staining method (*la reazione nera*) that facilitated the observation of the internal structures of nerve cells (Figure 4.2). Normal staining techniques failed to capture the detail of the thin filamentary extensions of neural cells, including the axon and the dendrites. Golgi's staining method labels a limited number of cells at random and in their entirety; dendrites and the cell soma are stained in brown and black, providing the required contrast for light microscopic observations. However, his ink pen drawings of the first observed internal reticular apparatus based on this staining method were met with skepticism regarding their very existence and significance in the function of the cell. Many scientists argued that the darkly stained networked structure was simply an artifact of the fixation or impregnation method that Golgi used, while others argued that the observed structure was an artifact composed of other organelles and cellular structures (Morré and Mollenhauer, 2009). It is clear nonetheless from Golgi's early work that he was convinced that the apparatus was not only a novel organelle but also that it played an essential role in cellular secretion processes.

Santiago Ramón y Cajal, a Spanish physician also studying the nervous system, took up the task of further characterizing the function and morphology of Golgi's reticular structure. While Golgi first identified the structure and

Plant Cells and their Organelles, First Edition. Edited by William V. Dashek and Gurbachan S. Miglani.
© 2017 John Wiley & Sons, Ltd. Published 2017 by John Wiley & Sons, Ltd.

Figure 4.1 Portrait of Camillo Golgi.

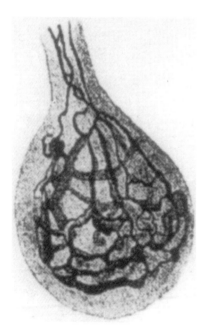

Figure 4.2 Original drawing by Camillo Golgi, illustrating the reticular network (then unknown Golgi apparatus) in a nerve cell.

developed the histological staining method that enabled the light microscopic detection, Cajal solidified the identity of the novel organelle with his seminal studies in a variety of cell types and cell stages. Specifically, Cajal was instrumental in describing changes that occurred to the organelle during cell differentiation and altered cellular metabolism—which taken together strongly supported the notion that the Golgi apparatus (GA) is involved in secretion. Cajal also demonstrated the presence of GA in a variety of cell types, providing further support for an important and universal role of the apparatus in the cell. Together, Golgi and Cajal laid the foundation for functional studies of the apparatus in the years to come (Morré and Mollenhauer, 2009).

In the early 1900s, as more and more researchers engaged in GA studies, a detailed picture of the function of the organelle was emerging. The organelle was shown to have a consistent association with secretory products. Key studies by Nassonov, in which he investigated the localization and movement of secretory granules through the reticular apparatus, conclusively demonstrated that the apparatus was involved in cellular secretion. He described the apparatus as having a specific role in the collection of secretory products and their subsequent discharge (Morré and Mollenhauer, 2009). Furthermore, this process was shown to be specific to different cell types, with unique and specific material being processed through the apparatus depending on the cell's identity.

By 1929, with the publishing of a cytology review of secretion processes, the GA had finally caught the interest of the cell biology community as a new organelle that plays a crucial role in the processes of secretion. However, despite the growing evidence of the importance of Golgi in cellular secretion, the challenging visualization of the organelle caused some scientists to still doubt its existence.

Until this point the GA had only been observed using light microscopy and until the advent of more sensitive microscopy methods, research on the new organelle proceeded with an air of doubt and skepticism. Numerous cytologists were convinced the GA was not a distinct new organelle but simply a section of fragment of a known organelle, perhaps undergoing some morphological change, or even an artifact of the histological silver or osmium staining and visualization technique. The difficulty of the organelle's visualization and isolation, in different tissues and organisms, also casted serious doubt on its presence in plant and non-neuronal animal cells. Given the approximate size of the Golgi of 2 μm, the observation using light microscopy was always a formidable challenge and remains so even with today's light microscopes. The existence of the equivalent of a Golgi apparatus in plants was a very controversial notion with some scientists arguing that there was no such organelle in plant cells (Morré and Mollenhauer, 2009).

There coexisted numerous explanations and theories during this time in the history of GA research—much of which were contradictory and controversial. In 1953, the first electron micrographs of the GA were published by Dalton and Felix, which ultimately and unequivocally demonstrated its existence in the cell. The reticular apparatus that Golgi first described more than 50 years earlier now had a clear morphological structure, one that included "flattened sacs and groups of vesicles" (Morré and Mollenhauer, 2009). With the technology for high-resolution visualization finally available, the controversy surrounding whether the Golgi apparatus was a new, distinct organelle ended.

The next wave of Golgi exploration focused on the elucidation of the bio-chemical function, the differentiation of the poles of the organelle, its significance, and the exact and detailed role of the apparatus in secretion.

Plant vs animal Golgi

The majority of research done after the organelle's discovery has been conducted within animal systems. Only relatively recently has GA research in plants begun to take on a life of its own. The field of Golgi research in both plants and animals are the basis of various industries. Plant secretion products in particular are an integral part in many biotechnologies today.

From the vast studies of the GA in mammalian systems and the more recent research in plants, we know that the organelle does indeed have a generally conserved form, which is somewhat surprising given its highly dynamic nature. However, this Golgi "blueprint" can vary not only between different kingdoms as in plants and animals, but also between cell types within the same organism.

General structure

The Golgi apparatus consists of stacked cisternae composed of flattened mem-brane discs; historically in plants, these were referred to as dictyosomes (Pavelka and Robinson, 2003). The Golgi complex is also polar, with membranes at one end of the stack differing in both composition and thickness from those at the other end. The entry point into the GA, called the *cis* face, is closely associated with the endoplasmic reticulum (ER) and acts as the receiving center of the Golgi. At the opposite end of the Golgi stack is the *trans* face, which acts as the shipping center from the Golgi apparatus. Dynamics of Golgi trafficking change according to different cellular cues which leads to the synthesis and secretion of specific Golgi products. Golgi trafficking changes in relation to the functions of other cel-lular components, environmental signals, cell stage, and overall needs of the cell. While the basic structure and function is the same between both plant cells and animal cells, there are a few key differences between the two. Ongoing research is establishing the detailed differences as well as their conserved properties.

Major differences

The differences between animal and plant Golgi include morphology, the cargo processed or carried through the organelle, behavior during cell division, and location within the cell (Pavelka and Robinson, 2003). In fact, skeptics cited these differences as an argument against the existence of a plant GA in the years prior to its definitive description (Morré and Mollenhauer, 2009). These distinct

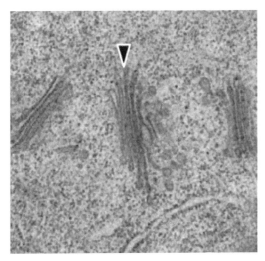

Figure 4.3 Electron micrograph demonstrating multiple Golgi in a plant cell. Section through a single *Arabidopsis* meristematic cell showing multiple immunolabeled Golgi stacks. A single Golgi stack is indicated by the arrowhead. Source: Adapted from Kang (2010, p. 275). Reproduced by permission of Elsevier Inc.

differences should not, however, be of much surprise considering the different functions, environments, and structures of plant and animal cells. One particular aspect of plant cells that is a major distinguishing factor is the cell wall, which consists in part of Golgi-secreted polysaccharides. Further notable differences are briefly outlined in the following text.

Subcellular position

The **position** of Golgi in plant and mammalian cells is an easily noticeable difference. Golgi in mammalian cells tends to reside near the nucleus, extending only as far as the perinuclear region in highly secreting cells. Golgi in plant cells however is polydispersed, with stacks distributed throughout the entire cell. While animals typically contain less than 10 Golgi stacks in the entire cell, plants contain many stacks, ranging from less than 25 to several hundred in each cell (Figure 4.3). Furthermore, in plants each individual Golgi stack acts in concert with the other stacks in the cell, forming a collective network and system of endomembrane trafficking between the ER and the plasma membrane (Pavelka and Robinson, 2003).

Cargo

The **material** that the Golgi modifies and sorts also differs between plant and animal cells. The plant cell requires extensive synthesis and secretion of noncellulosic polysaccharides from the GA in order to form the cell wall, which animal

cells do not (Albersheim *et al.*, 2010). During cell division, different needs mani-
fest themselves in unique Golgi behavior. In animal cells, the GA is disassembled into vesicular bodies as membrane trafficking ceases at the onset of
division. Conversely in plant cells, Golgi stacks persist throughout cellular division (Pavelka and Robinson, 2003; Faso *et al.*, 2009b). In fact, the number of
Golgi stacks in a dividing plant cell can double during metaphase/anaphase, presumably to meet the high demand for new polysaccharides at the plane of cell
division, termed the "cell plate," where the new cell wall is formed between the
two daughter cells (Pavelka and Robinson, 2003).

Cytoskeletal association

The **cytoskeletal association** has also been shown to be unique between
plants and mammals. These differences were established by chemical inhibition experiments, in which specific components of the cytoskeleton are
selectively disrupted (Hawes *et al.*, 2003). While plant Golgi moves along
actin filaments, animal Golgi is more associated with microtubules (Brandizzi
and Wasteneys, 2013). Upon disruption of microtubules, animal Golgi loses
its organization around the nucleus and disperses throughout the cytoplasm,
whereas identical treatment does not affect the organization of plant Golgi.
Conversely, actin depolymerization causes extensive spatial disorganization
of the plant Golgi, with much more subtle effects in animal cells (Egea *et al.*,
2013).

   While the reason for these differences has been speculated upon, no all-
encompassing hypothesis has yet emerged. There are many unanswered
questions surrounding the details of plant Golgi. Organelle organization,
interaction with other cellular components, cellular function, and observation and analysis techniques are some of the main focuses in plant Golgi
research.

Plant Golgi introduction

There are specific functions that the plant Golgi performs in the cell: it secretes
polysaccharides, proteins, and lipids, and it serves as a major site for polysaccharide synthesis, protein modification, and glycosylation, or the addition of sugar
components to proteins and lipids (Morré and Mollenhauer, 2009; Albersheim
et al., 2010). Product modifications occur sequentially through the cisterna via
enzymes within the Golgi lumen, specifically in the medial cisterna. Many molecules such as lipids, membrane proteins, and vacuolar enzymes undergo their
final modification in the Golgi and then leave it to go to their final destination in
the cell. One frequently used analogy describes the ER as a mailbox and the
Golgi as a post office, in that it collects and sorts material for shipment from the
ER to the final destination.

General structure and Golgi polarity

Structurally, the GA is composed of membrane tubules, referred to as cisternae, that are stacked on top of each other. The cisternae encase the lumen of the GA in which biochemical functions are carried out. The Golgi apparatus, usually composed of 6–8 cisternae per Golgi, is arranged with a distinct polarity—a gradient of different functions are carried out within the cisternae depending on their position in the overall stack. The plant Golgi is also characterized by a gradient in the cisternae lumen width, with the widest lumen occurring in the *cis* cisternae and the narrowest at the *trans*. The polarity of the GA is a fundamental aspect of the organelle and highlights the sequential modification and packaging of the secretory products (Morré and Mollenhauer, 2009).

A general description of the GA includes three main cisternae as they are defined by their relative position with respect to each other: *cis*, medial, and *trans*. The *cis* face of the Golgi stack is that closest to the ER; the medial is the middle region of the organelle; the *trans* face is at the far end of the Golgi, away from the ER. There is an additional domain beyond the *trans* face called the *trans*-Golgi network (TGN); this network of secretory vesicles is commonly considered a distinct cellular structure rather than a part of the cisternal GA (Figure 4.4) (Worden *et al.*, 2012).

The distinct cisternae of the GA are key to the characteristic gradient of its functionality. Material is transported to the Golgi bidirectionally, with material simultaneously being secreted, endocytosed, and recycled to the TGN (Figure 4.5). Anterograde transport originates from the ER and then proceeds through the Golgi in a *cis*-to-*trans* fashion, whereas retrograde transport occurs in the opposite direction. Similarly, while post-Golgi vesicles transfer material for secretion, endocytosed cargo is recycled back to the TGN. It is challenging to confidently assign resident enzymes and functions to specific cisternae

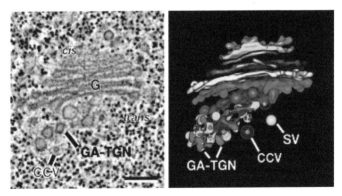

Figure 4.4 Plant Golgi structure as shown by electron microscopy and tomography. Electron tomography of Golgi apparatus (GA), exhibiting distinct cisternae, the *trans*-Golgi network (TGN), and specific vesicles types (clathrin-coated vesicle or CCV; SV, secretory vesicle). Source: Adapted from Kang *et al.* (2011, p. 317).

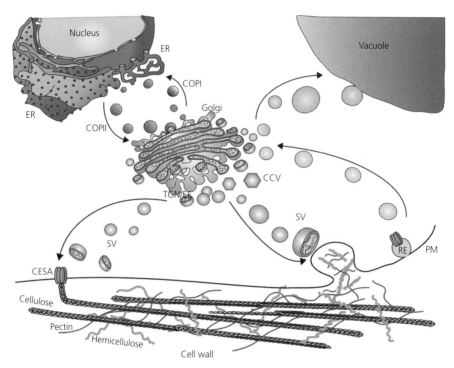

Figure 4.5 Golgi-mediated vesicular trafficking. Diagram of Golgi-mediated vesicular trafficking in the plant cell including bidirectionally trafficking between Golgi and ER with COP-coated vesicles, trafficking to the vacuole, recycling of membrane and proteins from the plasma membrane, and secretion of structural cell wall polysaccharides. CESA complexes embedded in the plasma membrane synthesizing cellulose are shown. CCV, clathrin-coated vesicle; CESA, cellulose synthases; PM, plasma membrane; RE, recycling endosome; SV, secretory vesicle; TGN/EE, *trans*-Golgi network/early endosome.

given the dynamic bidirectional budding and fusion of vesicles carrying different products at various levels of modification. The Golgi can be viewed as the organizational center for material movement throughout the cell. With the different regions of Golgi (*cis*, medial, *trans*) being involved in different, albeit overlapping functions, the enzyme/protein residency and therefore biochemical activities in the stacks are distinct. However, definitive functional assignment into separate divisions is challenging in contrast to morphological and structural identification (Glick, 2000; Pavelka and Robinson, 2003; Morré and Mollenhauer, 2009).

The *cis* face serves as the connection to the ER, with material being trafficked to and from the ER. Analogously, the *trans* face mediates the cargo traffic between the Golgi and the plasma membrane and/or other compartments of the cell via the TGN. Recently, it was established that the TGN is an early endosome (EE), that is, a compartment where endocytosed material is trafficked. The medial

Golgi resides between the two faces and is the location where the bulk of the biochemical modifications and synthesis occurs. Each directional process between the Golgi and other cellular components involves distinct vesicles. Among them, three easily identified vesicle types exist: clathrin-coated vesicles (CCVs) and coat protein complexes I and II coated vesicles (COPIs and COPIIs) (Aniento *et al.*, 2003).

CCVs are characterized by a cage-like structure composed of heavy and light clathrin chains that surround the vesicles and primarily bud from the TGN. They are also involved in endocytosis from the plasma membrane.

COPI and COPII vesicles are characterized by a unique complex of encapsulating coat proteins called "coatomer." The respective vesicle coats are formed through the sequential recruitment and interaction between a transmembrane protein embedded in the budding membrane, a small GTPase and the coatomer. Specifically, the COPI complex is composed of seven different protein subunits, while the COPII is made up of two protein complexes, both heterodimers. Both vesicle types are involved with ER-Golgi transport albeit in different directions. COPI vesicles are primarily involved with retrograde trafficking from the Golgi to the ER and between Golgi cisternae. Anterograde transport from the ER to the Golgi is carried out by COPII vesicles (Aniento *et al.*, 2003).

Structure and organization

Golgi-ER continuum

The plant GA is a very dynamic organelle which communicates intimately with the ER during protein secretion. However, it remains controversial whether these two organelles are physically linked and form a continuous tubular network or are solely associated through COPI/II vesicular traffic. While vesicular budding can be clearly observed on both the *trans* and the *cis* faces of the Golgi apparatus, much less experimental data exist supporting the budding of vesicles from the ER to the Golgi. COPII immunolabeling experiments with *Arabidopsis* root tip cells demonstrated the presence of COPII proteins in "vesicle-like structures" (Brandizzi and Barlowe, 2013). Additionally, research indicates that plant Golgi and ER are physically associated via a tethering matrix in the cell, even though the Golgi is actively moving through the cytoplasm at a relatively fast rate (Sparkes *et al.*, 2009; Lerich *et al.*, 2012). This is supported by evenly dispersed ER exit sites (ERES), or sites of COPII budding that are only present in ER locations continuously associated with the *cis*-Golgi (Brandizzi and Barlowe, 2013). Taken together, the ephemeral presence of COPII vesicles and the physical link between ER and Golgi are key challenges for COPII vesicle capture. Given the rapid release of the vesicles from specific ERES, which are already closely associated with Golgi cisternae, COPII budding from the ER is difficult to observe.

Connections between Golgi stacks

The rapid mobility of plant Golgi suggests the presence of tethering elements between cisternae that maintain proper stack organization. In mammalian cells, several protein complexes and coiled-coil proteins often called "golgins" are responsible for maintaining Golgi stack integrity. Golgins attach to the membrane in two ways: either through transmembrane domains or by interaction with other membrane proteins (peripheral) (Osterrieder, 2012). By forming homo- or heterodimers with other golgins and/or membrane proteins, the complex forms a "tether" between the two membranes. Plant Golgi have golgins located primarily at the *cis* and *trans* cisternae. They are thought to be involved with intercisternal tethering, tethering to the ER, and directing post-Golgi transport vesicles (Osterrieder, 2012).

There also exist elements between individual cisternae that appear to be unique to plant Golgi. Intercisternal filaments or fibers are arranged parallel to each other and are found midway between two cisternae. Interestingly, the number of filaments between cisternae follows a gradient: fewest between *cis*-side cisternae and most between *trans*-side cisternae. It has been speculated that these filaments are involved in the lumen width gradient specific to plant Golgi or stack stabilization; however, their precise nature and function has not yet been determined (Morré and Mollenhauer, 2009).

Plant Golgi and the cytoskeleton

Another distinguishing feature of plant Golgi organization lies in the unique relationship with the cytoskeleton. While mammalian Golgi are tethered to and move along microtubules, plant Golgi travel along actin filaments via myosin motors. The very organization of the Golgi stacks and TGN is primarily actin dependent. Actin is involved in the organization and movement of the organelle as a whole, and also in post-Golgi trafficking to the plasma membrane or vacuole. Movement between the Golgi and ER, however, does not appear to be cytoskeleton dependent, which stands in stark contrast to mammalian cells (Faso *et al.*, 2009b; Brandizzi and Wasteneys, 2013).

Historically, the actin cytoskeleton has been viewed as the main system by which Golgi bodies move in the plant cell; however, recently microtubules have also been shown to be involved with specific aspects of Golgi-derived vesicle trafficking (Brandizzi and Wasteneys, 2013). Golgi carrying cellulose synthase complexes (CSCs) pause at cortical microtubules during secretion to the plasma membrane. This pause has been associated with cellulose synthase (CESA) incorporation into the plasma membrane. In addition, microtubules have been associated with CESA-carrying vesicles, referred to as small CESA-containing compartments (SMACs) or microtubule-associated cellulose synthase compartments (MASCs) (Bashline *et al.*, 2014).

Models of Golgi organization

The overall organization and synthesis of Golgi stacks is currently described by two different models: the stationary cisternae/vesicular shuttle model and the maturation/cisternal progression model (Glick, 2000; Nebenführ, 2003).

The stationary cisternae model proposes that Golgi cisternae are stable, enduring components of the organelle. In this model, individual cisternae interact with other, distinct cisternae strictly by vesicular transport while retaining their characteristic cisternae-specific resident proteins. The maturation model, however, describes cisternae as "flowing" from the ER, with a new *cis* cisternae formed from fused ER vesicles. The new ER-derived cisternae gradually mature and take the place of the older cisternae. The stack is continuously being replaced with new cisternae, and older cisternae are used in secretory vesicles as they reach the *trans* end of the organelle. This model requires recycling of the resident enzymes from the older to the new cisternae in order to maintain the correct gradient of biochemical activities throughout the stack. Current opinion favors the maturation model of Golgi stack synthesis; however, the details of the timing and method of new cisternae delivery from the ER to the Golgi and correct recycling of resident enzymes remain unclear.

Golgi-mediated vesicular trafficking

The secretion process begins when vesicles coated with the specialized coat protein (COPII) and loaded with cargo, namely lipids and proteins, leave the ER (Marti *et al.*, 2010). These vesicles bud from specific locations of the ER, the ER exit sites, and traffic to the Golgi where they fuse and deliver their cargo. The materials are then shuttled through the Golgi stacks sequentially from *cis* to *trans* where they undergo various modifications. Post-Golgi-mediated trafficking transports secretory material to the apoplast or to the vacuole through intermediate compartments. Endocytosis and protein recycling involving CCVs originates at the plasma membrane and follows retrograde traffic to endosomes, including the EE. Movement of material occurs by different vesicle types, mostly determined by their characteristic coat proteins (Aniento *et al.*, 2003). However, the mechanisms by which these distinct vesicles fuse with other cellular components (i.e., the vacuole and the plasma membrane) are, in principle, similar. SNARE proteins are the major component responsible for vesicle fusion. They are embedded in the vesicle membrane and the acceptor compartment; with and their interactions bringing about the lipid membrane fusion. SNARE proteins or other molecular components such as GTPases involved in membrane fusion are also determinants of vesicle types.

Clathrin-coated vesicles

A clathrin coat, or cage as it is frequently described, consists of three chains of two different types (heavy/~190 kDa and light/~25 kDa chains) as well as adaptor proteins that mediate the interaction between the clathrin chains and the cargo (Pavelka and Robinson, 2003). When viewed under an electron microscope, CCVs appear spiny and are of regular geometric structure. Golgi transport via this vesicle type takes place at the *trans* face of the Golgi and TGN through a process that includes activation of the clathrin assembly prior to physical deformation of the membrane to form the vesicle. Specifically, ADP-ribosylation factor 1 (ARF1) is recruited to the location of the Golgi membrane which is identified by the localized cargo. Following ARF1 activation by GTPase, adaptor proteins and the clathrin chains are recruited and begin to assemble into a cage. This encourages vesicle budding as the increasing recruitment of clathrin chains form a cup-like structure which then forms a completely enclosed vesicle as the cage encapsulates the membrane (Pavelka and Robinson, 2003). Sorting of cargo in CCVs may be due to the binding of cytoplasmic tails of specific cargo receptors to the adaptors, which then aids the recruitment of clathrin. Therefore, the cargo plays a crucial role in the site at which clathrin-mediated budding occurs. A similar process occurs at the plasma membrane during endocytosis with CCVs budding and trafficking extracellular material into the cell.

COPII vesicles

COPII coats are composed of four proteins, dimers of Sec23-24 and Sec13-31 (Pavelka and Robinson, 2003; Marti *et al.*, 2010; Faso *et al.*, 2009a). They function between the ER and the Golgi by moving material from the ER to the *cis* face of the Golgi. COPII coat complexes take advantage of cytosolic and lumenal bonds between the coat and the cargo in the ER lumen. COPII coat formation is achieved by the concerted function of various proteins found both in the cytosol and on the ER membrane. Assembly of the COPII vesicle takes place at the ER; similarly to other vesicular pathways, it begins with the activation of a GTPase (Sar1, or secretion-associated Ras-related protein 1) in the cytosol by the Sec12 protein in the ER membrane (Marti *et al.*, 2010). This induces ER membrane association with Sar1, where the components of the COPII coat complex are then recruited. Following assembly of the COPII complex, it associates with Sar1, which distinguishes between secretion cargo and ER-resident proteins. The process is complete as the COPII complex causes a physical deformation, generating membrane curvature that induces vesicle formation and budding. COPII vesicles are formed at the ERES that are often oriented opposite the *cis*-Golgi. The tight spatial and temporal association between the two has given rise to the notion that they move together in the cell as a complete "secretory unit" held together by physical interactions (Sparkes *et al.*, 2009; Lerich *et al.*, 2012). Searches in the plant genome for tethering factors, similar to those found in the animal kingdom, established a few candidates that are potentially connecting the Golgi and ER. However, the highly dynamic nature of the process has been difficult to discern with certainty. Two main models currently exist describing the process: one involves the presence of a constant physical connection between ER and Golgi while the other favors an ephemeral nature in which only transient connections exist between the two organelles, the so-called kiss-and-go model (Marti *et al.*, 2010). Support exists for both models from experiments involving a variety of different plant systems, which lead to suggestions that the interaction is plant family- or plant species-specific. For example, in one experiment using an optical trap the GA was dragged through the cytoplasm, and it was demonstrated that ER tubules moved with the Golgi, supporting the hypothesis of a physical connection (Sparkes *et al.*, 2009). Definitive evidence of the nature of this connection in plants, the precise timing of COPII vesicle fusion with the Golgi (i.e., whether or not it occurs *during* Golgi movement or immediately after movement has ceased), has yet to be established.

COPI vesicles

While COPII-coated vesicles are more associated with the export of secretory material from the ER to the Golgi, COPI-coated vesicles are involved with transport of material from the Golgi to the ER as well as between Golgi stacks. The COPI coat complex is made up of two subunits that form a coat around budding vesicles by interacting with ARF1 (Pavelka and Robinson, 2003). Activated ARF1-GTP is attached to the membrane and recruits the coatomer proteins. Sorting of cargo into these vesicles is thought to be achieved by a family of transmembrane proteins, p24 and p23 (Pavelka and Robinson, 2003). These proteins are implicated in both the formation of the COPI coat and the recruitment of suitable cargo. The process has been well described for mammalian systems with proteins homologues existing in plants. The p24 and p23 proteins are thought to achieve coat formation by the presence of domains on either side of the budding vesicle membrane. The p24 proteins are able to act as receptors for cargo on the lumenal side of the Golgi membrane as well as bind to GTPases (like ARF1) and coatomers on the cytoplasmic side of the membrane. Analogous to COPII vesicle formation correlating with ERES, it is thought that COPI vesicle fusion occurs at specific ER import sites (ERIS); however, this has yet to be shown in plants (Faso *et al.*, 2009b).

Golgi movement

Cessation of Golgi movement in association with COPI budding and fusion with the ER suggests that fusion does not occur during Golgi movement in the cytoplasm but only with immobile Golgi stacks. Further evidence suggests that COPII transport occurs via a similar mechanism, in which ER-derived vesicles fuse with the Golgi only when its movement is paused (Lerich *et al.*, 2012). This further implies a role for tethering factors that keep the vesicles (COPI or COPII) near the Golgi so that the vesicles do not move away as the Golgi moves through the cytoplasm (Sparkes *et al.*, 2009). Taken together, this describes an interesting, plant-specific mode of transport between the Golgi and the ER—one that takes into account the fast mobility and close association of Golgi and ER import/export sites in plant cells (Osterrieder, 2012).

Plant Golgi-dependent cellular processes

Even though the various vesicle types and protein transport mechanisms were first discovered in animal and yeast systems, plants seem to follow similar mechanisms and possess homologous players associated with the Golgi. However, due to the sessile lifestyle of plants and the unique aspects of plant-specific structures, there exist several key differences in Golgi function and dynamics between plants and animal cells.

Polysaccharide synthesis and deposition

The cell wall is a unique component of plant life that is built up through the fusion of Golgi-derived vesicles and their cargo. The polysaccharides that make up the cell wall are weaved together to form a dynamic structure that protects the cell from various stresses and pathogenic assaults. Specifically, there are three main polysaccharides that make up the plant cell wall: cellulose, hemicellulose, and pectin (Albersheim *et al.*, 2010). The mechanisms by which hemicellulose and pectin reach the apoplast of the plant cell and are incorporated into the cell wall matrix are all related to and dependent on secretion from the Golgi apparatus. Cellulose microfibrils as the main structural "beams" in the wall are not synthesized in the Golgi cisterna as are the other polysaccharides (Driouich *et al.*, 2012). Instead, cellulose is extruded into microfibrils in the apoplast by cellulose synthases that are embedded in the plasma membrane. However, cellulose synthases reach the plasma membrane by traveling through the Golgi where they are modified and packaged for secretion (Albersheim *et al.*, 2010; Worden *et al.*, 2012). Hemicellulose, thought of as the "ties" or "cables" between the cellulose beams, and pectin as the embedding matrix are both synthesized in the Golgi (Scheller and Ulvskov, 2010). The biosynthesis of the polysaccharides occurs sequentially in a *cis*-to-*trans* fashion until they are packaged into vesicles that are transported to the plasma membrane, where they are released into the apoplast (Caffall and Mohnen, 2009; Driouich *et al.*, 2012). Structural proteins are also part of the cell wall, and their delivery to the apoplast is also Golgi mediated (Albersheim *et al.*, 2010; Oikawa *et al.*, 2013).

The buildup of new cell walls is particularly informative for the study of the Golgi function. With each new cell division, a new cell wall must be formed at the plane of division, or the midline of the parent cell. In contrast to animal cell division, plant cells divide from the interior of the cell outward. Cell wall polysaccharide deposition and membrane accumulation begins at the interior of the cell and grows outward radially to form the cell plate, which matures

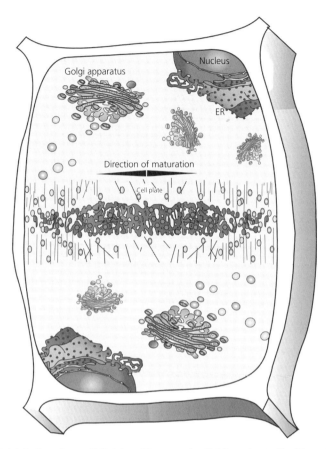

Figure 4.6 Golgi during plant cell division. Diagram of a dividing plant cell with multiple intact Golgi actively contributing membrane and polysaccharides to the formation of the cell plate. Hexagons in the center of the cell plate represent deposited polysaccharides during cell division.

and fuses with the previously existing cell wall, giving rise to two daughter plant cells (Figure 4.6) (Albersheim *et al.*, 2010). The Golgi secretes vesicles containing cell wall components within tight spatial and temporal limits to facilitate the formation of the cell plate. Plant Golgi are distinguished by the fact that they do not disintegrate during cell division as they do in animal cells; however, the mechanism(s) by which cell wall polysaccharide secretion occurs in synchrony with cell division is not yet completely understood. Specifically, the precise mechanisms controlling directionality and timing of deposition are not known. The TGN serves as the vehicle for cell wall components, whereas the stacked GA itself is more involved with the synthesis of the polysaccharides and modification of the proteins destined for the cell wall (Driouich *et al.* 2012; Worden *et al.*, 2012).

Glycosylation

Glycosylation, the addition of a sugar component, like glycan chains to proteins and lipids, is one of the main functions of the GA (Morré and Mollenhauer, 2009; Oikawa *et al.*, 2013). Modification of proteins by adding sugar residues aids in protein stability by preventing degradation as well as affecting protein folding and tertiary structure. This is a universal process in plants and animals; however, key differences in the type of glycosylation on the protein such as the absence of sialic acid in plant N-glycans exist (Morré and Mollenhauer, 2009). In fact, these differences are a debated concern in the field of food biotechnology and plant pharmaceutical production. These modifications affect the biochemical properties and physiological functions of proteins, and the differences between plant and animal Golgi glycosylation can potentially induce unforeseen and unexpected reactions in the human body once ingested (Ungar, 2009). Therefore, research focuses currently on deciphering what controls and affects glycosylation, how it affects the animal and human system once plant produce is consumed as food or medicine, and how we can manipulate plants to produce proteins with animal-"type" glycosylation properties.

Imaging and visualization

Electron microscopy (EM) conclusively established the existence of the GA over 50 years ago (Morré and Mollenhauer, 2009). The high resolution and level of detail obtained in electron micrographs remain useful today in studying the physical structure of the organelle. The orientation of stacks, overall dimensions of the Golgi, and even individual vesicles can be identified in electron micrographs (Han *et al.*, 2013). The process of Golgi preparation for EM is particularly challenging, given the difficulties in preserving Golgi structure and organization as the samples are being chemically and/or physically fixed. Current state-of-the-art EM preparation techniques such as high-pressure freezing (HPF) coupled with electron tomography yield high-quality, high-resolution three-dimensional (3D) images that reveal never-before-seen cellular structures (Kang, 2010; Han *et al.*, 2013). HPF is carried out by instantly stopping all cellular processes while preserving the cellular architecture by freezing it in a glass-like, vitreous form, sometimes referred to as "amorphous ice" (Han *et al.*, 2013). This method allows the detailed preservation of the cell without disrupting the very structure one wishes to observe. The native cellular organization is conserved by eliminating the disruptive dehydration step found in traditional, chemical fixation methods. In this method, the vitrified sample is fully hydrated in a highly viscous state. This method also precludes the need for secondary detection methods like staining since the cellular components maintain their natural densities.

There are multiple ways to vitrify samples, each with distinct benefits depending on the sample. Plunge freezing is highly suited to small cells, such a prokaryotes, since freezing occurs rapidly. For large eukaryotic cells, however, vitrification for Golgi visualization purposes is best done by HPF fixation. Proper vitrification requires a specific cooling rate to produce an even fixation of the sample. The high pressure makes fixation of thicker tissues easier. A similar method—the self-pressurized rapid freezing (SPRF)—introduced recently vitrifies samples by submerging the sample enclosed in a copper tube, in freezing cryogen (Kang, 2010).

While vitrification is particularly useful in the visualization of delicate organelles such as the Golgi apparatus, there are certain limitations of cryo-EM sectioning. The amorphous ice is sensitive to the highly energetic electron beam; therefore, the exposure time must be limited (Han *et al.*, 2013). During fixation in amorphous ice, extreme care must be taken to prevent cracks or disruptions in the tissue/ice, which could affect the resultant image. Additionally, subsequent preparations can be done with these vitrified samples to aid in the visualization and analysis; these include freeze substitution and embedding which allow ultrathin sectioning.

A balance in the sample preparation for visualization and maintenance of the integrity and clarity of the sample must be struck, sometimes compromising the obtainable resolution. However, noise reduction routines and tomographic reconstruction in transmission EM software and advanced imaging equipment can help produce highly detailed data. When done properly, individual structures at the protein complex level can be seen in vitrified samples, something which is impossible with resin-embedded samples. These techniques have aided in the visualization of protein complexes, protein interactions with the membrane, vesicle-type characterization, and 3D reconstruction of the Golgi structure (Han *et al.*, 2013). Identification of the macromolecular structure and organelles in relation to the whole cell has been referred to as "visual proteomics" (Han *et al.*, 2013). In conjunction with traditional molecular tags, this technology can help increase the information available toward understanding the specific role proteins play not only in relation to Golgi but also in the cell as a whole. However, protein discovery based on cellular localization studies is limited by the fact that the identity of the proteins observed is impossible to discern solely from an image. Advancement in the visualization of the Golgi can offer much needed insights in the physical interactions between Golgi and other organelles, its relation to the cytoskeleton, and behavior during different cellular stages, such as during division.

While our ability to visualize the GA has advanced incredibly since its first images in neural cells, the precise, native ultrastructure of an intact, functioning Golgi in a live cell remains elusive.

Golgi research is moving towards the characterization of proteins in different compartments or distinct cisternae using a combination of cellular and molecular techniques, including fluorescent confocal microscopy and proteomic analysis. Fluorescent proteins fused to Golgi proteins are also utilized for protein localization, dynamics, transport, and interaction studies and are particularly useful in mutant and gene identification. Confocal microscopy paired with fluorescent tagging techniques allows high-resolution imaging of live cells, invaluable in endomembrane trafficking studies. Fluorescent protein tags fused to proteins of interest are also used in protein dynamics studies, employing Förster resonance energy transfer (FRET) for protein interactions and fluorescence recovery after photobleaching (FRAP) in which monitoring of recovery rates can quantify the protein transport dynamics (Stefano *et al.*, 2013). Further, tracking of fluorescent reporters yields information on the directionality of protein transport through the endomembrane system.

Isolation and analysis

While there is a rich history of Golgi imaging techniques since the very discovery of the organelle by Camillo Golgi, there remain distinct challenges in the visualization and isolation of GA components, such as distinct cisterna, even today.

Current GA isolation techniques include employing a density gradient, free-flow fractionation, and localization of organelle proteins by isotope tagging (LOPIT) (Parsons *et al.*, 2013). Immunoisolation of tagged vesicles has also proven to be a useful tool to study specific components in the Golgi. Following isolation, proteomic analysis based on mass spectrometry (MS) can be used to identify proteins in specific components (Figure 4.7) (Parsons *et al.*, 2012).

Density gradients

Isolation based on density gradient centrifugation involves homogenization of plant material to break the cell wall and plasma membrane and free the subcellular components. Afterward, the plant homogenate is applied to various stratified aqueous media layers that separate along their different densities, such as sucrose. During centrifugation, the components move through the layers and collect on top of the layer with a density slightly higher than the component itself, thereby separating the cellular components by their individual densities. The approach is suitable for the enrichment of Golgi and TGN vesicles; however; isolation of pure vesicles is compromised by co-fractionation of compartments with similar density (Parsons *et al.*, 2013).

Free-flow electrophoresis

Membranes of different organelles have distinct protein and lipid compositions, which results in specific surface charge signatures, which in turn can be employed for isolation by FFE. Subcellular compartments move through an electric field.

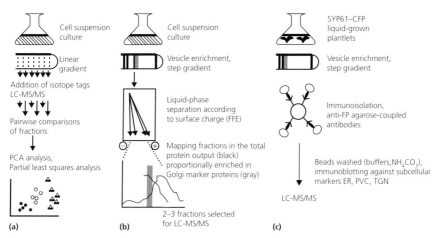

Figure 4.7 Golgi and TGN vesicle isolation and proteomic analysis techniques. (a) LOPIT analysis of the Golgi proteome based on co-clustering of proteins. (b) Free-flow electrophoresis (FFE) purification and identification of Golgi proteins based on Golgi surface charge and mass spectrometry. (c) Immunoisolation of TGN vesicles (SYP61 for example) by first enriching vesicles by density centrifugation and vesicle isolation using antibody-coupled beads specific for the bait tag. Mass spectrometry is subsequently carried out to dissect the proteins in the specific vesicle isolated. Source: Adapted from Parsons *et al.* (2013, p. 2). Reproduced by permission of the original authors.

Specifically, the process involves adding the raw cellular extracts to a chamber through which they move by laminar flow, with a perpendicular electric field separating the components of the sample. This technique has been successfully used in the isolation of tonoplast (vacuole membrane), plasma membrane, and the separation of whole organelles like the mitochondria (Parsons *et al.*, 2012).

Immunoisolation

Direct vesicle isolation is very challenging due to the large similarity of *physicochemical* properties between different vesicle populations. Further complicating the dissection of the highly dynamic vesicular network are the multidirectional movements of vesicles and the coexistence of resident proteins and transient cargo. However, recent pioneering efforts have demonstrated that isolation of TGN compartments in their native state, paired with a quantitative identification of their cargo, is possible and offers a viable and practicable approach for the dissection of endomembrane trafficking pathways. Immunoisolation can be used to separate vesicles featuring specific membrane surface-exposed target proteins (Parsons *et al.*, 2012). SNARE proteins, such as Syntaxin of Plants 61, (SYP61) localized at TGN compartments can be employed for vesicle isolation. The proteome of isolated SYP61 vesicles was characterized providing information about their cargo (Drakakaki *et al.*, 2012). The purity of the isolation is of key importance for subsequent proteomic analysis, in which contaminants from other components prevent characterization of a single organelle.

Proteomic analysis

Proteomics research on Golgi depends on the ability to isolate Golgi and its specific components. A detailed analysis of the GA proteome can be conducted using MS. Experimentally, the proteins are enzymatically digested, ionized, and separated based on their mass-to-charge ratio; peptide fragments are detected in the form of a spectra and, with the help of databases, assigned to specific proteins/peptides (Parsons *et al.*, 2012).

Based on gene annotations of model systems, such as *Arabidopsis*, we can categorize the proteins in the plant Golgi proteome by their purported function in the cell. Proteins involved in a wide variety of cellular processes have been identified in this way. For example, recent proteomic studies have identified proteins involved in endomembrane trafficking, protein modification and transport, and polysaccharide biosynthesis using gene annotations and functional predictions (Parsons et al., 2013). The breath of cellular processes is not altogether surprising, given the involvement of secretion in nearly every aspect of cellular life. Given the intimate relationship of the Golgi with other components in the cell, identifying the core resident versus transient proteins seems the most challenging part of dissecting the Golgi proteome.

There are numerous methods of capturing and analyzing the proteins in the plant Golgi. Depending on the method used, different but partially overlapping, proteomes have been established. The two most prominent methods for Golgi proteomic analysis include FFE and LOPIT. LOPIT employs MS to separate isotope tagged proteins separated by sucrose gradients. During this process, cell homogenate is separated by gradient centrifugation. Comparison of different fractions allows for determination of abundance ratios of isotope masses calculated for each protein. The assumption is that proteins located in the same compartment will have the same ratios, and thus they will cluster together (Parsons *et al.*, 2012).

To date, 452 proteins have been assigned to the Golgi and 145 to the TGN (Parsons *et al.*, 2012). The major protein categories represented by the Golgi proteome include proteins involved in sugar metabolism (specifically, glycosyltransferases involved with polysaccharide biosynthesis make up 20% of the proteome), transporters (12%), and transferases (12%). Transient proteins and contaminants also make up a sizable portion (~12% each) of the investigated proteome (Parsons *et al.*, 2012). There are also a number of proteins involved in trafficking, like GTPases, and protein modification such as glycosylation factors present. The high proportion of the plant Golgi proteome containing polysaccharide synthesis/modification proteins emphasizes the intimate connection and unique role of plant Golgi in forming and maintaining the cell wall.

Interestingly, the proteome of the TGN isolated SYP61 vesicles revealed the presence of many trafficking proteins. In addition, components of the cellulose biosynthesis machinery (e.g., CESA3 and CESA6) and other proteins involved in cell wall development were also identified in the SYP61 vesicle proteome, suggesting that SYP61 vesicles are involved in the trafficking of cell wall components (Drakakaki *et al.*, 2012). Colocalization experiments validated the results with proteins including ECHIDNA, the mutants of which show perturbed cell elongation and cell wall deficiencies, emphasizing the role of SYP61 vesicles in trafficking of cell wall biosynthetic machinery (Boutte *et al.*, 2013; McFarlane *et al.*, 2013). As a major juncture in post-Golgi trafficking, TGN proteomic research compared to Golgi, plasma membrane and apoplast proteomes could offer valuable insights into the secretory pathway and the formation of the cell wall.

Golgi genetics and genomics

Many genes with Golgi-specific functions have been identified using genomics and genetics approaches. There exists a wealth of information on plant specific proteomes, genomes, and transcriptomes, in particular for the model plant *Arabidopsis*. This in combination with forward and reverse genetics allows for the identification and characterization of Golgi proteins.

In principle, genes found using bioinformatics pipeline strategies can be studied using reverse genetics (using mutants for the corresponding genes), cell biology, and biochemistry. Through specific phenotypes and biochemical analysis, the biological role of the corresponding gene products can be deduced.

Recently, the use of forward genetics has led to the identification of new plant Golgi-associated proteins. Chemical treatment can induce random mutations in the plant genome. The mutant plants can then be screened based on the localization of fluorescent-tagged reporter Golgi-associated proteins (Faso *et al.*, 2009b; Stefano *et al.*, 2013). Outlined in the following text are several examples of genetic approaches for studying plant Golgi processes.

Cell wall biosynthetic genes

There are several gene products specific to cell wall polysaccharide biosynthesis and other cell wall-related functions that reside in Golgi, such as cellulose synthase complexes and glycosyltransferases. Proteomic analysis in combination with bioinformatics and reverse genetics has revealed the function of many cell wall glycosyltransferases involved in the biosynthesis of hemicellulose and pectins. The overall complex structure of pectin, with its highly branched structure and extensive modifications, is likely to require many genes for its biosynthesis and modification. Examples of using genetics for elucidating the function of pectin biosynthetic genes include the homogalacturonan biosynthesis gene *QUASIMODO1* (*QUA1*) and several *GALACTERONOSYL-TRANSFERASES* (*GAUTs*). *QUA1* was discovered using a mutant screen in *Arabidopsis* in which *qua1* exhibited reduced levels of homogalacturonan (Bouton *et al.*, 2002).

In contrast to pectin, many gene products involved in the biosynthesis of *Arabidopsis* hemicellulose (xyloglucan) have been characterized by reverse genetics, including the xyloglucan xylosyltransferase (XXT) mutants. Furthermore, studies using the *katamari1* (*kam1*)/*mur3* mutant revealed an interesting intersection between trafficking and cell wall polysaccharide synthesis. MUR3 encodes a galactosyltransferase, which adds a galactosyl residue to a growing hemicellulose polymer. MUR3 is allelic to KAM1 which is involved in endomembrane–actin microfilament interactions. The *kam1*/*mur3* mutant results in an altered endomembrane phenotype, whereas the lack of galactosyltransferase activity alters hemicellulose structure (Madson *et al.*, 2003). It has been suggested that KAM1 is involved in providing feedback from the cell wall to the cytoskeleton to control Golgi-mediated vesicle trafficking. Although these glycosyltranferase mutants have been characterized regarding their biochemical activity in polysaccharide biosynthesis, a detailed structural analysis of Golgi and TGN in these mutants can provide more insights on how they affect Golgi structure.

Golgi organization and structure

Searches in the plant Golgi proteome for proteins involved in Golgi organization and structure have identified plant homologues with those found in mammalian cells. For example, mammalian golgins form tethering complexes between cisternae that are responsible for the overall structure of the Golgi. Plant homologues to mammalian golgins are responsible for tethering between cisternae, tethering between Golgi and ER, and are involved in directing post-Golgi vesicle trafficking. Further, a GFP fusion to a candidate plant golgin (AtGRIP), which shares a 50% sequence identity to the mammalian version, exhibited Golgi localization in tobacco cells. However, while there are several homologous golgin complex proteins, the Golgi reassembly and stacking proteins (GRASPs) family is interestingly absent from the plant genome. In mammalian cells, GRASPs are involved in Golgi reassembly following mitosis (Osterrieder, 2012). The fact that plant Golgi do not disassemble during cell division suggests that plants either developed a new method of maintaining Golgi organization or lost the ability to disassemble.

In eukaryotic cells, there are golgins and other multisubunit proteins involved in Golgi tethering. The conserved oligomeric Golgi (COG) complex, for example, is involved with intra-Golgi transport: transport between Golgi and ER and Golgi integrity. While characterization of COG in plants is not complete, there are *Arabidopsis* homologues for each of the eight mammalian COG subunits. A mutant in COG7 was identified by the yellow coloration and aberrant morphology of *Arabidopsis* embryos and initially referred to as *embryo yellow* (*eye*). The *eye* mutation caused mislocalization of a Golgi protein, changed the size of

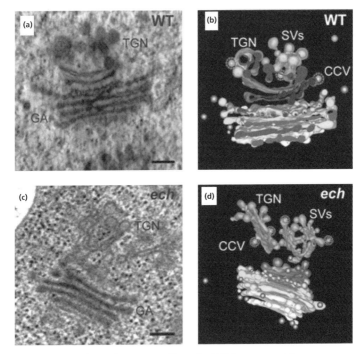

Figure 4.8 *Arabidopsis echidna* (*ech*) mutant affects the TGN structure. Electron tomography of *ech* mutant Golgi and TGN compared to wild-type (WT) Golgi (d and b, respectively). (a) and (c) Single electron micrographs of WT and *ech*, Golgi and TGN respectively. The *ech* mutant Golgi exhibits a more tubular TGN structure and fewer secretory vesicles (SVs) present compared to WT, indicating a role of ECHIDNA is post-Golgi trafficking. Source: Adapted from Boutte *et al.* (2013, p. 16263). Reproduced by permission of Proceedings of the National Academy of Sciences.

the GA and was proposed to be involved in Golgi structure maintenance (Ishikawa *et al.*, 2008).

During the past decade, studies using mutants, which show altered vesicular trafficking to and from Golgi, have provided a wealth of information on Golgi-dependent transport processes. Prominent examples are ECHIDNA, a TGN-localized protein involved in post-Golgi secretion, and MAIGO5, a COPII vesicle coat protein. ECHIDNA was identified in a mutant screen based on an aberrant hypocotyl growth phenotype and was also present in the SYP61 proteome. Recently, *echidna* (*ech*) has been associated with TGN structural defects in that it interferes with post-Golgi trafficking of auxin transporters to the plasma membrane and the trafficking of cell wall polysaccharides, particularly pectins, to the apoplast (Figure 4.8) (Boutte *et al.*, 2013; McFarlane *et al.*, 2013).

A mutation in the *Arabidopsis* Sec16 subunit of COPII, named *maigo5* (*mag5*) (which is an ortholog to the Sec16 in yeast and humans), leads to aberrant ER structures and disrupted ER export. AtSec16 was found to be vital for protein transport from ERES and COPII coat recycling between the Golgi and ER (Takagi *et al.*, 2013).

Overall, genetic approaches to identify and characterize Golgi proteins utilize various processes from fluorescence tagging in conjunction with fluorescence confocal microscopy, mutant screening, and genome searches for homologues to mammalian Golgi components. Using genomic tools to study plant Golgi constitutes a complementary approach with the potential of identifying players in the Golgi structural maintenance and function.

Chemical genomics

An alternative to random genetic mutations to perturb, study, and control protein function in the cell is chemical genomics. This approach, similar to pharmacological drug and herbicide discovery, uses exposure to small molecules to rapidly, reversibly, and importantly conditionally (dose-dependent response, location and temporal) alter the function of specific proteins. Using chemicals to study protein function rather than the traditional mutant analysis avoids potential lethality issues when gene products are essential (Drakakaki *et al.*, 2011).

Brefeldin A has been the major pharmacological diagnostic compound for Golgi-dependent trafficking. In recent screens, a new set of promising compounds affecting endomembrane trafficking in plants were identified. Several of these chemicals have demonstrated they can arrest dynamic endomembrane processes. For example, Endosidin 1 (ES1) was used to demonstrate recycling of the plasma membrane protein PIN2 to the SYP61 compartment. An exhaustive confocal microscopy-based screen identified a library of compounds affecting endomembrane trafficking *in vivo*. This screen discovered a different small molecule, Endosidin 7 (ES7) as a useful tool for studying cell plate maturation (Drakakaki *et al.*, 2011; Worden *et al.*, 2012).

Significance

The Golgi apparatus, being the hub of endomembrane trafficking particularly for the cell wall, vacuole, and other compartments, has biotechnological importance. Proteins and molecules of significance in food and biomass production and medicine are synthesized, modified, and trafficked through the Golgi (Xu *et al.*, 2012).

With the exponential population growth and globally rising living standards, the world's energy demands have grown. This has led to the exploration of alternative energy sources beyond fossil fuels including biofuels (Pauly and Keegstra, 2010). Plant cell walls constitute a virtually unlimited source of renewable biomass which can be converted into biofuel. The fact that biofuels are relatively carbon neutral, that is, not adding to the CO_2 in the atmosphere, is also an attractive aspect when considering global warming. However, the deconstruction of cell wall polysaccharides to yield fermentable monosaccharides is rather inefficient, and this is currently still a bottleneck in the development of sustainable and economically viable biofuels. Understanding cell wall degradation requires deeper insights into cell wall biosynthesis in which Golgi plays an essential role. Once better understood, the cell wall can potentially be modified to yield better biomass conversion rates and add to the economic viability of biofuels.

Further, the cell wall provides food/feed sources and protects plants from external adverse influences such as pathogens or abiotic stresses (Albersheim *et al.*, 2010). Understanding Golgi-mediated processes involved in cell wall biosynthesis can be used to design effective strategies for cell wall manipulation to enhance the productive yield and robustness of plants (Pauly and Keegstra, 2010).

Overall, plants have a vast potential as "green factories" for fuels, medicine, raw proteins, and food/feed (Pauly and Keegstra, 2010; Xu *et al.*, 2012). While each of these processes involves a variety of different pathways, they each intersect at the Golgi apparatus, where the material produced by the plant cell is modified and then transported and deposited at its final destination.

One emerging concern, however, lies in the fact that plant glycosylation patterns differ from those found in mammals. The biological significance of these differences, specifically whether they cause allergen responses in humans, is a current area of research in plant biotechnology (Gomord *et al.*, 2010). Understanding plant-based glycosylation in detail can assist in the design of desirable glycan modifications of plant-produced proteins.

By understanding the specific processes governing biosynthesis and cargo modification in the Golgi, we can better direct our efforts regarding plant improvements and utilization to various ends.

Acknowledgment

We acknowledge the NSF/IOS-1258135 grant to Georgia Drakakaki.

References

Albersheim, S.D.A., Roberts, K., Sederoff, R. and Staehelin, A. (2010) *Plant Cell Walls*, Garland Science, New York.

Aniento, F., Bernd Helms, J. and Memon, A.R. (2003) How to make a vesicle: coat protein-membrane interactions. The Golgi apparatus and the plant secretory pathway. *Annual Plant Review*, **9**, 36–54.

Bashline, L., Li, S. and Gu, Y. (2014) The trafficking of the cellulose synthase complex in higher plants. *Annals of Botany*, **114**, 1059–1067.

Bouton, S., Leboeuf, E., Mouille, G. *et al.* (2002) QUASIMODO1 encodes a putative membrane-bound glycosyltransferase required for normal pectin synthesis and cell adhesion in *Arabidopsis. The Plant Cell*, **14**, 2577–2590.

Boutte, Y., Jonsson, K., Mcfarlane, H.E., *et al.* (2013) ECHIDNA-mediated post-Golgi trafficking of auxin carriers for differential cell elongation. *Proceedings of the National Academy of Sciences of the United States of America*, **110**, 16259–16264.

Brandizzi, F. and Barlowe, C. (2013) Organization of the ER-Golgi interface for membrane traffic control. *Molecular Cell Biology*, **14**, 382–392.

Brandizzi, F. and Wasteneys, G.O. (2013) Cytoskeleton-dependent endomembrane organization in plant cells: an emerging role for microtubules. *The Plant Journal*, **75**, 339–349.

Caffall, K.H. and Mohnen, D. (2009) The structure, function, and biosynthesis of plant cell wall polysaccharides. *Carbohydrate Research*, **344**, 1879–1900.

Drakakaki, G., Robert, S., Szatmari, A.M. *et al.* (2011) Clusters of bioactive compounds target dynamic endomembrane networks *in vivo*. *Proceedings of the National Academy of Sciences of the United States of America*, **108**, 17850–17855.

Drakakaki, G., Van de Ven, W. *et al.* (2012) Isolation and proteomic analysis of the SYP61 compartment reveal its role in exocytic trafficking in *Arabidopsis*. *Cell Research*, **22**, 413–424.

Driouich, A., Follet-Gueye, M.L., Bernard, S. *et al.* (2012) Golgi-mediated synthesis and secretion of matrix polysaccharides of the primary cell wall of higher plants. *Frontiers in Plant Science*, **3**, 79.

Egea, G., Serra-Peinado, C., Salcedo-Sicilia, L. and Gutierrez-Martinez, E. (2013) Actin acting at the Golgi. *Histochemistry and Cell Biology*, **140**, 347–360.

Faso, C., Chen, Y.N., Tamura, K. *et al.* (2009a) A missense mutation in the *Arabidopsis* COPII coat protein Sec24A induces the formation of clusters of the endoplasmic reticulum and Golgi apparatus. *The Plant Cell*, **21**, 3655–3671.

Faso, C.B.A., Boulaflous, A. and Brandizzi, F. (2009b) The plant Golgi apparatus: last 10 years of answered and open questions. *FEBS Letters*, **583**, 3752–3757.

Glick, B.S. (2000) Organization of the Golgi apparatus. *Current Opinion in Plant Biology*, **12**, 450–456.

Gomord, V., Fitchette, A.-C., Menu-Bouaouiche, L., Saint-Jore-Dupas, C., Plasson, C., Michaud, D. and Faye, L. (2010) Plant-specific glycosylation patterns in the context of therapeutic protein production. *Plant Biotechnology Journal* **8**: 564–587.

Han, H.M., Bouchet-Marquis, C., Huebinger, J. and Grabenbauer, M. (2013) Golgi apparatus analyzed by cryo-electron microscopy. *Histochemical Cell Biology*, **140**, 369–381.

Hawes, C., Saint-Jore, C. and Brandizzi, F. (2003) The Golgi apparatus and the plant secretory pathway, *Annual Plant Review*, **9**, 63–73.

Ishikawa, T., Machida, C., Yoshioka, Y., Ueda, T., Nakano, A. and Machida, Y. (2008) EMBRYO YELLOW gene, encoding a subunit of the conserved oligomeric Golgi complex, is required for appropriate cell expansion and meristem organization in *Arabidopsis thaliana*. *Genes Cells*, **13** (6), 521–535.

Kang, B.-H. (2010) Electron microscopy and high-pressure freezing of *Arabidopsis*. *Methods Cell Biology*, **96**, 259–283.

Kang, B.-H., Nielsen, E., Preuss, M.L., Mastronarde, D. and Staehelin, L.A. (2011) Electron tomography of RabA4b- and PI-4Kβ1-labeled trans Golgi network compartments in *Arabidopsis*. *Traffic*, **12**, 313–329.

Lerich, A., Hillmer, S., Langhans, M., Scheuring, D., Van Bentum, P. and Robinson, D.G. (2012) ER import sites and their relationship to ER exit sites: a new model for bidirectional ER-Golgi transport in higher plants. *Frontiers in Plant Science*, **3**, 143.

Madson, M., Dunana, C., Li, X., Verma, R., Vanzin, G.F., Caplan, J. and Rieter, W.D., (2003) The *MUR3* gene of *Arabidopsis* encodes a xyloglucan galactosyltransferase that is evolutionarily related to animal exostosins. *The Plant Cell*, **15**, 1662–1670.

Marti, L.F.S., Renna, L., Stefano, G. and Brandizzi, F. (2010) COPII-mediated traffic in plants. *Trends in Plant Science*, **15**, 522–528.

McFarlane, H. E., Watanabe, Y., Gendre, D. *et al.* (2013) Cell wall polysaccharides are mislocalized to the vacuole in echidna mutants. *Plant Cell Physiology*, **54**, 1867–1880.

Morré, D.J. and Mollenhauer, H.H. (2009) *The Golgi Apparatus: The First 100 Years*, Springer, New York.

Nebenführ, A. (2003) Intra-Golgi transport: escalator or bucket brigade? The Golgi apparatus and the plant secretory pathway. *Annual Plant Review*, **9**, 76–87.

Oikawa, A., Lund, C.H., Sakuragi, Y. and Scheller, H.V. (2013) Golgi-localized enzyme complexes for plant cell wall biosynthesis. *Trends in Plant Science*, **18**, 49–58.

Osterrieder, A. (2012) Tales of tethers and tentacles: golgins in plants. *Journal of Microscopy*, **247**, 68–77.

Parsons, H.T. I., J., Ito, J., Park, E. *et al.* (2012) The current state of the Golgi proteomes. In: Heazlewood, J. (ed.), *Proteomic Applications in Biology*, InTech. Available at: http://www.intechopen.com/books/proteomic-applications-in-biology/the-current-state-of-the-golgi-proteomes (accessed May 5, 2016).

Parsons, H.T.D., Drakakaki, G. and Heazlewood, J.L. (2013) Proteomic dissection of the *Arabidopsis* Golgi and trans-Golgi network. *Frontiers in Plant Science*, **3**, 1–7.

Pauly, M. and Keegstra, K. (2010) Plant cell wall polymers as precursors for biofuels. *Current Opinion in Plant Biology*, **13**, 305–312.

Pavelka, M. and Robinson, D.G. (2003) The Golgi apparatus in mammalian and higher plant cells: a comparison. The Golgi apparatus and the plant secretory pathway. *Annual Plant Review*, **9**, 16–31.

Scheller, H.V. and Ulvskov, P. (2010) Hemicelluloses. *Annual Review of Plant Biology*, **61**, 263–289.

Sparkes, I.A., Ketelaar, T., de Ruijter, N.C. and Hawes, C. (2009) Grab a Golgi: laser trapping of Golgi bodies reveals *in vivo* interactions with the endoplasmic reticulum. *Traffic*, **10**, 567–571.

Stefano, G.O., A., Hawes, C. and Brandizzi, F. (2013) Endomembrane and Golgi traffick in plant cells. *Methods Cell Biology*, **118**, 69–83.

Takagi, J., Renna, L., Takahashi, H. *et al.* (2013) MAIGO5 functions in protein export from Golgi-associated endoplasmic reticulum exit sites in *Arabidopsis*. *The Plant Cell*, **25**, 4658–4675.

Ungar, D. (2009) Golgi linked protein glycosylation and associated diseases. *Seminar in Cell Development and Biology*, **20**, 762–769.

Worden, N., Park, E. and Drakakaki, G. (2012) Trans-Golgi network—an intersection of trafficking cell wall components. *Journal of Integrative Plant Biology*, **54**, 875–886.

Xu, J., Dolan, M.C., Medrano, G., Cramer, C.L. and Weathers, P.J. (2012) Green factory: plants as bioproduction platforms for recombinant proteins. *Biotechnological Advances*, **30**, 1171–1184.

Further reading

Grangeon, R., Agbeci, M., Chen, J., Grondin, G., Zheng, H. and Laliberté, J. (2012) Impact on the endoplasmic reticulum and Golgi apparatus of turnip mosaic virus infection. *Journal of Virology*, **86** (17), 9255–9265.

Kelly, E.E., Giordano, F., Horgan, C.P., Jollivet, F., Raposo, G. and McCaffrey, M.W. (2012) Rab30 is required for the morphological integrity of the Golgi apparatus. *Biology of Cell*, **104**, 84–101.

Pantazopoulou, A. (2016) The Golgi apparatus: insights from filamentous fungi. *Mycologia*, **108** (3), 603–622.

Rios, R.M. (2014) The centrosome–Golgi apparatus nexus. *Philosophical Transactions of the Royal Society B*, **369**, 1650, DOI:10.1098/rstb.2013.0462.

Wilson, C., Venditti, R., Rega, L.R., Colanzi, A., D'Angelo, G. and De Matteis, M.A. (2011) The Golgi apparatus: an organelle with multiple complex functions. *Biochemical Journal*, **433** (1), 1–9.

Xu, H., Su, W., Cai, M., Jiang, J., Zeng, X. and H. Wang. (2013) The asymmetrical structure of Golgi apparatus membranes revealed by *in situ* atomic force microscope. *PLoS ONE*, **8** (4), e61596, doi:10.1371/journal.pone.0061596.

CHAPTER 5

Microbodies

Robert Donaldson

Department of Biological Sciences, George Washington University, Washington, DC, USA

Introducing peroxisomes

Oxidases that use O_2 to take hydrogen from substrates such as glycolate or fatty acids produce hydrogen peroxide (H_2O_2). H_2O_2 is a dangerous molecule that can damage proteins, lipids and nucleic acids. **Catalases** are responsible for removing the H_2O_2 as it is produced. Thus, housing oxidases along with catalase in a membrane enclosure is a strategy that evolved in eukaryotic cells. This type of enclosure was christened by de Duve as the **peroxisome** (de Duve *et al.*, 1960). Peroxisomes are found in all eukaryotes, plants, animals, yeasts, fungi, etc.

Peroxisomal oxidases are alternatives to mitochondrial dehydrogenases. In the mitochondria the electrons and protons (hydrogens) contribute to the electron transport chain and the generation of adenosine triphosphate (ATP) and do not combine with oxygen until much of the energy has been recovered. However, the direct transfer of hydrogens to oxygen in peroxisomes results in energy being discharged as heat. This allows metabolic processes to take place without necessarily being coupled to ATP generation. That is, a cell (or tissue) can conduct peroxisomal oxidations even if it has minimal energy needs for its own purposes. This will become apparent in the case of peroxisomal metabolism (photorespiration) in photosynthetic cells and in the oil-storing cells of seeds and pollen (fatty acid β-oxidation).

Peroxisomes were first recognized as biochemical units associated with microscopically visible structures (**microbodies**) in cells in the late 1960s when biochemists realized that coherent groups of enzymes, which comprised metabolic sequences or pathways, behaved as physical entities (de Duve *et al.*, 1960). Experimentally, these entities moved together in centrifugation of tissue lysates. The experiments depended on techniques that broke open the cells in a tissue (leaves, germinating seedlings, roots) in ways that did not disrupt the membrane enclosed organelles. Thus, plant biochemistry labs in the 1970 came to resemble juice bars. The plant tissue could be chopped with razor blades or with brief

Plant Cells and their Organelles, First Edition. Edited by William V. Dashek and Gurbachan S. Miglani.
© 2017 John Wiley & Sons, Ltd. Published 2017 by John Wiley & Sons, Ltd.

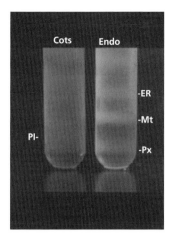

Figure 5.1 Density gradient separation of organelles. Peroxisomes (Px), plastids (Pl), mitochondria (Mt) and ER were separated from homogenates of castor bean (*Ricinus communis*) cotyledon (Cots) or endosperm (Endo) by centrifugation through a sucrose gradient, 60% (w/w) from the bottom to 20% (w/w) at the top. Peroxisomes equilibrate at a density of approximately 1.25 g/ml in the centrifugal field (Donaldson, unpublished).

pulses of a blender, resulting in a green juice from leaves or a creamy juice from sprouting seeds. The buffer medium was designed to protect the integrity of the membrane enclosures and would consist of isosmotic sucrose or mannitol and biocompatible buffer ions and salts. The organelles could then be separated from each other based on their size, mass or density by centrifugation of the lysate. Larger, more massive organelles such as nuclei or starch grains would sediment faster. Mitochondria and chloroplasts required more centrifugal force and time to collect as a pellet at the bottom of a centrifuge tube, while smaller things such as membrane fragments of the endoplasmic reticulum or ER (microsomes) needed even greater forces to collect. Peroxisomes are in the same size range as mitochondria and chloroplasts, about 1 μm in diameter. It was discovered that peroxisomes could be separated from other organelles by centrifugation through a density gradient of sucrose solutions because the peroxisomes were dense (Figure 5.1). Once the peroxisomes were resolved from the other organelles their metabolic capacities could be more thoroughly addressed, and they could be identified as visible structures in cells. A specific stain for catalase activity allowed the peroxisomes to be identified as the **microbodies** seen in electron micrographs (Frederick and Newcomb, 1969a, 1969b).

Leaf peroxisomes

Investigators working on photosynthetic carbon metabolism soon found that enzymes associated with photorespiration were clustered in peroxisomes isolated from spinach. This included glycolate oxidase, which consumes oxygen and produces H_2O_2 and glyoxylate. Catalase was logically there to take care of the H_2O_2. Also present was an aminotransferase that converted the glyoxylate to glycine. The enzymes that processed the glycine were found in the mitochondria where two molecules of glycine are combined to produce a molecule of serine,

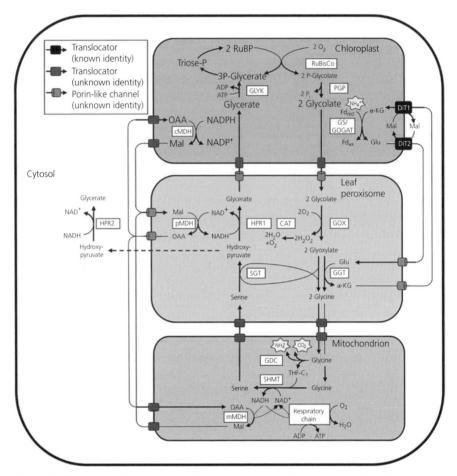

Figure 5.2 Photorespiration is mediated by peroxisomes working between chloroplasts and mitochondria. It begins with RuBisCo taking in O_2 in place of CO_2 resulting in phospho-glycolate which is dephosphorylated by a phosphatase (PGP) and becomes the substrate for the peroxisomal glycolate oxidase (GO). The glyoxylate produced is converted to glycine by either a serine glyoxylate aminotransferase (SGT) or a glutamate glyoxylate aminotransferase (GGT). Pairs of glycines are converted to serine in the mitochondria by glycine decarboxylase (GDC) and serine hydroxymethyl transferase (SHMT). In the peroxisome serine is deamidated by the SGT, and the resulting hydroxyl pyruvate goes on to glycerate via the hydroxypyruvate reductase (HPR1). Glycerate is returned to the reductive pentose cycle in the chloroplast by the glycerate kinase (GLYK). Organelle-specific isoforms of malate dehydrogenase (cMDH, pMDH, mMDH) support the exchange of reducing equivalents (NADH, NADPH). Source: Adapted from Hu *et al.* (2012), used with permission.

releasing CO_2 and NH_3. The serine is converted to glycerate by peroxisomal enzymes. It took some time to understand where the glycolate was coming from. It was known to be a by-product of the chloroplasts' carbon cycle and eventually glycolate was found to be a direct product of the ribulose bisphosphate oxygenase (Figure 5.2).

In a simple sense this photorespiratory process appears to be a distraction from photosynthesis. It is counterproductive since O_2 is incorporated in place of CO_2; and to make matters worse, CO_2 is being released; it is photorespiration. However, it allows a plant to close its stomata to limit water loss and to deal with the consequences of diminished CO_2 concentrations and elevated O_2 within the leaf tissue. Thus peroxisomal photorespiration is an adaptation for plants' water use efficiency. The plant can close its stomata and the CO_2 being released can be used by the chloroplasts as they continue to use light to drive the electron transport activities. The ATP and NADPH produced by the light reactions are used to convert the recycled CO_2 to sugars. When the stomata are closed to the extent that the rate of CO_2 being released is equal to or less than the rate of CO_2 being incorporated, there will be no net accumulation of sugar.

The biochemical characterization of peroxisomes made it possible for microscopists to visualize the organelles in the crowded cytoplasm of plant cells. A histochemical stain was devised to visualize catalase (De Duve *et al.*, 1960). Antibodies made to recognize peroxisomal enzymes could be tagged with gold particles that were visible in electron micrographs. Thus, organelles that microscopists had originally characterized as **microbodies**, for lack of a functional definition, were now recognized as **peroxsiomes**. What was particularly interesting about the micrographs was the appearance of catalase in crystalline structures within some peroxisomes (http://www.cellimagelibrary.org/images/35989). The mystery remains as to whether these crystals have functional significance or if they are simply a consequence of the concentrations of the protein that accumulate in the organelles.

Peroxisomes in oil seeds and pollen

Peroxisomal fatty acid oxidation

Seeds and pollen often contain oil (triglycerides) in addition to starch as a stored energy source to be used to fuel growth and development during germination. Starch in seeds is easy to deal with as the sugars released are readily translocated from cotyledons or endosperm into the expanding tissues of the seedling. Triglycerides contain more energy than starch, but they are not transportable in the plant vascular system; plants lack the circulating lipoproteins found in animals. Thus, plants convert stored triglycerides to sugars for export. This conversion begins with the hydrolysis of triglycerides in the oil bodies triglycerides to release fatty acids, followed by the oxidation of the fatty acids to yield acetyl-CoA, which is then converted to 4-carbon dicarboxylic acids. These dicarboxylic acids are used to create hexoses and transportable disaccharides (Kaur *et al.*, 2009).

This creation of sugars from triglycerides is like the alchemy of transforming lead into gold. The metabolic magic that creates the sugars involves four different cellular compartments: oil bodies, glyoxysomes, cytosol and mitochondria. The oils are stored as triglycerides in oil bodies in cotyledons, endosperm or pollen. These oil bodies are discrete organelles surrounded by a membrane, a monolayer of phosphoglyceride. A bilayer is not necessary because the fatty acid tails of the phosphoglycerides can intermingle with the hydrophobic triglycerides within the oil bodies. The membrane monolayer also includes an oil body-specific protein, an oleosin which may function to stabilize the oil bodies during dehydration of the seed or pollen. During rehydration lipases embedded in the oil body membrane hydrolyze the triglycerides within and release the fatty acids.

The fatty acids emanating from the oil bodies enter glyoxysomes, which are specialized peroxisomes that house fatty acid oxidizing (β-oxidation) enzymes along with glyoxylate cycle enzymes to accomplish the conversion to 4-carbon dicarboxylic acids. The fatty acids are brought into the glyoxysomes by an ABC transport protein in the glyoxysomal membrane, perhaps facilitated by physical contact between the two organelles. Once inside the glyoxysomes the fatty acids are connected to Coenzyme-A by acyl-activating enzymes that require ATP. The glyoxysomal membrane also has transport proteins specifically designed to bring in ATP from the cytosol since glyoxysomes and peroxisomes do not have the capacity to generate ATP. Quite the contrary, glyoxysomal and peroxisomal metabolisms bypass ATP generation as we will see in the next event. The resulting acyl-CoAs are subjected to oxidases, which are typical peroxisomal flavin oxidases, that transfer H's directly to oxygen via FAD. In animals most fatty acid oxidation takes place in mitochondria where the acyl-CoAs are processed by dehydrogenases that transfer electrons directly into the mitochondrial electron transport chain leading to ATP generation. The peroxisomal oxidases skip that and the energy is dissipated as heat. Since this process depends on molecular O_2 oil-storing seeds cannot germinate anaerobically unlike starch-storing seeds such as rice.

There are multiple Acyl-CoA oxidases within glyoxysomes to accommodate the different types of fatty acids encountered, different numbers of carbons (C16–C22) and double bonds. These form multienzyme complexes with the other proteins needed to complete the β-oxidation process and the release of Acetyl-CoA units (two carbons) from the fatty acids. The oxidase creates a temporary double bond in the fatty acid which is then hydrated followed by a dehydrogenase that produces NADH and thiolysis by CoA that yields an Acetyl-CoA. Momentarily, we will consider the fate of this NADH along with the NADH produced in the subsequent glyoxylate cycle.

The glyoxylate cycle

The conversion of fatty acid to carbohydrate involves a special type of peroxisome, the glyoxoysome (Figure 5.3). The glyoxylate cycle enzymes in the glyoxysomes in oil-metabolizing cells (in seeds, pollen) combine pairs of acetyl-CoA

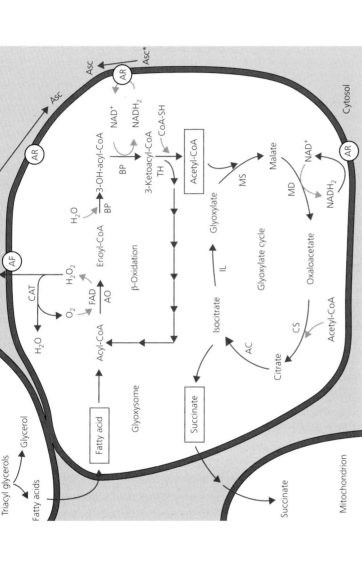

Figure 5.3 β-Oxidation, glyoxylate cycle and antioxidant enzymes in glyoxysomes. The β-oxidation enzymes are acyl-CoA oxidase (AO), enoyl-CoA hydratase combined with 3-hydroxy acyl-CoA dehydrogenase in the bifunctional protein (BP) and 3-ketoacyl-CoA thiolase (TH). Glyoxylate cycle enzymes are isocitrate lyase (IL), malate synthase (MS), citrate synthase (CS), aconitase (AC) and malate dehydrogenase (MD). The glyoxylate cycle uses the acetyl-CoA from β-oxidation to construct succinate which is eventually converted to hexose and exportable sucrose through the mitochondria and cytosol. Membrane enzymes include ascorbate peroxidase (AP) and ascorbate free radical reductase (AR). Both catalase (CAT) and AP consume hydrogen peroxide. AP and AR work together to consume the NADH along with the H_2O_2 produced by β-oxidation. Asc, ascorbate; Asc·, ascorbate free radical. Source: Reproduced with permission of Donaldson *et al.*, (2008).

units to produce 4-carbon dicarboxylic acids (oxaloacetate, malate and succinate). Citrate synthase combines an acetate with oxaloacetate to make citrate (C6), which is converted to isocitrate by aconitase out in the cytosol. This means that the citrate needs to exit the glyoxysome and the isocitrate returns. Isocitrate lyase releases succinate (C4) and glyoxylate (C2) from the citrate. Malate synthase combines the glyoxylate with another acetate, and the resulting malate is either converted to oxaloacetate by a glyoxysomal NAD-malate dehydrogenase or a cytosolic dehydrogenase, the consequence being the production of NADH in the glyoxysome or outside in the cytosol. NADH in the glyoxysome can be used to support the detoxification of the H_2O_2 as discussed later. The glyoxysomal membrane has an electron transport system that has the capacity to oxidize the NADH at the rate that it is produced by β-oxidation (Alani *et al.*, 1990; Donaldson and Fang, 1987; Luster and Donaldson, 1987, Luster *et al.*, 1988). Otherwise, to allow the oxidation of fatty acids and the glyoxylate cycle to continue the NADH produced needs to be dispatched in the cytosol or the mitochondria. Malate can carry the reducing equivalents out into the cytosol and into mitochondria. If the cells do not require as much ATP as might be produced by the mitochondria, there are ways that mitochondria can dissipate the energy (through the alternative oxidase discussed in Chapter 7). All of this implies a certain amount of metabolic traffic through the membrane enclosing the organelle (Kunze and Hartig, 2013).

Peroxisomes in other plant cell types conduct β-oxidation-like processes. This includes the synthesis of jasmonates, which are regulatory substances involved in growth, development and defences. Also, included are the oxidations of polyamines and branched chain amino acids (isoleucine, leucine and valine) as mentioned earlier.

Nitrogen metabolism in peroxisomes

A variety of nitrogen-containing compounds are subject to peroxisomal metabolism including ureides, polyamines and branched chain amino acids. Urate metabolism is especially important in legumes where ureides produced in root nodules transport nitrogen to leaves and developing seeds. This process begins with uricase that oxidizes urate, produces H_2O_2 and ultimately leads to the conversion of the urate to allantoin. The polyamines, spermine, spermidine and putrescine can all be metabolized by peroxisomal oxidases and converted to γ-aminobutyric acid (GABA). Branched-chain amino acids (isoleucine, leucine and valine) are also catabolized in peroxisomes by β-oxidation-type activities.

Defence molecules from peroxisomes

Plants respond to pathogens and chewing insects by producing signaling substances and protective molecules, some of which are derived from peroxisomes. A fascinating observation is the clustering of peroxisomes around fungal penetration sites (Lipka *et al.*, 2005) along with the report that photorespiratory enzymes are upregulated by and confer resistance to fungal infection. The H_2O_2 produced by the peroxisomal glycolate oxidase appears to provide protection from pathogens. Also, peroxisomes house PEN proteins, which function as glycosyl and glucosinolate hydrolases releasing compounds that inhibit fungi or that function as signals for deposition of callose, an amorphous glucan that serves as a barrier to pathogen penetration and as a matrix for protective substances.

Jasmonates (jasmonic acid, methyl jasmonate and related oxylipins) are produced by plants in response to wounding and fungal infection. They are derived from unsaturated fatty acids such as linolenate (18:3) by a process that is initiated by lipoxygenase in the chloroplasts. Eventually, peroxisomal β-oxidation completes the synthesis. Jasmonates are volatile regulatory substances that induce pathogen resistance in remote uninfected tissues by inducing the expression of defence-related genes (e.g., genes coding for protease inhibitors that interfere with insect digestive enzymes). Jasmonates also have regulatory roles in flowering, seed germination and root growth.

Antioxidant processes in peroxisomes

Hydrogen peroxide produced in peroxisomes has the potential to be detrimental to proteins, lipids and nucleic acids. As we have seen, many of the processes that take place in peroxisomes generate H_2O_2 including photorespiration, fatty acid metabolism, ureide metabolism and jasmonate synthesis. Thus, proteins contained in peroxisomes and bathed in H_2O_2 would seem to be especially vulnerable. Peroxisomal proteins can be oxidized by H_2O_2 exposure *in vitro* or *in vivo* and mass spectroscopy analysis of the proteins shows that particular regions of the proteins are subject to oxidation (Anand *et al.*, 2009). Some peroxisomal proteins are more vulnerable than others (Nguyen and Donaldson, 2005). For example, isocitrate lyase (a glyoxylate cycle enzyme) is particularly sensitive to H_2O_2, as is its product, glyoxylate (Yanik and Donaldson, 2005), while the peroxisomal malate dehydrogenase is less sensitive. The isocitrate lyase activity is protected by a physical association of the protein with catalase.

Catalases

Peroxisomes house a variety of antioxidant defences beginning with catalase, the tetrameric heme-protein that converts two molecules of H_2O_2 to two H_2O molecules and one O_2 molecule (Mhamdi *et al.*, 2012). Thus it recycles half of the oxygen consumed by the oxidases. Catalase has a very high turnover number, that it processes H_2O_2 very rapidly. However, it has a low affinity for H_2O_2 and is not so effective at low concentrations of H_2O_2. Not surprisingly, catalase is one of the most abundant enzymes in peroxisomes and, in some cases, accumulates to the extent that it forms crystals within the peroxisomes. Plants are capable of producing different isoforms of the catalase protein that are expressed under different conditions. *Arabidopsis*, for example has three different isoforms of catalase, CAT2 associated with photorespiration in leaf mesophyll cells and imbibed seed, CAT1 associated with β-oxidation in pollen but not in imbibed seed and CAT3 expressed in senescent leaves. Interestingly, plant catalases often have a calmodulin binding domain which means that the activity can be regulated by Ca^{++} concentration. Another interesting capability for catalase is its peroxidative action on substrates such as alcohols. Plant tissues produce ethanol under hypoxic conditions and that ethanol can be metabolized by catalase or alcohol dehydrogenase when oxygen becomes available.

Ascorbate peroxidases

Ascorbate peroxidases (APs) use ascorbate to detoxify H_2O_2, producing H_2O and monodehydroascorbate (Figure 5.3). APs have a higher affinity for H_2O_2 than catalase and thus can scavenge H_2O_2 at concentrations lower than catalase. The monodehydroascorbate is converted back to ascorbate by a monodehydroascorbate reductase (AR) which uses NADH as a source of reductant. This is one way in which the NADH produced by peroxisomal dehydrogenases can be reoxidized. Conveniently AP and AR are associated with the peroxisomal membrane, an appropriate location for a system that can detoxify H_2O_2 that escapes catalase (Karyotou and Donaldson, 2005)

Ascorbate may also be oxidized to dehydroascorbate in the presence of H_2O_2 or superoxide ($O_2^{\bullet-}$). Peroxisomal dehydroascorbate reductase then uses glutathione to revive the ascorbate and the oxidized glutathione re-reduced by NADPH-glutathione reductase. NADPH can be generated by an NADP-isocitrate dehydrogenase within the organelle. Thus, the ascorbate- and glutathione-dependent antioxidant systems are sustained by the metabolic dehydrogenases within the peroxisomes (del Rio *et al.*, 2006).

Other antioxidant proteins

Other antioxidant proteins in peroxisomes include superoxide dismutase (SOD), peroxiredoxins and thioredoxins. SOD converts superoxide to H_2O_2. Peroxiredoxins can be oxidized by H_2O_2 and re-reduced by thioredoxins.

Peroxisomal movements

Peroxisomes move around in cells as do chloroplasts and mitochondria. Plant peroxisomes move by being attached to myosin motors which then pull the organelles along actin filaments powered by ATP. This is in contrast to animal peroxisomes that use dynein or kinesin motors to move on microtubules. The same myosin motors that move peroxisomes also move other organelles. The myosin motors apparently have specific sites for tethering each type of organelle. Also, there is a particular protein on the peroxisomal membrane that attaches it to the myosin. There are several different myosins that are capable of propelling peroxisomes including myosins XI-2, XI-I, XI-E, and XI-K in *Arabidopsis*.

Plant peroxisomes move faster on their actin tracks around 5 µm/s when compared to animal cell peroxisomes that move on microtubules around 1 µm/s. This was observed in living cells using a peroxisomal protein tagged with a yellow or green fluorescent protein expressed in *Arabidopsis* (epidermal, root hair, trichomes) or onion epidermal cells (Jedd and Chua, 2002; Mathur *et al.*, 2002; Figure 5.2). The details of these movements are fascinating. Peroxisomes tend to gather at a particular location in a cell and then move off in different directions at different rates. Some peroxisomes remain immobile, while other move different distances (20–70 µm) in a series of jumps. They can change directions and velocities independently. All of these observations indicate that peroxisomes are not casually swept along by cytoplasmic streaming.

Why do peroxisomes – and other organelles – move around in cells? One possibility is metabolic mixing, to move the organelles among sources of substrates and to allow them to distribute their metabolic products throughout the cytoplasm. Peroxisomal metabolism involves interactions with chloroplasts and mitochondria, and these interactions could be transient. Also, peroxisomes import their protein contents from the cytosol and the movements may facilitate the import process. Peroxisomes would also need to be dispersed after undergoing division, fission. Of course, cell division involves distributing organelles into progeny cells, which would involve moving the organelles to appropriate locations prior to cytokinesis.

Peroxisomal division

As cells divide so must their organelles. A peroxisome divides by first elongating, then its membrane is constricted and then fission occurs. One of the peroxisomal membrane proteins, PEROXIN-11 (PEX11), is responsible for the initial elongation phase. PEX11 homologs are also found in yeast and animals. There are multiple forms of this protein in plants, five in *Arabidopsis*, for example. Some PEX11 proteins are embedded in the membrane such that both the N- and C-terminal ends are exposed to the cytosol, while in others the C-terminus extends into the interior matrix of the organelle. These multiple PEX11 proteins are apparently functionally redundant since mutations or diminished expression of any one of these does not have obvious consequences.

The constriction of a peroxisome prior to fission is brought about by Fission-1 (FIS1) and dynamin-related proteins (DRPs). FIS1 proteins, two isoforms in *Arabidopsis*, are present on both mitochondrial and peroxisomes, anchored in the membranes by the C-terminal end. FIS1 proteins attach to DRPs which form a ring around the elongated peroxisome. DRPs exert GTPase activity as they tighten around the site of constriction.

Cyanidioschyzon merolae

This is a unicellular red alga that provides a dramatic visual record of the process of peroxisomal fission (Imoto *et al.*, 2013). *Cyanidioschyzon merolae* has only one chloroplast, one mitochondrion, a nucleus and one peroxisome. The organelles undergo fission in that sequence. The genome of *C. merolae* codes for only two dynamin proteins (Dnm1 and Dnm2), Dnm2 being responsible for chloroplast fission. Dnm1 resides in the cytosol during the G1 phase, interphase, of the cell cycle. During the process of cell division, Dnm1 molecules are first recruited to the mitochondrion and participate in its fission. Then, Dnm1 molecules migrate to the peroxisome to form a constriction ring. Meanwhile there is a filamentous assemblage inside the organelle at the site of constriction that somehow works together with the dynamin to complete the fission of the peroxisome. There are no FIS1 homologs in *C. merolae*. The remaining mystery in peroxisomal fission is what molecules identify the constriction site.

Peroxisomal growth

The enlargement and division of peroxisomes requires that they acquire additional proteins and membrane material. Membrane lipids can be supplied by the ER (Donaldson, 1976). Peroxisomes lack DNA and ribosomes so that all of their protein must be imported from the outside. Since ribosomes are not associated with the organelle, protein import is post translational rather than co-translational. As is the case for the import of proteins into any particular organelle, the proteins have specific targeting information among their amino acid sequences. Since the targeting of proteins to peroxisomes is post translational the information could be located anywhere in the sequence, at N-terminus, the C-terminus or in between, but it is often at the C-terminus. A peculiar circumstance is that peroxisomal proteins are translocated into the organelle, through the membrane, in a folded state. This being the case requires that the targeting information, wherever it is located in the protein structure, must be accessible to the receptor protein that initiates the process of delivery from the cytosol into the organelle (Kaur *et al.*, 2009; Lanyon-Hogg *et al.*, 2010).

Since peroxisomal proteins are imported fully folded the appropriate cofactors are most likely included in the structures. For example, each monomer of the catalase tetramer contains a heme-iron cofactor. The oxidases contain flavin adenine dinucleotide (FAD). Other cofactors that transiently interact with the enzymes such as ATP and NAD^+ are probably imported into peroxisomes separate from proteins.

Peroxisomal targeting signals

Peroxisomal targeting signals (PTS) are sequences of particular amino acids in the newly synthesized peroxisomal proteins that provide the information for delivery from the cytosol into the peroxisomes. The most common sequence, the PTS1 sequence, is located at the C-terminus of the protein and consists of the three residues, such as serine, lysine and leucine (−SKL) or the common alternative, alanine, arginine and methionine (−ARM). Thus the PTS1 consists of a small side chain, a positively charged side chain and a hydrophobic side chain with a free carboxyl group that is negatively charged. There is considerable freedom in these choices; the S may be replaced by A, P, C, G or T; the R may be

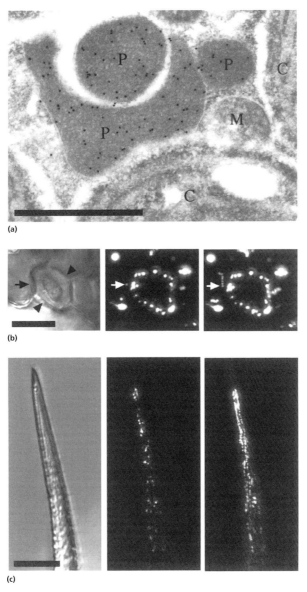

(a)

(b)

(c)

Figure 5.4 Peroxisomal movements. (a) GFP-PTS1 protein is seen in leaf mesophyll cell peroxisomes, P, by immunoelectron microscopy using anti-GFP antibodies with 10-nm gold particles. The GFP-PTS1 protein is not found in mitochondria, M, or chloroplasts, C. 1 μm bar. (b) Peroxisomes move in mesophyll cells but not in leaf guard cells. The left panel is a visible light microscope image of guard cells surrounded by mesophyll cells. The center panel is a fluorescent microscope image of GFP-PTS1 in the guard cells, and the arrow points to a GFP-peroxisome in a mesophyll cell. The right panel is a composite of seven consecutive images of that GFP-PTS1 peroxisome taken at 1 s intervals showing the movement of that peroxisome. The bar is 25 μm. (c) Peroxisomes in a trichome. The right image is a composite of seven fluorescence images taken 1 s apart, while the middle panel is one of the series, and on the left a visible light image. Peroxisomes are propelled by myosin motors on actin tracks. Source: Adapted from Jedd and Chua (2002), used with permission of *Oxford Journals*.

replaced by K, N, S, M, H or L; and the L may be replaced by M, I, V, F or Y. This short sequence is sufficient to direct a protein from the cytosol into the peroxisomes. However, the efficiency of import can be affected by other nearby residues. The experiments that demonstrated this involved engineering green fluorescent tagged proteins (GFPs) with and without a PTS1 sequence and using fluorescence microscopy to see the distribution of the proteins in the cytosol and peroxisomes (Figure 5.4). Catalase identifies the peroxisomes. The PTS1 sequence is conserved in all eukaryotes, animals, fungi, yeasts and plants, indicating that peroxisomes were an early innovation in eukaryotic evolution.

The PTS's of catalases deviate from the canonical PTS1. The C-terminus of a typical plant catalase would consist of LASRLNVRPSM. Experiments have shown that many of these residues are important for import including the L's, S's, R's, P and M. The efficiency of catalase import is weaker than most other peroxisomal protein. Perhaps because it is so abundant, it is designed to minimize competition with the import of other PTS1 proteins.

Less often proteins destined for peroxisomes have a PTS2 near the N-terminus, usually within the first 30 amino acids. Typically this looks like RLXXXXXHL with some variation of the L's and X's often being basic but rarely acidic. The PTS2 is often cleaved after import, in contrast to the PTS1's which are not removed after import.

Receptors for peroxisomal protein import

Peroxisomal proteins are initially synthesized on ribosomes in the cytosol. The import process depends on a host of proteins known collectively as peroxins (Pex). These were identified as mutations in genes that were necessary for peroxisomal biogenesis in yeast. A defect in any one of the 26 *PEX* genes resulted in an absence of peroxisomes. Many of the Pex proteins, which have homologs in plants, have defined roles in the import process. Proteins that are destined for delivery to peroxisomes are synthesized on cytosolic ribosomes rather than on ER-bound ribosomes. This is also the case for proteins destined for chloroplasts or mitochondria, at least those transcribed and translated from genes in the nucleus. It may be recalled that chloroplasts and mitochondria contain DNA that codes for some of the proteins resident in those organelles. However, peroxisomes have no DNA, and thus all of their proteins are transcribed from genes in the nucleus and translated through cytosolic ribosomes.

Since the targeting information, the PTS1, is often located at the carboxy terminus the proteins are completely translated and released from the ribosomes before the PTS1 is recognizable. Prior to recognition by the import receptor and delivery to the peroxisomes, the proteins are folded along with their requisite cofactors. In contrast, mitochondrial and chloroplast proteins do not assume their ultimate folded forms until after they have been threaded through the double membranes enclosing those organelles. Next, the folded peroxisomal protein is recognized by a cytosolic receptor protein, Pex5p recognizing the PTS1

C- Terminal PTS1

Glycolate oxidase
Arab   ...RTHIKTDWDTPHYLS<u>AKL</u>
Rice   ...RAHIYTDAERLARPF<u>PRL</u>
Soyb   ...IVTDWDQPRILPRAL<u>PRL</u>

Hydroxypyruvate reductase
Arab   ...PNASPSIVNSKALGLPV<u>SKL</u>
Rice   ...PAACPSIVNAKQLGLPS<u>SKL</u>
Soyb   ...PAACPSIVNAKALGLPT<u>SKL</u>

Isocitrate lyase
Arab   ...RPGADGMGEGTSLVVAK<u>SRM</u>
Rice   ...SWTGPGSESSSHVLAK<u>SRMM</u>
Soyb   ...TRSGAVNIDRGSIVVAK<u>ARM</u>

Catalase 2
Arab   ...QADKSLGQK<u>LASRLNVRPSI</u>
Rice   ...QADRSLGQK<u>LASRLSAKPSM</u>
Soyb   ...QADRSLGQK<u>IASHLNLKPSI</u>

N-terminal PTS2

Acyl-CoA oxidase 3
Arab   MESRREKNPMTEEESDGLIAAR<u>RIQRLSLHL</u>SPS...
Rice   MDPSYPPSATAR<u>RAAAIARHL</u>AGLS...
Soyb   MQTPNCEAER<u>RIQRLTLHL</u>NPT...

Figure 5.5 Peroxisomal targeting sequences that direct proteins from the cytosol into plant peroxisomes. The PTS1 sequences underlined are at the C-terminal end of the proteins. Typically it is the last three amino acid residues, a polar residue, a positively charged basic residue and the carboxy terminal residue with a non-polar side chain. The PTS1 amino acids are not removed after import into the peroxisomes. Catalases are atypical in requiring more residues to effect targeting. PTS2 sequences (underlined) are near the N-terminal end of a protein. As shown, the sequence is commonly something like RIXXXXXHL...,RLXXXXXHL... or QLXXXXXHL... having more basic and hydrophobic residues and fewer acidic residues within the sequence. This N-terminal region of the protein is usually removed after import. Source: Adapted from Hu *et al.* (2012), used with permission.

proteins or the receptor Pex7p recognizing the PTS2 proteins (Figure 5.5). This means that the targeting information is exposed even though the protein is folded. The recognition domain of Pex5p consists of seven tetratricopeptide repeats that form α-helices around a crevice that accommodates the PTS1. Asparagine residues in these α-helices are positioned to interact with the PTS1 residues. When the Pex7p receptor picks up a PTS2 protein, it associates with Pex5p prior to docking on a peroxisome. The receptor proteins, Pex5p and Pex7p, are at large in the cytosol until they pick up some cargo (Figure 5.6).

Receptor docking, discharging and recycling
Once the receptor protein has taken on a cargo molecule, for example a Pex5p receptor has recognized the C-terminal LASRLNVRPSM of a catalase monomer folded around its heme-iron, the Pex5p changes shape and reveals a binding site for docking on the membrane surface of peroxisomes. The peroxisome, as may be remembered, could be tethered by a myosin motor and moving along an

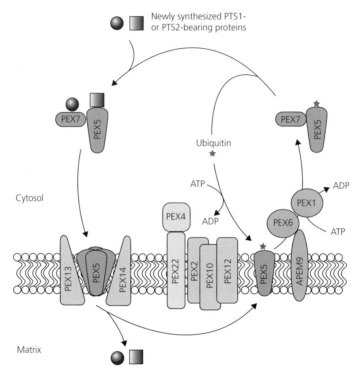

Figure 5.6 The peroxisomal import apparatus. Peroxisomally destined proteins are translated from cytosolic ribosomes and their targeting sequences, PTS1 or PTS2 are recognized by PEX5 or PEX7 receptors which then dock on the PEX13/14 complex in the membrane. The cargo proteins are released into the interior of the peroxisomes and the receptors returned to the cytosol through interactions with the PEX 4, 22, 2, 10, 12, 6 and 1. In the process PEX5 is ubiquitinated by the PEX4. Source: Hu *et al.* (2012), used with permission.

actin filament. The Pex5p docks on a pair of peroxisomal membrane proteins, Pex14 and Pex13. These are transmembrane proteins that are exposed to the cytosol as well as the interior matrix of the peroxisome. The Pex5p binds to a hydrophobic region of the Pex14 protein that is not exposed to the cytosol, possibly on the matrix side of the membrane such that the cargo is now on the matrix side, but still attached to the Pex5p. At this point there is some mystery as to how the folded cargo protein and the Pex5p receptor can squeeze through the membrane lipid bilayer. The process is not at all like that of mitochondrial and chloroplast protein import where the proteins are threaded through the membranes in a linear, unfolded state and then fold once they have entered the organelle. The process is similar to the import of proteins into the nucleus. However, the nuclear envelope has relatively large portals that allow larger protein assemblages to pass through.

   At this point we need to assume that the cargo protein, for example a catalase molecule, is now on the matrix side of the peroxisomal membrane but still attached to its receptor protein, the Pex5p. The cargo must be released into the matrix, but it is not clear how this happens and what triggers the detachment of the cargo from the receptor protein. Also, it is not clear if the receptor protein actually enters the matrix before releasing the cargo or if the receptor remains in the membrane as the cargo is unloaded. In any case, once the receptor is free of cargo it is returned to the cytosol with the assistance of several other Pex proteins in the membrane. First, the receptor interacts with a RING-finger complex consisting of Pex2, Pex10 and Pex12. Then the receptor encounters Pex1 and Pex6, which are ATPases required for recycling of the receptor proteins. ATP hydrolysis helps to drive the receptor retrotranslocation, but the job is not finished without the participation of another Pex pair, namely Pex22 and Pex4. Pex22 is another membrane protein that serves as an anchor for Pex4 which is a ubiquitin ligase. The addition of monoubiquitin to a Pex5 cysteine residue allows the receptor to be released into the cytosol to search for additional PTS1 cargo proteins. However, if a chain of several ubiquitin peptides is attached to a particular Pex5 lysine, then the protein is sent to the proteasome for degradation. This may be necessary if the receptor protein is damaged or if it is no longer needed as the import of peroxisomal proteins becomes less active in a maturing cell. How this is regulated is not known.

Proteomics, genomics and bioinformatics

Of course we would really like to know about all of the various proteins and functions going on inside of peroxisomes. Certainly the functions would depend on the type of cell and the physiological state of the cell. One approach is to use the computational tools of bioinformatics to search genomic (DNA) databases for genes that might code for proteins having peroxisomal targeting information, PTS1 or PTS2 in their amino acid sequences. However, the presence of this information in the DNA does not necessarily mean that the genes are transcribed into RNA and translated as protein. Another approach, the proteomic approach, is to isolate the peroxisomes and then use various protein separation techniques followed by mass spectroscopy to identify proteins. Neither approach is entirely satisfactory. Bioinformatic analyses predict 542 PTS1 and 110 PTS2 proteins for the *Arabidopsis* genome (Bussell *et al.*, 2013). Proteomic analyses identify 204 proteins in peroxisomes obtained from *Arabidopsis*. However, only 97 proteins are shared between these approaches. Bioinformatics would seem to overestimate and proteomics underestimate. Both approaches probably include proteins that do not belong to peroxisomes. Not surprisingly the peroxisomal proteome is simpler than that of chloroplasts (>6000) or mitochondria (>4000).

Bioinformatic predictions can be improved by showing that the genes are actually transcribed as indicated in transcriptome databases and that transcripts are expressed under the same conditions as known peroxisomal genes, such as those for β-oxidation, glyoxylate cycle or photorespiration. Also, fusions can be created with fluorescent proteins attached to putative targeting sequences to show that the proteins are located in peroxisomes when expressed in plant cells. The peroxisomes are identified with a known protein, such as catalase, made to fluoresce at a different wavelength. Thus, peroxisomal residence is verified when the two fluorescent labels co-locate.

Proteomic analysis is appealing because it is applied to tangible, isolated peroxisomes which are isolated from particular tissues, such as leaves, cotyledons or cultured cells. It is potentially an analysis of genes that have actually been transcribed and translated, with the caveat that some low-abundance proteins may not be detected and yet may have important functions. Membrane proteins would be much less abundant and thus less likely to be represented in proteomic analysis. The basic approaches to purifying peroxisomes are refinements of the techniques used originally to characterize them. Peroxisomes may be isolated from leaves, cotyledons, endosperm, whole plants, non-green plants grown in the dark or cultured cells realizing that the organelles from tissues such as a leaves may come from different cell types. Also, cultured cells may not produce the same types of peroxisomes and protein contents as found in tissues from normal plants. For example, leaf tissue would include epidermal cells, stomatal guard cells and the more abundant mesophyll cells. Peroxisomes in the different cell types almost certainly have different functions and different protein compositions. Despite these uncertainties let's follow a proteomic protocol to see where it leads us.

First, the organelles are released from the cells by breaking through the cell walls and membranes. This is accomplished with a mortar and pestle or with various mechanical devices with blades that cut through the cells. While the cells are being broken open the goal is to keep the organelles intact and at the same time disconnect the organelles from each other. The composition of the solutions used as media during the isolation process is also important; the pH, osmoticum and ion concentrations should mimic the cytoplasm. Protease inhibitors are often included since the process of lysing cells can expose the organelles to protein-degrading enzymes that would have been sequestered in the intact cells. The cell lysate/tissue homogenate is then subjected to centrifugations to sediment out larger organelles, nucelii, chloroplasts and mitochondria. The fraction enriched in peroxisomes is then centrifuged through a density gradient made up of different concentrations of sucrose and/or Percoll, a colloidal material that is not osmotically active. Peroxisomes sediment into more dense solutions than the other organelles and appear as a white, cloudy layer near the bottom of the density gradient; brownish mitochondria are above that; and yellow or green plastids further up. The peroxisomes can be further purified by free-flow electrophoresis that separates organelles by their surface charge characteristics (Figure 5.1).

The purity of the organelles can be assessed by measuring the content of potential contaminants, such as mitochondrial cytochrome c oxidase, or chlorophyll along with measurements of peroxisomal-specific enzyme like catalase.

The protein components of the purified organelles are separated and analysed by increasingly sophisticated techniques, the objective being to identify proteins based on amino acid sequences determined by mass spectroscopy. Often the proteins are dissociated from each other with detergents and separated by two-dimensional electrophoresis or resolved by liquid chromatography. Two-dimensional electrophoresis involves first subjecting the protein mixture to isoelectric focusing which separates proteins according to their net charges imparted by amino acid residues that bear positive charges (histidine, lysine, arginine) or negative charges (aspartate, glutamate). (An example of such a 2D gel can been seen at http://www.pnas.org/content/104/27/11501/F3.expansion.html.) Liquid chromatography distributes proteins based on their different hydrophobicities as a consequence of non-polar residues (leucine, valine, phenylalanine, etc.). Any of these approaches can favour certain types of proteins over others and, for example, can miss membrane proteins. A simpler approach is to separate the proteins by mass in a one-dimensional SDS-gel to cut the gel into approximately 20 segments or to go directly to mass spectroscopy (the shotgun approach). These more direct techniques complicate the end analysis, but the computational tools available make it possible.

At this point we have proteins that have been partially separated from each other but not necessarily purified. We may have clues about their molecular sizes and charges which suggest identities of the proteins. Mass spectrometry brings us closer to specific identifications. The protein samples from a 1D or 2D gel, liquid chromatography or the shotgun extract can be digested with trypsin, a proteolytic enzyme that cuts a protein at the positively charged residues, lysine and arginine. This produces peptide fragments of discrete sizes, perhaps 40 or more fragments. The mass spectrometer tells us the exact mass of these peptides, usually not all of them, maybe only 25% of them. But this can be enough to identify the protein that produces the peptides. If this is not enough, then a second mass spectrometer takes a few peptides and fragments each one of them at random producing many peptides consisting of various numbers of residues. Computational tools make it possible to determine which residues are in these fragments and to define the sequence of amino acids in the original peptide. Now the computational software can go to the genome of that organism and find out what DNA sequences would produce those amino acid sequences and peptides of those masses. In many cases that tells us what the protein is, a catalase, an oxidase or something that has not been previously attributed to peroxisomes. Thus we would want to verify that the protein is really present in peroxisomes.

There are a couple of things that we can do to verify peroxisomal location of a protein identified in the proteomic analysis. One, we can look to see if it has a PTS1 or PTS2 sequence at the C-terminal and near the N-terminal region as described earlier. Two, we can take that sequence information from the end of the protein and attach it to a fluorescent reporter protein such as GFP or YFP, express it in cultured plant cells and use fluorescence microscopy to see if the protein construct co-localizes with a peroxisomal protein (e.g., catalase). At the time of this writing there were about 163 proteins that have survived this gauntlet of criteria (Bussell *et al.*, 2013). These include enzymes associated with photorespiration, fatty acid oxidation and the glyoxylate cycle. Many of the enzymes occur in multiple isoforms. For example, *Arabidopsis* and rice have five different Acyl-CoA oxidases with specificities for different chain-length fatty acids. The proteomic analysis has also revealed protein kinases and proteases. The kinases would be involved in phosphorylating proteins, regulating their activities, something that remains to be characterized. The proteases may be responsible for removing the PTS2 segments as mentioned earlier. Also, proteases may be there to degrade unused proteins such as the β-oxidation and glyoxylate cycle enzymes in an expanding and greening cotyledon that is transitioning from nutrient mobilization to photosynthesis and photorespiration.

Now that we have identified a protein we might wonder how the protein fits into the scheme of life in a particular type of cell. However, you may recall that we may not know what type of cell that protein is associated with, depending on the tissue that was used to obtain the peroxisomes to begin with. The protein may have the characteristics of a protease or kinase. If so, what are the substrates for the protein? The proteomics analysis does not inform that. Does the protein have sites that resemble sites for phosphorylation; and if so, under what conditions is the protein phosphorylated? If the protein appears to be a part of a metabolic pathway that is thought to occur elsewhere in cells, we would like to know about the flow of substrate intermediates into and out of peroxisomes. These are issues that have yet to be resolved.

Whence peroxisomes?

Peroxisomal biologists have pondered this question since the discovery of the organelle in the late 1960s. There are two questions implied here. (i) How do cells create new peroxisomes as the cells divide and grow or when a physiological condition calls for more peroxisomes with particular contents? (ii) How did peroxisomes evolve? The answer to the second question is not so easy as it is for chloroplasts or mitochondria. Peroxisomes lack DNA and RNA that would have been indications of an endosymbiotic origin. This also provides challenges for answering the first question. Cells can create more peroxisomes by fission as described earlier. We learned that peroxisomes can grow by importing proteins from the cytosol. But this does not explain how the membrane is expanded. The evidence points to the ER as the source of membrane lipids and some membrane proteins (Donaldson, 1976). Also, there are indications that the ER can produce peroxisomal buds and create new peroxisomes, *de novo*. The connections between ER and peroxisomes remain controversial (Barton *et al.*, 2013).

Peroxisomes may have evolved before or after mitochondria in eukaryotic cells. Both organelles depend on oxygen which became abundant in the atmosphere about two billion years ago. Peroxisomes could have been an early protobacterial symbiont whose genetic information completely migrated to the nucleus of the host. The fission process and the import of proteins from the cytosol are evidence for an endosymbiotic origin. However, phylogenetic analysis of peroxisomal proteins suggests that the organelles were a later innovation in eucaryotes. The evidence for this comes from the similarities of many of the peroxin proteins to those involved in the import apparatus for proteins involved in the ER-associated degradation (ERDA) pathway. Misfolded proteins are translocated out of the ER by the ERDA into the cytosol for proteosomal degradation. This implies that peroxisomes evolved from the ER which mirrors the idea that peroxisomes may be created *de novo* from ER buds. The conservation of PTS1, PTS2 and peroxins among all eukaryotes, yeasts, animals and plants also suggests that peroxisomes evolved early in the eukaryotic lineage.

Plant peroxisomal coda

Peroxisomes are peculiar organelles. They are metabolic organelles, but unlike mitochondria or chloroplasts they are not dedicated to conserving energy. Peroxisomal oxidases by-pass energy conservation by transferring reducing equivalents directly to oxygen. Peroxisomes in leaves participate in photorespiration, a process that appears to be the result of an evolutionary accident. Ribulose bisphosphate carboxylase (RuBisCo) being the victim of plants being extremely effective at increasing the concentration of O_2 in the earth's atmosphere. That O_2 reached a level that competes with CO_2 in the RuBisCo reaction.

Peroxisomes lack DNA which means that they have to import all of their components from the cytoplasm. Other organelles such as mitochondria and chloroplasts import some of their constituents from the cytoplasm, but they contain some genetic information to build on and the guide their replications. Also, chloroplasts and mitochondria bring in proteins from the cytosol by threading them in an unfolded state through their double membranes. Peroxisomes allow folded, multimeric proteins to pass through their single membranes. How that happens is not understood yet.

The absence of DNA also lends to the mystery of the evolutionary origins of peroxisome. Chloroplasts and mitochondria have many clues in their DNA that point to their origins. Peroxisomes have left us guessing.

References

Alani, A.A., Luster, D.G. and Donaldson, R.P. (1990) Development of endoplasmic reticulum and glyoxysomal membrane redox activities during castor bean germination. *Plant Physiology*, **94**, 1842–1848.

Anand, P., Kwak, Y., Simha, R. and Donaldson, R.P. (2009) Hydrogen peroxide induced oxidation of peroxisomal malate synthase and catalase. *Archives of Biochemistry and Biophysics*, **491**, 25–31.

Barton, K., Mathur, N. and Mathur, J. (2013) Simultaneous live-imaging of peroxisomes and the ER in plant cells suggests contiguity but no luminal continuity between the two organelles. *Frontiers in Physiology*, **4**, pp. 196.

Bussell, J.D., Behrens, C., Ecke, W. and Eubel, H. (2013) *Arabidopsis* peroxisome proteomics. *Frontiers in Plant Science*, **4**, pp. 101.

De Duve, C., Beaufay, H., Jacques, P. *et al.* (1960) Intracellular localization of catalase and of some oxidases in rat liver. *Biochimica et Biophysica Acta*, **40**, 186–187.

del Rio, L.A., Sandalio, L.M., Corpas, F.J. *et al.* (2006) Reactive oxygen species and reactive nitrogen species in peroxisomes. Production, scavenging, and role in cell signaling. *Plant Physiology*, **141**, 330–335.

Donaldson, R.P. (1976) Membrane lipid metabolism in germinating castor bean endosperm. *Plant Physiology*, **57**, 510–515.

Donaldson, R.P. and Fang, T.K. (1987) β-Oxidation and glyoxylate cycle coupled to NADH: cytochrome c and ferricyanide reductases in glyoxysomes. *Plant Physiology*, **85**, 792–795.

Donaldson, R.P., Kwak, Y. and Yanik, T. (2008) Plant peroxisomes and glyoxysomes in Hetherington, A.M. (ed.) *Encyclopedia of Life Sciences*, John Wiley & Sons, Ltd, Chichester.

Frederick, S.E. and Newcomb, E.H. (1969a) Cytochemical localization of catalase in leaf microbodies (peroxisomes). *Journal of Cell Biology*, **43**, 343–353.

Frederick, S.E. and Newcomb, E.H. (1969b) Microbody-like organelles in leaf cells. *Science*, **163**, 1353–1355.

Hu, J., Baker, A., Bartel, B. *et al.* (2012) Plant peroxisomes: biogenesis and function. *The Plant Cell*, **24**, 2279–2303.

Imoto, Y., Kuroiwa, H., Yoshida, Y. *et al.* (2013) Single-membrane-bounded peroxisome division revealed by isolation of dynamin-based machinery. *Proceedings of the National Academy of Sciences of the United States of America*, **110**, 9583–9588.

Jedd, G. and Chua, N.H. (2002) Visualization of peroxisomes in living plant cells reveals acto-myosin-dependent cytoplasmic streaming and peroxisome budding. *Plant and Cell Physiology*, **43**, 384–392.

Karyotou, K. and Donaldson, R.P. (2005) Ascorbate peroxidase, a scavenger of hydrogen peroxide in glyoxysomal membranes. *Archives of Biochemistry and Biophysics*, **434**, 248–257.

Kaur, N., Reumann, S. and Hu, J. (2009) Peroxisome biogenesis and function. *The Arabidopsis Book/American Society of Plant Biologists*, **7**, pp. e0123.

Kunze, M. and Hartig, A. (2013) Permeability of the peroxisomal membrane: lessons from the glyoxylate cycle. *Frontiers in Physiology*, **4**, 204.

Lanyon-Hogg, T., Warriner, S.L. and Baker, A. (2010) Getting a camel through the eye of a needle: the import of folded proteins by peroxisomes. *Biology of the Cell*, **102**, 245–263.

Lipka, V., Dittgen, J., Bednarek, P. *et al.* (2005) Pre- and postinvasion defenses both contribute to nonhost resistance in *Arabidopsis*. *Science*, **310**, 1180–1183.

Luster, D.G. and Donaldson, R.P. (1987) Orientation of electron transport activities in the membrane of intact glyoxysomes isolated from castor bean endosperm. *Plant Physiology*, **85**, 796–800.

Luster, D.G., Bowditch, M.I., Eldridge, K.M. and Donaldson, R.P. (1988) Characterization of membrane-bound electron transport enzymes from castor bean glyoxysomes and endoplasmic reticulum. *Archives of Biochemistry and Biophysics*, **265**, 50–61.

Mathur, J., Mathur, N. and Hülskamp, M. (2002) Simultaneous visualization of peroxisomes and cytoskeletal elements reveals actin and not microtubule-based peroxisome motility in plants. *Plant Physiology*, **128**, 1031–1045.

Mhamdi, A., Noctor, G. and Baker, A. (2012) Plant catalases: peroxisomal redox guardians. *Archives of Biochemistry and Biophysics*, **525**, 181–194.

Nguyen, A.T. and Donaldson, R.P. (2005) Metal-catalyzed oxidation induces carbonylation of peroxisomal proteins and loss of enzymatic activities. *Archives of Biochemistry and Biophysics*, **439**, 25–31.

Yanik, T. and Donaldson, R.P. (2005) A protective association between catalase and isocitrate lyase in peroxisomes. *Archives of Biochemistry and Biophysics*, **435**, 243–252.

Further reading

Acton, A. (2013) *Microbodies—Advances in Research and Application*, Scholarly Editions, ScholarlyBrief, Atlanta, Georgia.

Goto-Yamada, S., Mano, S., K., Okawa, K. *et al.* (2015) Dynamics of the light-dependent transition of plant peroxisomes. *Plant and Cell Physiology*, **56**, 1264–1271.

Hruban, Z. and Rechcigl, M. (2013) *Microbodies and Related Particles: Morphology, Biochemistry, and Physiology*, Academic Press, New York, NY.

Lauersen, K.J., Willamme, R., Coosemans, N., Joris, M., Kruse, O. and Remacle, C. (2016) Peroxisomal microbodies are at the crossroads of acetate assimilation in the green microalga *Chlamydomonas reinhardtii*. *Algal Research*, **16**, 266–274.

Lee, H.N., Kim, J. and Chung, T. (2014) Degradation of plant peroxisoimes by autophagy, *The Plant Science*, **5**, 139.

Moyer, A.E., Zheng, W., Johnson, E.A. et al. (2014) Melanosomes or microbes: Testing an alternative hypothesis for the origin of microbodies in fossil feathers. Scientific Reports, 4, Article number 4233, doi:10.1038/ srep04233.

CHAPTER 6

Microtubules, intermediate filaments, and actin filaments

William V. Dashek

Retired Faculty, Adult Degree Program, Mary Baldwin College, Staunton, VA, USA

Microtubules

Structure

Microtubules are part of the cytoskeleton and can assemble into bundles, for example, the cortical microtubules (Hyams and Lloyd, 1994; Gunning and Steer, 1996; Lloyd and Chan, 2004; Eckardt, 2008), the prophase band (Wasteneys, 2002; Karahara *et al.*, 2009; Smirnova *et al.*, 1998), the spindle (Pickett-Heaps and Forer, 2009), and the phragmoplast (Guo *et al.*, 2009). Cortical microtubules can appear in proximity to the plasmalemma and may be attached to it via cross bridges (Lloyd and Chan, 2004). These microtubules seem to be involved in orienting cellulosic **microfibrils** in the wall (Wasteneys and Ambrose, 2009). The prophase band originates during late G2 phase and delineates the future site of division (Müller *et al.*, 2010). The phragmoplast promotes the synthesis of the new cell plate (Pastuglia *et al.*, 2006).

Transmission electron micrographs revealed microtubules to be **elongated hollow** tube-like structures (Figure 6.1). However, microtubules can be curved. Negative staining (see **Chapter** 2) showed them to be composed of 13 protofilaments (**a linear row of tubulin dimers**) along a microtubule's circumference (Kwiatkowska, 2006). Microtubules in cross section appear as circular structures (~25 mm in diameter) with a less dense electron core (Figure 6.1; Chang-Jie and Sonobe, 1993). These tubules can be from 0.2 to 2.5 cm in length.

Chemical composition

Biochemistry demonstrated that the protofilaments are composed of tubulin dimers (Fosket and Morejohn, 1992; Donhauser et al., 2010) formed by the polymerization of α- and β-tubulins (Hyams and Lloyd, 1994). This polymerization requires α-tubulin (Murata *et al.*, 2005) and a chaperonin for folding α- and β-tubulins (Gao *et al.*, 1993). The balance of α- and β-tubulins has been reported to be controlled by a *KIS* gene (Kirik *et al.*, 2002). Alpha tubulin can

Plant Cells and their Organelles, First Edition. Edited by William V. Dashek and Gurbachan S. Miglani.
© 2017 John Wiley & Sons, Ltd. Published 2017 by John Wiley & Sons, Ltd.

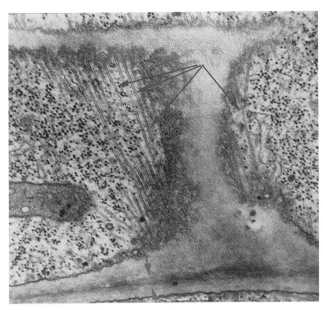

Figure 6.1 Microtubules in the middle lamella (arrows, microtubules). Source: Courtesy of Eldon Newcomb, University of Wisconsin. An electron micrograph illustrating both cross and longitudinal sections of microtubules in *Ornithogalum umbellatum occurs in* Kwiatkowska (2006), with permission from Folia Histochemica and Cytobiologica.

be posttranslationally modified by tyrosine carboxypeptidase and tubulin ligase to yield nitrotyrosine α-tubulin (Jovanovic *et al.*, 2010).

Microtubules exhibit a polarity: the plus end and the minus. The terminal subunit of the less dynamic end is α-tubulin, and β-tubulin terminates the faster growing end (Guo *et al.*, 2009).

Plant α-tubulin is localized with all microtubule arrays (Wasteneys and Brandizzi, 2013). These investigators suggested that it plays a role in nucleation.

In addition to structural proteins, microtubule-associated proteins (MAPs) can regulate microtubule organization (Vantard *et al.*, 1993; Amos and Schlieper, 2005; Mao *et al.*, 2005). One of the MAPs is kinesin, a protein dimer that binds to microtubules through an adenosine triphosphate (ATP) hydrolysis-driven motor domain. This binding permits kinesis to traffic vesicles (tethering), organelles, and proteins from the negative end to the positive end of a microtubule (Reddy, 2001). A calmodulin-domain kinesin can link actin filaments and microtubules in rice (Frey *et al.*, 2009). Some examples of kinesins are kinesin-like calmodulin-binding protein (KCBP) from *Arabidopsis* (Vos *et al.*, 2000), MAP spc 98p (Kaloriti *et al.*, 2007), and *Arabidopsis* ATKI (Marcus *et al.*, 2001, 2003). A hypothetical diagram exhibiting the locations of MAPs along the length of a microtubule occurs in Kaloriti *et al.* (2007). Whether dynein, another MAP, occurs in plants is still not proven with certitude. Other MAPs are M8RI, CLASP, MAP65s, and katanins (Wasteneys and Ambrose, 2009).

Finally, microtubule-interacting proteins are those that bind to microtubules (Mandelkow and Mandelkow, 1995). Quilichini and Munch stated that the binding of these proteins can concentrate proteins and/or regulate protein activity.

Biogenesis

Microtubules can polymerize and depolymerize at both the plus (β-tubulin) and minus (α-tubulin) ends, a process termed "dynamic instability" (Moore et al., 1997; Cassimeris, 2007; Nick, 2008; Guo et al., 2009). Tubulin dimers, which carry guanosine triphosphate (GTP) molecules, can add at both ends but preferentially add at the positive end and depolymerize at the negative end.

Certain plants lack distinct mobile organizing centers or **dynein** (Vaughn and Harper, 1998; Wasteneys, 2002). In these plants, kinesins may organize microtubules (Ambrose et al., 2005). In other plants, microtubules originate from small, dispersed nucleation centers (Murata et al., 2005; Kaloriti et al., 2007). Murata et al. demonstrated that new microtubules branch off preexisting microtubules in the center of certain plant cells. Similar observations were made by Chan et al. (2009) for the cortical microtubules of Arabidopsis. Pastuglia et al. (2006) provided evidence that γ-tubulin plays a role in the synthesis and organization of Arabidopsis microtubule arrays. This conclusion was based on reverse genetics. Finally, the biogenesis of microtubules can be affected by anti-microtubule dimers, for example, propyzamide, oryzalin, trifluralin (Marcus et al., 2001), and benomyl (deAndrade-Montieri and Matinez-Rossi, 1999).

Functions

The functions of plant microtubules are copious (Table 6.1). One of the best known is spindle formation (Kumagai and Hasezawa, 2001; Wasteneys, 2004), which has been used for investigations of microtubule origin (Wittmann and Waterman-Storer, 2001). The microtubules play a significant role in cell plate formation (Figure 6.2). Hepler and Jackson (1968) carried out a fine structure investigation of phragmoplast and developing cell plate origin in cultured endosperm cells of Haemanthus katherinae. The phragmoplast consisted of an accumulation of microtubules positioned at right angles to the developing plate (Smirnova et al., 1998). The 200–240 Å diameter microtubules occurred in small clusters located at 0.2–0.3 µm intervals along the plate. Vesicles derived from a cytoplasmic matrix aggregate between clusters of microtubules and coalesce to yield the plate (Seguí-Simarro et al., 2004). It is generally accepted that cortical microtubules in the cell periphery regulate the alignment of cellulose microfibrils in the cell wall (the alignment hypothesis— see Baskin, 2001; Dixit and Cytr, 2004). Baskin (2001) proposed a template incorporation model whereby nascent microfibrils are incorporated into the cell wall via binding to a scaffold. The role of cortical microtubules is proposed

Table 6.1 Functions of microtubules in plants.*

Function	Reference
Construction of the mitotic spindle for separation of chromosomes, an antiparallel array of microtubules	Whittmann *et al.* (2001)
	Zhong *et al.* (2002)
Oriented deposition of cellulose microfibrils and cell wall strength	Marcus *et al.* (2001)
Stomatal development and guard and cell function	Lucas *et al.* (2005)
Cytoplasmic streaming	Reddy and Reddy (2004)
Organelle transport, e.g., mitochondria	Boute *et al.* (2001)
Plasmalemma orientation	Reddy (2001)
Directional growth	Lloyd and Chan (2004)
Cytoskeleton dynamics	Reddy (2001)
Polarized growth involvement	Reddy (2001)
Possible role in perception of touch and gravity signals	Bisgrove (2008)
Modulating signaling pathways	Wasteneys (2004)

* The functioning of plant microtubules may depend upon polymerization (including phosphorylation of tyrosine residues).

to bind and orient scaffold components at the plasma membrane. Of interest is the possibility that gibberellins may regulate microtubule orientation (Shibaoka, 1994; Locascio *et al.*, 2013). The orientation appears to involve a c-chaperone required for tubulin folding. Nick (2013) reported that plant microtubules may also function as sensors for the perception of mechanical membrane stress. The author stated that the sensory function depends on dynamic instability. Microtubules were proposed "as elements of a sensory hub that decodes stress-related signal signatures, with phospholipase D as an important player."

Recent research has implicated microtubules in programmed cell death (Smertenko and Franklin Tong, 2011; Nakaba *et al.*, 2013). The later investigators noted the disappearance of microtubules during cell death of ray parenchyma in *Abies sachalinensis*. Phospholipase D, which is involved in cell death, has been reported to be associated with microtubules in plant cells (Dhonukshe *et al.*, 2003).

Intermediate filaments

Structure

The higher plant cytoskeleton appears to contain 10 nm intermediate filaments (Yang *et al.*, 1992; De La Cruz and Ostap, 2004). According to Cruz and Moreno Diaz de la Espina (2009), the plant cytoskeleton is composed of

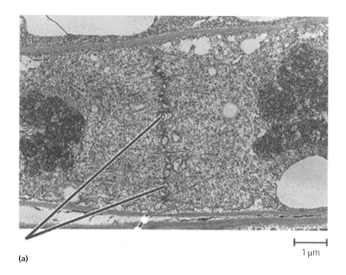

(a)

1 μm

Microtubules

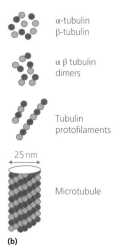

α-tubulin
β-tubulin

- Microtubules are assembled by polymerization of α-tubulin and β-tubulin molecules.

α β tubulin dimers

- α and β tubulin unite to form heterodimers.

- α - β tubulin dimers assemble into tubulin protofilaments, each consisting of alternating α- and β-units.

Tubulin protofilaments

- Tubulin protofilaments are assembled into cylindrical structures, the microtubules.

25 nm

Microtubule

- There are 13 protofilaments along the circumference of microtubule. The protofilaments are arranged longitudinally giving the appearance of alternating spirals of α- and β-units.

- Microtubules have a diameter of 25 nm.

(b)

Figure 6.2 (a) TEM of microtubules during cell plate formation. Source: http://files/quia/ users/lmcgeemitosis/cell-plate-picture.gif. Another figure (not shown here) is at http:// publishing.cdlib.org/ucpressebooks/data/13030/n2/ft796nb4n2/figures/ft796nb4n2_00007.jpg (b) Microtubule structure. Source: Cuscheri (2009).

coiled-coil filaments. These are rope-like structures composed of fibrous subunits (Figure 6.3b). This yields an extended λ-helical rod domain with head and tail regions (Liu, 2010). Su *et al.* (1990) reported keratin-like intermediate filaments in maize protoplasts as did Hao *et al.* (1992) for *Allium cepa*. McNulty and Saunders (1992) reported the isolation of possible lamins from

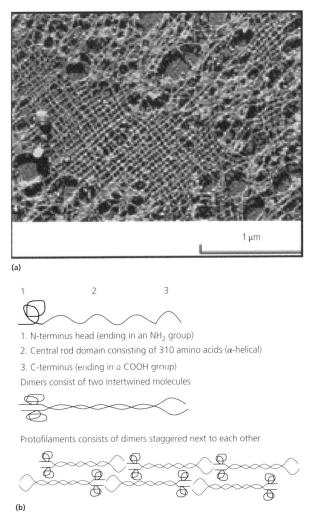

1. N-terminus head (ending in an NH_2 group)
2. Central rod domain consisting of 310 amino acids (α-helical)
3. C-terminus (ending in a COOH group)
Dimers consist of two intertwined molecules

Protofilaments consists of dimers staggered next to each other

(b)

Figure 6.3 Intermediate filaments: (a) micrograph of intermediate filaments. Source: http://cytochemistry.net/cell-biology/intermediate_filaments.htm. (b) 10-nm intermediate filaments showing three domains of IF proteins, dimmers, and protofilaments. Source: Cuscheri (2009).

pea nuclei. Then, Minguez and Moreno Diaz de la Espina (1993) provided an immunological characterization of lamins for the onion cell nuclear matrix. Recently, Ciska and Moreno (2013) stated that the plant protein family NMCP/LINC/CRWN may be plant lamin analogues based on functional analysis of NMCP/LINC mutants. However, consensus thinking about whether the occurrence of lamin intermediates in plants is similar to that in animals (Goldman *et al.*, 2008) is lacking (Meier, 2007).

Chemical composition and biogenesis

There could be at least three putative nuclear intermediate filament proteins, 65, 60, and 54 ka (Yang *et al.*, 1992), and multiple isoelectric forms (Blumenthal *et al.*, 2004). Yang *et al.* (1992) demonstrated that intermediate filaments could be assembled in a cell-free system. These filaments were reported to be indistinguishable from native intermediate filaments in both protein composition and morphology. Min *et al.* (1999) used scanning tunneling microscopy and transmission electron microscopy (TEM) to reveal that plant acidic and basic keratins can assemble into dimers followed by 10 nm filaments *in vitro*. In animal cells, intermediate filament polymerization begins with a monomer that coils about another monomer yielding a dimer. Then, the latter forms a tetramer: the basic subunit of the intermediate filament. Whether these events occur in plants yielding a nuclear filament remains to be determined (Fiserova *et al.*, 2009).

Actin filaments (microfilaments)

Structure

Another component of the cytoskeleton is actin (Figure 6.4), an abundant protein (Sheterline *et al.*, 1998; Volkmann and Baluska, 1999) although there are actin monomers (4SKDa). TEM has revealed polymerized filaments (Staiger *et al.*, 2000; Higaki *et al.*, 2007; Pollard and Cooper, 2009). These filaments, 7 nm in diameter, are created via the polymerization of the G-actin monomers (Figure 6.4). The possibility of actin-organizing centers at the cell periphery exists (Volkmann and Baluska, 1999). Branched actin filaments can be formed with the aid of actin-mediated 2/3 protein. Localized increases in polymerized actin can occur via mechanical wounding of post-harvest tubers (Morelli *et al.*, 1998). Finally, actin filament function can be visualized by fluorescent fusion protein 60S (Schenkel *et al.*, 2008) and fluorescent phallotoxins (Lloyd, 1998).

Chemical composition and biogenesis

There are extended F-actin networks that are composed of F-actin bundles (Volkmann and Baluska, 1999). Also, there can be networks of short F-actin oligomers. In addition, G-actin can assemble as actin oligomers consisting of a few actin molecules. Actin monomers can assemble into polymeric filaments (Staiger *et al.*, 2000). The polymerization is usually coupled to ATP hydrolysis. Actin-binding proteins (Table 6.2) control the size and activity of the monomer pool (Staiger *et al.*, 2000). The other functions of actin-binding proteins are presented in Table 6.2. Disruptors of microtubules and actin filaments are presented in Table 6.3.

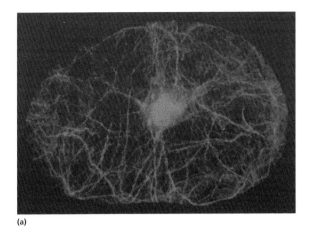

(a)

Molecular structure of actin filaments

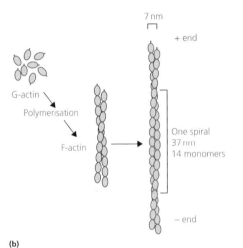

7 nm

+ end

G-actin

Polymerisation

F-actin

One spiral
37 nm
14 monomers

– end

(b)

- Actin filaments (known as F-actin) are assembled from globular molecules of G-actin.

- Actin filaments consist of double strands of G-actin arranged in a spiral.

- Each spiral turn spans a length of 37 nm and contains about 14 monomers (7 per strand).

- Each filament has a diameter of 7 nm.

- Actin filaments have polarity indicated as + and – ends.

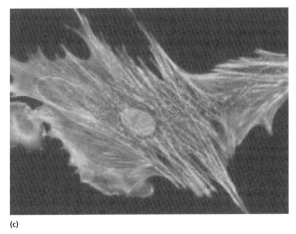

(c)

Figure 6.4 Actin filaments: (a) actin filament networks—fluorescent image. Source: Courtesy R. Meagher—UGA genetics. (b) Molecular structure of actin filament. Source: Cuscheri (2009). (c) Actin filaments—fluorescent image. Source: http://micro.magnet.fsu.edu/primer/techniques/fluorescence/gallery/cells/apm/apmcellsexlarge4.html.

Table 6.2 Actin-binding proteins (ABPs) and their possible functions.*

ABP	Function	Reference
Profilin Actin-binding protein	Promotes depolymerization of actin filaments by binding G-actins	de Ruijter and Emmons (1999)
Proper cell morphogenesis	Regulates the pool of monomers available for polymerization, promotes actin assembly, **sequesters G-actin monomer**	Harries *et al.* (2005)
Actin-depolymerizing factor	Prevents actin filament elongation or loss of monomers from the ends of filaments	de Ruijter and Emmons (1999)
Spectrin family	Produce bundles to networks	de Ruijter and Emmons (1999)

* Cofilin binds filamentous F-actin.

Table 6.3 Disruptors of microtubules, microfilaments, and intermediate filaments.*

Subcellular structure	Disruptor	References
Microtubule	Ansamitocin P-3	Vanghateri *et al.* (2013)
	Dolastatin	Müller *et al.* (2014)
	Myoseverin	Rosanic *et al.* (2000)
	Nocodazole	Vasquez *et al.* (1999)
	TN-16	Arai *et al.* (1983)
	Lubeluzole	Burkhurt *et al.* (1999)
	Plant alkaloids	Weaver (2014)
	Cytochalasin	Shevchenko (2011)
	Oryzalin	Seagull (1990)
Microfilaments	Cytochalasin	Shevchenko (2011)
	Oryzalin	Seagull (1990)
	Latrunculin B	Collings *et al.* (2006)
	2,3,5-Triiodobenzoic acid	Kojo *et al.* (2013)
	Jasplakinolide	Dhaliwal (2013)
	Mycalolide B	Takakuwa *et al.* (2000)
	Wiskostatin	Cheng *et al.* (2011)
Intermediate Filaments	Acrylamide	Eckert (1986)

* Note paclitaxel appears to be a microtubule stabilizer.

Functions

The functions of actin filaments are diverse (Table 6.4). Of recent research interest is the perception–signaling roles of plant cells (Drøbak *et al.*, 2004). Volkmann and Baluska (1999) suggested that signaling is involved in guard cell movements, mechano- and gravity-sensing host–pathogen interactions, and wound-healing.

Table 6.4 Possible functions of actin filaments.

Function	Reference
Molecular tracks for cytoplasmic streaming and organelle movements	Staiger *et al.* (2009)
	Jeng and Welch (2001)
Chloroplast orientation	Kong and Wada (2011)
Guidance of the cell plate to the division site	Staiger *et al.* (2000)
Support molecule vesicle trafficking	Hussey *et al.* (2006)
Control of actin cytoskeleton	Staiger *et al.* (2000)
Coordinate cytoplasmic responses to extra- and intracellular signals	Cárdenas *et al.* (1998)
Response to pathogen stimulation	Deeks and Hussey (2009)
Control of cell shape	Pollard and Cooper (2009)
Control of growth movement	Hussey *et al.* (2006)

References

Ambrose, J.C., Li, W., Marcus, A. *et al.* (2005) A minus-end-directed kinesin with plus-end tracking protein activity is involved in spindle morphogenesis. *Molecular Biology of the Cell*, **16**, 1584–1592.

Amos, L.A. and Schlieper, D. (2005) Microtubules and maps. *Advances in Protein Chemistry*, **71**, 257–298.

Arai, T. (1983) Inhibition of microtubule assembly *in vitro* by TN-16, a synthetic antitumor drug. *FEBS Letters*, **155** (2), 273–276.

Baskin, T.I. (2001) On the alignment of cellulose microfibrils by cortical microtubules: a review and a model. *Protoplasma*, **215**, 1–4, 150–171.

Bisgrove, S.R. (2008) The roles of microtubules in tropisms. *The Plant Science*, **175**, 747–755.

Blumenthal, S.S., Clark, G.B. and Roux, S.J. (2004) Biochemical and immunological characterization of pea nuclear intermediate filament proteins. *Planta*, **218**, 965–975.

Boute, N., Pernet, K. and Issad, T. (2001) Monitoring the activation state of the insulin receptor using bioluminescence resonance energy transfer. *Molecular Pharmacology*, **60**, 640–645.

Burkhart, K.K., Beard, D.C., Lehman, R.A. and Billingsley, M.L. (1998) Alterations in tau phosphorylation in rat and human neocortical brain slices following hypoxia and glucose deprivation. *Experimental Neurology*, **154** (2), 464–472.

Cárdenas, L., Vidali, L., Domínguez, J. *et al.* (1998) Rearrangement of actin microfilaments in plant root hairs responding to *Rhizobium etli* nodulation signals. *Plant Physiology*, **116**, 871–877.

Cassimeris, L. (2007) Tubulin delivery: polymerization chaperones for microtubule assembly? *Developmental Cell*, **13**, 4, 455–456.

Chan, J., Sambade, A., Calder, G. and Lloyd, C. (2009) Arabidopsis cortical microtubules are initiated along, as well as branching from, existing microtubules. *The Plant Cell*, **21**, 2298–2306.

Chang-Jie, J. and Sonobe, S. (1993) Identification and preliminary characterization of a 65 kDa higher-plant microtubule-associated protein. *Journal of Cell Science*, **105**, 891–901.

Cheng, C.Y., Wong, E.W.P., Lie, P.P.Y. *et al.* (2011) Environmental toxicants and male reproductive function. *Spermatogenesis*, **1** (1), 2–13.

Ciska, M. and Moreno, S. (2013) NMCP/LINC protein: putative lamins in plants. *Plant Signal Behavior*, **15**, 8–11.

Collings, D.A., Lill, A.W., Himmelspach, R. and Wasteneys, G.O. (2006) Hypersensitivity to cytoskeletal antagonists demonstrates microtubule–microfilament cross-talk in the control of root elongation in *Arabidopsis thaliana. New Phytologist*, **170**, 2, 275–290.

Cruz, J.R. and Moreno Diaz de la Espina, S. (2009) Subnuclear compartmentalization and function of actin and nuclear myosin I in plants. *Chromosoma*, **118**, 193–207.

Cuscheri, A. (2009). The motile machinery and cytoskeleton of the cells. http://Staff.um.edu.mt/acus1/07Motile.pdf (accessed March 18, 2016).

De La Cruz, E.M. and Ostap, E.M. (2004) Relating biochemistry and function in the myosin superfamily. *Current Opinion in Cell Biology*, **16**, 61–67.

deAndrade-Montieri, C. and Matinez-Rossi, N. (1999) The influence of microtubules in *Asperglilus nidulans. Genetics and Molecular Biology*, **22**, 309–313.

Deeks, M.I. and Hussey, P.J. (2009) *Plant Actin Biology* (ELS). Wiley, New York.

de Ruijter, N.C.A. and Emons, A.M.C. (1999) Actin-binding proteins in plant cells. *Plant Biology*, **1**, 26–35.

Dhaliwal, A. (2013) Activators and inhibitors in cell biology research. *Materials and Methods*, **3**, 185. http://dx.doi.org/10.13070/mm.en.3.185.

Dhonukshe, P., Laxalt, A.M., Goedhart, J. *et al.* (2003) Phospholipase D activation correlates with microtubule reorganization in living plant cells. *The Plant Cell*, **15**, 2666–2679.

Dixit, R. and Cytr, R. (2004) The cortical microtubule array: from dynamics to organization. *The Plant Cell*, **16**, 2456–2552.

Donhauser, Z.J., Jobs, W.B. and Binka, E.C. (2010) Mechanics of microtubules: effects of protofilament orientation. *Biophysics Journal*, **99** (5), 1668–1675.

Drøbak, B.K., Franklin-Tong, V.E. and Staiger, C.J. (2004) The role of the actin cytoskeleton in plant cell signaling. *New Phytologist*, **163** (1), 13–30.

Eckardt, N.A. (2008) High-resolution imaging of cortical microtubule arrays. *The Plant Cell*, **20**, 817–819.

Eckert, B. (1986) Alteration of the distribution of intermediate filaments in PtK1 cells by acrylamide II: Effect on the cytoplasmic organelles. *Cell Motility and the Cytoskeleton*, **6** (1), 15–24.

Fiserova, J., Kiseleva, E. and Goldberg, M.W. (2009) Nuclear envelope and nuclear pore complex structure and organization in tobacco BY-2 cells. *The Plant Journal*, **59**, 243–255.

Fosket, D.E. and Morejohn, L.C. (1992) Structural and functional organization of tubulin. *Annual Review of Plant Physiology and Plant Molecular Biology*, **43**, 201–240.

Frey, N., Klot, J., and Nick, P. (2009) Dynamic bridges—a calponin-domain from rice links actin filaments and microtubules in both cycling and non-cycling cells. *Plant and Cell Physiology*, **50**, 1493–1506.

Gao, Y., Valnberg, I.E., Chow, R.L. and Cowan, N.J. (1993) Two cofactors and cytoplasmic chaperonin are required for the folding of α- and β-tubulin. *Molecular and Cellular Biology*, **13**, 2478–2485.

Goldman, R.D., Grin, B., Mendez, M.G., and Kuczmanniski, E.R. (2008) Intermediate filaments: versatile building blocks of cell structure. *Current Opinion in Cell Biology*, **20**, 28–34.

Gunning, B.E.S. and Steer, M.W. (1996) *Plant Cell Biology. Structure and Function*, John and Bartlett Publishers, Sudbury, MA.

Guo, L., Ho, C.K., Kong, Z. *et al.* (2009) Evaluating the microtubule cytoskeleton and its interacting proteins in monocots by mining the rice genome. *Annals of Botany*, **103**, 387–402.

Hao, S., Ha, A., Dezhag, J. *et al.* (1992) Nuclear lamina-like filaments and nuclear matrix in *Allium cepa* as revealed by scanning electron microscopy. *Cell Research*, **2**, 153–163.

Harries, P.A., Pan, A. and Quatrano, R.S. (2005) Actin-related protein2/3 complex component ARPC1 is required for proper cell morphogenesis and polarized cell growth in *Physcomitrella patens. The Plant Cell*, **17**, 2327–2339.

Hepler, K.B.P.K. and Jackson, W.T. (1968) Microtubules and early stages of cell-plate formation in the endosperm of *Haemanthus katherinae* Baker. *Journal of Cell Biology*, **38** (2), 4437–446.

Herrman, H., Strelkov, S.V., Burkhard, P., and Aebi, U. (2009) Intermediate filaments, primary determinants of cell architecture and plasticity. *Journal of Chemical Investigation*, **119**, 1772–1783.

Higaki, T., Sano, T. and Hasezawa, S. (2007) Actin microfilament dynamics and actin-side binding proteins in plants. *Current Opinion Plant Biology*, **7**, 651–660.

Hussey, P.I., Keteloar, T., Deeks, M.J. *et al.* (2006) Control of the actin cytoskeleton in plant cell growth. *Annual Review Plant Biology*, **57**, 109–125.

Hyams, J.S. and Lloyd, C.W. (1994) *Microtubules*. Wiley, New York, NY.

Jeng, R.L. and Welch, M.D. (2001) Cytoskeleton actin and endocytosis no longer the weakest link. *Current Biology*, **11**, R691–R694.

Jovanovic, A.M., Durst, S., and Nick, P. (2010) Plant cell division is specifically affected by nitro-tyrosine. *Journal of Experimental Botany*, **61**, 901–909.

Kaloriti, D., Galva, C., Parupalli, C. *et al.* (2007) Microtubule associated proteins in plants and the processes they manage. *Journal of Integrative Plant Biology*, **49**, 1164–1173.

Karahara, I., Suda, J., Tahara, H. *et al.* (2009) The preprophase band is a localized center of clathrin-mediated endocytosis in late prophase cells of the onion cotyledon epidermis. *The Plant Journal*, **57**, 819–831.

Kirik, V., Grini, P.E., Mathur, J. *et al.* (2002) The Arabidopsis TUBULIN-FOLDING COFACTOR A gene is involved in the control of the alpha/beta-tubulin monomer balance. *The Plant Cell*, **14** (9), 2265–2276.

Kojo, K.H., Higaki, T., Kutsuna, N. *et al.* (2013) Roles of cortical actin microfilament patterning in division plane orientation in plants. *Plant and Cell Physiology*, **54** (9), 1491–1503.

Kong, S. and Wada, M. (2011) New insights into dynamic actin-based chloroplast photorelocation movement. *Molecular Plant*, **4** (5), 771–781.

Kumagai, F. and Hasezawa, S. (2001) Dynamic organization of microtubules and microfilaments during cell cycle progression in higher plant cells. *Plant Biology Stuttgart*, **3**, 4–16.

Kwiatkowska, D. (2006) Flower primordium formation at the *Arabidopsis* shoot apex: quantitative analysis of surface geometry and growth. *Journal of Experimental Botany*, **57**, 571–580.

Liu, B. (2010) *The Plant Cytoskeleton (Advances in Plant Biology)*, Springer, New York, NY.

Lloyd, C. (1998) Actin in plants. *Journal of Cell Science*, **90**, 185–188.

Lloyd, C. and Chan, J. (2004) Microtubules and the shape of plants to come. *Nature Reviews. Molecular Cell Biology*, **5**, 13–23.

Locascio, A., Blázquez, M.A. and Alabadí, D. (2013) Dynamic regulation of cortical microtubule organization through prefoldin-DELLA interaction. *Current Biology*, **23** (9), 804–809.

Lucas, J.R., Nadeau, J.A. and Sack, F.D. (2005) Microtubule arrays and 'Arabidopsis' stomatal development. *Journal of Experimental Botany*, **57**, 71–79.

Mandelkow, E. and Mandelkow, E.M. (1995) Microtubules and microtubule-associated proteins. *Current Opinion Cell Biology*, **7**, 72–81.

Mao, T., Jin, L., Li, H. *et al.* (2005) Two microtubule-associated proteins of the *Arabidopsis* MAP65 family function differently on microtubules. *Plant Physiology*, **138**, 654–662.

Marcus, A.I., Moore, R.C. and Cyr, R.J. (2001) The role of microtubules in guard cell function. *Plant Physiology*, **125**, 387–395.

Marcus, A.I., Li, W., Ma, H. and Cyr, R.J. (2003) A kinesin mutant with an atypical bipolar spindle undergoes normal mitosis. *Molecular Biology of the Cell*, **14**, 1717–1726.

McNulty, A.K. and Saunders, M.J. (1992) Purification and immunological detection of pea nuclear intermediate filaments: evidence for plant nuclear lamins. *Journal of Cell Science*, **103**, Pt 2, 407–414.

Meier, I. (2007) Composition of the plant nuclear envelope: theme and variations. *Journal of Experimental Botany*, **58** (1), 27–34.

Min, G., Yang, C., Tong, X. and Zhai, Z. (1999) Assembly characteristics of plant keratin intermediate filaments *in vitro*. *Science in China Series C: Life Sciences*, **42** (5), 485–493.

Minguez, A. and Moreno Diaz de la Espina, S. (1993) Immunological characterization of lamins in the nuclear matrix of onion cells. *Jounal of Cell Science*, **106**, 431–439.

Moore, R.C., Zhang, M., Cassimeris, L. and Cyr, R.J. (1997) *In vitro* assembled plant microtubules exhibit a high state of dynamic instability. *Cell Motility and the Cytoskeleton*, **38**, 278–286.

Morelli, J., Zhou, W., Yu, J. *et al.* (1998) Actin deploymerization affects stress-induced translational activity of potato tuber tissue. *Plant Physiology*, **116**, 1227–1237.

Müller, J., Beck, M., Mettbach, U. *et al.* (2010) *Arabidopsis* MPK6 is involved in cell division plane control during early root development, and localizes to the pre-prophase band, phragmoplast, trans-Golgi network and plasma membrane. *The Plant Journal*, **61**, 234–248.

Müller, P., Martin, K., Theurich, S. *et al.* (2014) Microtubule-depolymerizing agents used in antibody–drug conjugates induce antitumor immunity by stimulation of dendritic cells. *Cancer Immunological Research*, **74**, 10.1158/2326-6066.CIR-13-0198.

Murata, T., Sonobe, S., Baskin, T.I. *et al.* (2005) Microtubule-dependent microtubules nucleation based on recruitment of tubulin in higher plants. *Nature Cell Biology*, **7**, 961–968.

Nakaba, S., Sano, Y. and Funada, R. (2013) Disappearance of microtubules, nuclei and starch during cell death of ray parenchyma in abies sachalinensis. *IAWA Journal*, **34** (2), 135–146.

Nick, P. (2008) *Plant Microtubules Development and Flexibility*, Springer, New York, NY.

Nick, P. (2013) Microtubules, signalling and abiotic stress. *The Plant Journal*, **75** (2), 309–323.

Pastuglia, M., Azimzadeh, J., Goussot, M. *et al.* (2006) γ-tubulin is essential for microtubule organization and development in *Arabidopsis*. *The Plant Cell*, **18**, 1412–1425.

Pickett-Heaps, J. and Forer, A. (2009) Mitosis: spindle evolution and the matrix model. *Protoplasma*, **235** (1–4), 91–99.

Pollard, T.D. and Cooper, J.A. (2009) Actin, a central player in cell shape and movement. *Science*, **326**, 1208–1212.

Reddy, A.S. (2001) Molecular motors and their functions in plants. *International Review of Cytology*, **204**, 97–178.

Reddy, V.S. and Reddy, A.S.N. (2004) Proteomics of calcium-signaling components in plants. *Phytochemistry*, **65**, 1745–1776.

Rosania, G.R., Chang, Y.T., Perez, O., *et al.* (2000) Myoseverin, a microtubule-binding molecule with novel cellular effects. *Nature Biotechnology*, **18** (3), 304–308.

Schenkel, M., Sinclair, A.M., Johnstone, D. *et al.* (2008) Visualizing the actin cytoskeleton in living plant cells using a photo-convertible mEos::FABD-mTn fluorescent fusion protein. *Plant Methods*, **21**, 10.1186/1746-4811-4-21.

Seagull, R.W. (1990) The effects of microtubule and microfilament disrupting agents on cytoskeletal arrays and wall deposition in developing cotton fibers. *Protoplasma*, **159** (1), 44–59.

Seguí-Simarro, J.M., Austin II, J.R., White, E.A. and Staehelin, L.A. (2004) Electron tomographic analysis of somatic cell plate formation in meristematic cells of *Arabidopsis* preserved by high-pressure freezing. *The Plant Cell*, **16**, 4, 836–856.

Sheterline, P., Clayton, J. and Sparrow, J.C. (1998) *Actin: Protein Profile*, 4th edn. Oxford University Press Inc., New York.

Shevchenko, G. (2011) Actin microfilament organization in the transition one of 'Arabidopsis'—ABD2-GFP roots under clinorotation. *Microgravity Science Technology*, **24**, 427–433.

Shibaoka, H. (1994) Plant hormone-induced changes in the orientation of cortical microtubules: alterations in the cross-linking between microtubules and the plasma membrane. *Annual Review of Plant Physiology and Plant Molecular Biology*, **45**, 527–544.

Smertenko, A. and Franklin Tong, V. E. (2011) Organization and regulation of the cytoskeleton in plant programmed cell death. *Cell Death Differentiation*, **18**, 1263–1270.

Smirnova, E.A., Reddy, A.S.N., Bowser, J. and Bajer, A.S. (1998) Minus end-directed kinesin-like motor protein, Kcbp, localizes to anaphase spindle poles in Haemanthus endosperm. *Cell Motility and the Cytoskeleton*, **41**, 271–280.

Staiger, C.J., Baluska, F., Volkmann, D. and Barlow, P. (2000) *Actin: A Dynamic Framework for Multiple Plant Cell Functions* (*Developments in Plant and Soil Sciences*), Springer, New York, NY.

Staiger, C.J., Sheahan, M.B., Khurana, P. *et al.* (2009) Actin filament dynamics are dominated by rapid growth and severing activity in 'Arabidopsis' cortical array. *Journal of Cell Biology*, **184**, 269–280.

Su, F., Gu, W., and Zhai, Z. (1990) The keratin intermediate filament-like system in maize protoplasm identified by using immunogold labeling. *Cell Research*, **1**, 11–16.

Takakuwa, R., Kokai, Y., Kojima, T. *et al.* (2000) Uncoupling of gate and fence functions of MDCK cells by the actin-depolymerizing reagent Mycalolide B. *Experimental Cell Research*, **257** (2), 238–244.

Vantard, M., Schellenbaum, P., Peter, C. and Lambert, A.M. (1993) Higher plant microtubule-associated proteins: '*in vitro*' functional assays. *Biochemie*, **75**, 725–730.

Vasquez, R.J., Howell, B., Yvon, A.M., Wadsworth, P. and Cassimeris, L. (1997) Nanomolar concentrations of nocodazole alter microtubule dynamic instability *in vivo* and *in vitro*. *Molecular Biology of Cell*, **8** (6), 973–985.

Vaughn, K.C. and Harper, J.D. (1998) Microtubule-organizing centers and nucleating sites in land plants. *International Review of Cytology*, **181**, 75–149.

Venghateri, J.B., Gupta, T.K., Verma, P.J., Kunwar, A. and Panda, D. (2013) Ansamitocin P3 depolymerizes microtubules and induces apoptosis by binding to tubulin at the vinblastine site. *PLoS ONE*, **8**, 10, e75182, doi:10.1371/journal.pone.0075182.

Volkmann, D. and Baluska, F.B. (1999) Actin cytoskeleton in plant: from transport network in signaling networks. *Microscopy Research and Technique*, **47**, 135–154.

Vos, J.W., Safadi, F., Reddy, A.N., and Helper, P.K. (2000) The kinesin-like calmodulin binding protein is differentially involved in cell division. *The Plant Cell*, **12**, 979–990.

Wasteneys, G.O. (2002) Microtubule organization in the green Kingdom: chaos or self-order. *Journal of Cell Science*, **115**, 1345–1354.

Wasteneys, G.O. (2004) Progress in understanding the role of microtubules in plant cells. *Current Opinion Plant Biology*, **7**, 651–660.

Wasteneys, G.O. and Ambrose, J.C. (2009) Spatial organization of plant cortical microtubules: close encounters of the 2-D kind. *Trends in Cell Biology*, **19**, 62–71.

Wasteneys, G.O. and Brandizzi, F. (2013) A glorious half-century of microtubules. *The Plant Journal*, **75** (2), 185–188.

Weaver, B.A. (2014) How Taxol/paclitaxel kills cancer cells. *Molecular Biology of the Cell*, **25** (18), 2677–2681.

Whittmann, T., Hyman, A. and Desai, A. (2001) The spindle: a dynamic assembly of microtubules and motors. *Nature Cell Biology*, **3**, E-28.

Wittmann, T. and Waterman-Storer, C.M. (2001) Cell motility: can Rho GTPases and microtubules point the way? *Journal of Cell Science*, **114**, Pt 21, 3795–3803.

Yang, C., Xing, L. and Zhay, Z. (1992) Intermediate filaments in higher plant cells and their assembly in a cell-free system. *Protoplasma*, **171**, 44–54.

Zhong, R., Burk, D.H., Morrison, III, W.H., and Ye, Z. (2002) A kinesin-like protein is essential for oriented deposition of cellulose microfibrils and cell wall strength. *The Plant Cell*, **14**, 3101–3117.

Further reading

Buchnik, I., Abu-Abied, M. and Sadot, E. (2014) Role of plant myosins in motile organelles: is a direct interaction required? *Journal of Integrative Plant Biology*, **57**, 23–30.

Hamada, T. (2014). Microtubule organization and microtubule-associated proteins in plant cells. *International Review of Cell and Molecular Biology*, **312**, 1–52.

Hashimoto, T. (2015) Microtubules in plants. *The Arabidopsis Book*, **13**: e0179.

Klotz, J. and Nick, P. (2012). A novel actin-microtubule cross-linking kinesin, ntkch, functions in cell expansion and division. *New Phytology*, **193**, 576–589.

Li, S., Sun, T. and Ren, H. (2015). The functions of the cytoskeleton and associated proteins during mitosis and cytokinesis in plant cells. *Frontiers in Plant Science*, **6**, 282, doi: 10.3389/fpls.2015.00282.

McMichael, C.M. and Bednarek, S.Y. (2013) Cytoskeletal and membrane dynamics during higher plant cytokinesis. *New Phytology*, **197**, 1039–1057.

Nick, P. (2014) *Plant Microtubules*, Springer, New York, NY.

Schneider, R. and Persson, S. (2015). Connecting two arrays: the emerging role of actin-microtubule cross-linking motor proteins. *Frontiers in Plant Science*, **6**, 415, doi:10.3389/fpls.2015.00415.

Takeuchi, M., Karahara, I., Kajimura, Kajimurio, N. *et al.* (2016) Single microfilaments mediate the early steps of microtubule bundling during preprophase band formation in onion cotyledon epidermal cells. *Molecular Biology of the Cell*, E15-12-0820.

CHAPTER 7

The mitochondrion

Ray J. Rose[1], Terence W.-Y. Tiew[1], and William V. Dashek[2]

[1]Center of Excellence for Integrative Legume Research, School of Environmental and Life Sciences, The University of Newcastle, Callaghan, New South Wales, Australia
[2]Retired Faculty, Adult Degree Program, Mary Baldwin College, Staunton, VA, USA

Structure and dynamics

Mitochondria (Figure 7.1) are membrane-bound organelles which are characteristic of eukaryotic cells (Day *et al.*, 2004; Logan, 2007; Millar *et al.*, 2008; Scheffler, 2011, Rose and Sheahan, 2012). Plant mitochondria were described from electron micrographs as small, oval organelles 1–2 μm long and 0.5–1 μm wide. Particularly with the advent of mitochondrially targeted green fluorescent protein and confocal microscopy, mitochondria visualized *in vivo* have, however, been shown to be highly pleomorphic (Figure 7.2). They can be small and spherical but also long and worm-like up to 30 μm long.

The number of mitochondria may vary from a few hundreds to a few thousands, depending not only on the cell type but also on the balance between fusion and fission (Logan, 2010). Mitochondrial division by fission was part of the developing understanding that mitochondria did not develop *de novo* but from pre-existing mitochondria (Kuroiwa *et al.*, 1998), while fusion of mitochondria was demonstrated more recently when suitable technology became available. Arimura and co-workers used co-localization of red and green forms of the photoconvertible fluorescent protein Kaede targeted to mitochondria to demonstrate fusion. More dramatically, as shown by Sheahan and co-workers when cultured mesophyll protoplasts are stimulated to initiate division (ultimately developing into whole plants) mitochondria undergo massive mitochondrial fusion, producing extremely long mitochondria. Mitochondrial fusion was confirmed experimentally by protoplast fusion where mitochondria of one cell expressed mitochondria-targeted green fluorescent protein (mtGFP) and mitochondria of the other cell was labeled with mitotracker; subsequent fusion produced a yellow signal due to co-localization of the GFP and mitotracker (Rose and Sheahan, 2012).

Plant Cells and their Organelles, First Edition. Edited by William V. Dashek and Gurbachan S. Miglani.

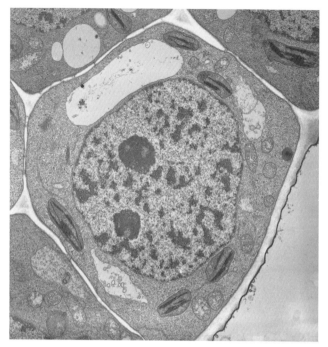

Figure 7.1 Electron micrograph of young plant cell with prominent nucleus and several mitochondria (e.g., see bottom right with two mitochondria located between two chloroplasts and cell wall).

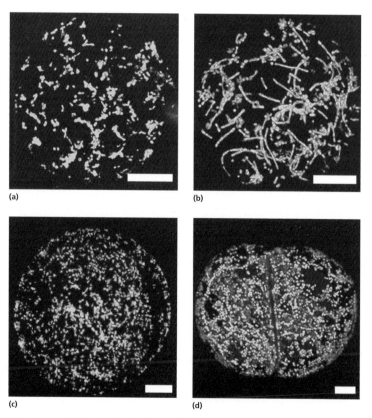

(a)

(b)

(c)

(d)

Figure 7.2 Fusion and fission in isolated protoplasts entering cell division and visualized using green fluorescent protein (GFP). The mitochondria (a) of isolated plant protoplast when cultured undergo fusion (b) then fission (c) prior to cell division where mitochondria distribute into daughter cells (d). Bar = 10 μm.

The dynamin-related and functionally redundant GTPase proteins DRP3A and DRP3B are key factors in mitochondrial fission. Together with their organelle anchor FISSION 1 (FIS1), they form a DRP3–FIS1 complex (Logan, 2010). *Arabidopsis* contains two homologs of FIS1: FIS1A (BIGYIN) and FIS1B. ELONGATED MITOCHONDRIA1 (ELM1) recruits DRP3A/B to the division site. PEROXISOMAL and MITOCHONDRIAL DIVISION FACTOR 1 and 2 (PMD1 and PMD2) are integral membrane proteins of the outer mitochondrial membrane that do not interact with the DRP3–FIS1 complex and likely also have a role in mitochondrial proliferation, though other functions are possible. Less is known about the mechanism of mitochondrial fusion in plants. In animal cells the mechanism of mitochondrial fusion is much better understood where again it is clear that dynamin family members (mitofusins) are key. Mitofusins interact with outer membrane proteins, and together with other proteins facilitate the fusion process. What we do know in plants is the apparent absence of homologs of the mammalian outer (Mfn) and inner (Opa I) membrane GTPases but dependence on an inner membrane electrical potential as in mammalian mitochondrial fusion (Rose and Sheahan, 2012).

The actin cytoskeleton is involved in mitochondrial transport, distribution and inheritance at cell division. However, it is not involved in the fusion process but with microtubules playing a role. Somewhat surprising however is the recent demonstration of actin within the mitochondria, but it could provide explanations for a dynamic mitochondrial morphology and nucleoid segregation in mitochondrial division.

As indicated, mitochondria interact with the cytoskeleton, but they may also interact with other organelles. This has not been critically investigated in plant cells, but recent investigations in animal cells suggest that mitochondrial division takes place at mitochondrial contact sites with the endoplasmic reticulum.

Traditional transmission electron microscopy (TEM) image of the mitochondrion shows it to consist of a smooth outer membrane and a convoluted inner membrane with finger-like projections called 'cristae'. Negative staining of isolated mitochondria demonstrated protrusions of the inner membrane consisting of a stalk and rounded termination (Figure 7.3). These knobs are the H^+-ATP synthases consisting of the F_1 'knob' and the F_0 'stalk'. Six distinct compartments can be recognized in the mitochondrion: the outer membrane, inner membrane, intermembrane space, crystal membranes, intercristal space and matrix (Logan, 2006).

In the electron microscope, within the mitochondrial matrix ribosomes as well as DNA fibrils can be visualized. Mitochondrial DNA in the mitochondrion is organized into nucleoids which can be visualised in the fluorescence microscope with the fluorescent dye DAPI. Nucleoids consist of mtDNA molecules complexed with protein, but there is little knowledge of the molecular organization. Mitochondrial DNA and RNA are discussed in more detail later.

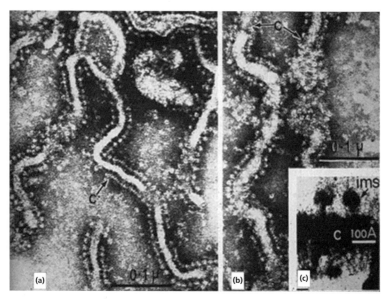

Figure 7.3 Negatively stained mitochondrion cristae with arrays of H^+-ATP synthase F_1 'knobs' (a and b). The F_0 'stalks' and F_1 'knobs' are shown at higher magnification (c). c, Cristae; ims, inner mitochondrial space (mitochondrial matrix).

The mitochondrial genome

The plant mitochondrial genome varies enormously in size from 208 kb to 2.9 Mb, which is substantially larger than other eukaryote mitochondrial genomes. Most plant mitochondrial genomes map as a single circular master chromosome (Figure 7.4) and a collection of sub-genomic circles due to recombination across direct repeats (Schuster and Brennicke, 1994). However, Alverson and co-workers have shown that the cucumber mitochondrial genome maps to three largely autonomous circular molecules, one very large and two smaller ones (1.56 Mb, 83 and 44 kb) in addition to the usual direct–repeat intragenomic recombination. The mitochondrial genome of *Arabidopsis* is 367 kb and the maize mitochondrial genome is 570 kb (Clifton *et al.*, 2004). Despite this large variation in size with different mitochondrial genomes, the number of coding genes remains similar. It is likely that the origins of large mitochondrial genomes reflects increased repeats, expansion of existing introns and acquisition of sequences from diverse sources (chloroplast, nuclear, viruses and bacteria). Isolated mitochondrial DNA is present in circular and other forms. Newer techniques that prevent shearing establish that organellar DNA is present as a mixture of monomers and head–tail concatemers of circular and linear forms; in addition, there are various branched structures. This structural variability can be accounted for by replication and recombination intermediates. Recombination utilizing the large number of repeats is involved in DNA repair, DNA replication

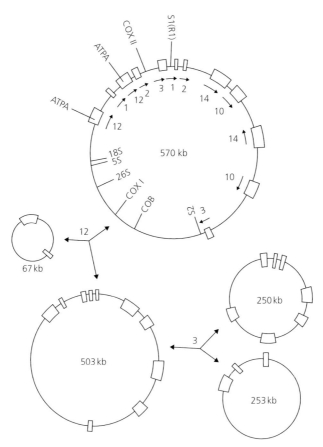

Figure 7.4 The maize 'master' chromosome and recombinants. Examples of intra-molecular recombination using 3 and 12 kb direct repeats of the 'master' chromosome are shown.

and evolution of mitochondrial genomes. These mitochondrial genome dynamics appear necessary for mitochondrial genome stability of the maternally inherited mtDNA, while facilitating a certain amount of variability for evolutionary purposes.

The mitochondrial genome then exists in a multipartite state with subgenomes that exist in circular and linear forms. Such genomic organization means that individual mitochondria can lack a complete genome (Preuten *et al.*, 2010). Mitochondrial fusion can overcome this problem of varying gene content in different mitochondria by ensuring that individual mitochondria can be part of a discontinuous whole. Mitochondrial fusion also accounts for the recombination of mitochondrial genomes between species with different mitochondrial genotypes, in somatic hybridization (Rose *et al.*, 1990).

Lynn Margulis' endosymbiont theory of the origin of mitochondria and their genome is now widely accepted (Gray, 2012). Mitochondria likely

Table 7.1 Mitochondrial-encoded genes.

Mitochondria component	Mitochondria encoded genes	Number of nuclear-encoded genes
Electron transport chain		
Complex I	*nad1, 2, 3, 4, 4L, 5, 6, 7, 9*	42
Complex III	*cob*	9
Complex IV	*cox1, 2, 3*	11
Complex V (H⁺-ATP synthase)	*atp1, 4, 6, 8, 9*	10
Translation machinery		
rRNA ribosomal large subunit	*rrn5, rrn26*	
rRNA ribosomal small subunit	*rrn18*	
Ribosomal proteins	*rps3, 4, 7, 12 rpl 2, 5, 16*	~73
Native tRNAs	*11*	~20
Chloroplast-like tRNAs	*4*	
Other		
Cytochrome c biogenesis	*ccmB, ccmC, ccmFC, ccmFN(2)*	
Protein translocator	*mttB*	
Intronic ORF (open reading frame) in the fourth intron of *nad1*	*mat-R*	

Genes encoded by the mitochondrial genome and the co-operation of the nuclear genome in forming the electron transport chain and translation machinery of the mitochondrion. Data largely based on *Arabidopsis*.

originated by the endosymbiosis of an α-proteobacterium by a primitive eukaryote or archaebacterium between 0.8 and 1.5 billion years ago (Martin and Mentel, 2010). However, the mitochondrion though maintaining some genes depends predominantly on nuclear-encoded genes as over long periods of time endosymbiont genes were transferred to the nucleus (Gray, 2012).

Plant mitochondrial DNA codes for about 32 proteins that comprise components of the electron transport chain as well as the translational apparatus (Gillham, 1994; Table 7.1). The electron transport chain (ETC) components that include mitochondria-encoded proteins are complex I (NADH – ubiquinone oxidoreductase), complex III (ubiquininone – cytochrome c oxidoreductase), complex IV (cytochrome c oxidase) and complex V H⁺-ATP synthase. Most of the proteins in these complexes are actually nuclear encoded so co-operativity, and cross-talk between nuclear and mitochondrial genomes are essential. The coding of the ETC and other components of plant mtDNA are shown in Table 7.1. Again we see with the ribosomal proteins that to build a ribosome, cooperativity between nuclear and mitochondrial genes is required. The tRNAs are also of special interest in that some tRNAs have to be imported from the nucleus and as a result of evolution we have endosymbiont, chloroplast and nuclear origins of tRNAs.

Mitochondrial genes may contain introns and, therefore, require splicing and processing to produce the mature transcript. Some mitochondrial genes may require editing (post-transcriptional modification of a cytosine to uracil residue). Plant mitochondrial transcription, splicing, processing and editing are considered in detail in a number of review chapters in Kempken's consideration of *Plant Mitochondria* (Kempken, 2010).

Comparison of the mitochondrial genome with chloroplast and nuclear genomes

As can be seen in Table 7.2, the mitochondrial genome contributes a small number of genes to the plant cell and fewer genes than does the smaller chloroplast genome, but they are critical genes (Table 7.1). There are multiple copies of the mitochondrial genome in a cell, as for the chloroplast genome, but absolute numbers will depend on the cell type and developmental stage. Copy number per mitochondrion is variable, as discussed before in relation to the requirement for mitochondrial fusion. As reported in *Arabidopsis* the coding region of a mitochondrial gene has fewer than half the nucleotides of a nuclear gene. Interestingly some mitochondrial genes have introns, a characteristic of eukaryotes indicating the dynamic evolutionary interactions between different genome compartments. The protein synthesizing machinery of the cytoplasm, mitochondria and chloroplast is unique for each compartment (Gillham, 1994; Table 7.2).

Table 7.2 Comparison of the three plant genomes.

	Nucleus	Plastid	Mitochondrion
Genome size	125 Mb	154 kb	367 kb
Protein coding genes	27 500	79	32
Genome equivalents	2	560	26
Gene order	Variable, but syntenic	Conserved	Variable
Average coding length (nt)	1900	900	860
Genes with introns (%)	79	18	12
Ribosome	80S	70S	78S
Ribosome LS and SS	60S, 40S	50S, 30S	60S, 44S
Ribosomal RNAs	25S, 17S, 5.8S, 5S	23S, 16S, 5S, 4.5S	26S, 18S, 5S

Comparison of the *Arabidopsis* mitochondrial, plastid and nuclear genomes.
S, sedimentation velocity.

The mitochondrial proteome and protein import

By far, most mitochondrial proteins are encoded by the nucleus and targeted to the mitochondrion (Millar *et al.*, 2005; Gray, 2012). About 1% of mitochondrial proteins are synthesized inside the organelle. In 2004, Heazlewood and co-workers identified 416 proteins using liquid chromatography-tandem mass spectrometry (LC-MS/MS). This number had increased to 471 distinct proteins by 2011 (Klodman *et al.*, 2011). It has been suggested that there may be as many as 1500 gene products targeted to the mitochondrion, based on targeting prediction software applied to the *Arabidopsis* genome. The 2011 study used an initial separation on two-dimensional blue native (BN)/SDS-PAGE (as opposed to 2D isoelectric focusing/SDS–PAGE) combined with MS/MS with improvements in mass spectrometry, computer technology and software. As can be seen from the ETC complexes (Table 7.1), mitochondrial proteins commonly form protein complexes, and 35 different complexes were identified in this latter study by Klodman and co-workers.

Given that most mitochondrial proteins are nuclear-encoded, synthesized on cytoplasmic ribosomes and imported from the cytosol, a sophisticated import apparatus is required. There are at least four different pathways involved in the importing and sorting of proteins targeted to mitochondria, requiring energy in the form of adenosine triphosphate (ATP), a transmembrane potential ($\Delta\Psi$) or redox reactions (Chacinska *et al.*, 2009). Nucleus-encoded proteins are mostly targeted to the mitochondria by a peptide signal sequence located at the N-terminal end of the protein cleaved after translocation, or by an 'internal' signal which is not cleaved after transport. Mitochondrial protein import typically involves unfolding of the preprotein by the Hsp70 chaperone, before docking with the preprotein import machinery (Figure 7.5), where Hsp70 acts as a molecular motor pulling the precursor proteins across the mitochondrial membranes. The translocase of the outer membrane (TOM) complex is the gateway for the import of most mitochondrial proteins (Carrie *et al.*, 2010; Figure 7.5). In *Arabidopsis* the current interpretation is that the TOM complex has seven subunits: two receptors located on the outer face of the outer mitochondrial membrane (Tom20 and OM64) and five import pore components (Tom40, Tom9, Tom7, Tom6 and Tom5). The insertion of β-barrel and α-helical proteins into the outer membrane is carried out by the sorting and assembly machinery of the outer mitochondrial membrane (SAM) complex. In plants it is known that the receptor protein is metaxin, and it is complexed with a protein called SAM50. Translocases of the inner membrane (TIM) complexes facilitate protein transfer to the mitochondrial matrix (the TIM17:23 complex in association with the pre-sequence translocase-associated motor PAM) or insertion into the inner membrane with TIM22. In addition to the TIM17:23 complex and TIM22, there is the mitochondrial intermembrane space import and assembly (MIA) complex associated with the import of proteins destined for the inner

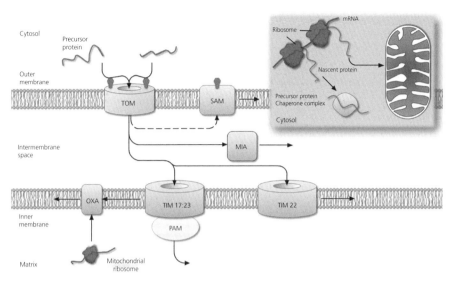

Figure 7.5 Mitochondrial import machinery. The diagram shows how proteins from the cytosol (and some proteins encoded by mitochondrial DNA) are transported to specific locations in the mitochondrion. Refer to the text for components.

membrane and matrix. Inner membrane proteins synthesized on mitochondrial ribosomes are inserted using the oxidase assembly (OXA) complex (Carrie *et al.,* 2010; Figure 7.5).

Respiratory metabolite transporters

Movement of metabolites and small molecules (up to ~1 kDa) across the outer mitochondrial membrane is facilitated by voltage-dependent anion channels (VDACs) or porins that form protein channels through which molecules can diffuse. Transport of molecules such as organic acids, amino acids and adenine di- and trinucleotides across the inner mitochondrial membrane is accomplished by a family of carriers that shuttle molecules between the matrix and the intermembrane space (Millar *et al.,* 2011).

The electron transport chain and oxidative phosphorylation

Generation of pyruvate for the tricarboxylic acid cycle

The mitochondrial matrix houses the tricarboxylic acid (TCA) or Krebs cycle which is the centrepiece of the carbon metabolizing machinery of plant mitochondria. Pyruvate provides the substrate for the cycle and is generated in the cytoplasm primarily by glycolysis, though pyruvate can also be generated in the mitochondrial matrix by malic enzyme (Millar *et al.*, 2011).

There are many detailed considerations of glycolysis in plants (e.g. Bowlby, 2006). The translocated sugar in plants is sucrose and is the substrate for glycolysis in most plants. Sucrose is converted into hexose phosphates. An outline of the flow of reactions from glucose-6-phosphate to pyruvate is shown in Table 7.3 and Figure 7.6a. In the glycolysis process four molecules of substrate-level ATP is produced, in addition to two molecules of NADH as part of the respiration process. There is evidence in *Arabidopsis* that enzymes of glycolysis are associated with the mitochondrial surface as well as the cytosol. It is suggested that the glycolytic enzymes associate dynamically with the mitochondrial surface to support substrate channelling and restrict the use of intermediates from competing metabolic pathways. As can be seen in Figure 7.6a, glucose-6-phosphate can be utilised by the pentose phosphate pathway to provide a source of carbon skeletons and intermediates, as required, for biosynthetic processes. Eventually triose phosphates can return to glycolysis and pyruvate formation.

There is another source of pyruvate generated within the mitochondrion from malate which can enter the mitochondrion and be converted to pyruvate by malic enzyme (Table 7.4). The release of CO_2 by the malic enzyme reaction in the mitochondria is also important in C_4 metabolism of certain C_4 Plants.

The TCA cycle

The pyruvate dehydrogenase complex in the mitochondrion links glycolysis to the TCA cycle. Oxidative decarboxylation of pyruvate to acetyl CoA, CO_2 and NADH provides acetyl CoA. The TCA cycle then continues to turn by the action of citrate synthase which catalyzes the condensation of acetyl CoA and oxaloacetate to produce citric acid. The subsequent enzymic reactions and the TCA cycle components are shown in Table 7.4 and Figure 7.6b. The oxidation of pyruvate in the cycle reduces four molecules of NAD^+ to NADH and one molecule of FAD to $FADH_2$ and oxaloacetate is regenerated. One molecule of ATP is produced by substrate-level phosphorylation during the reaction catalysed by succinyl-CoA synthetase. It is the reduced electron carriers NADH and $FADH_2$ that produce most of the ATP in the oxidative phosphorylation process by the ETC in the inner membrane of the mitochondrion. All the enzymes of the TCA cycle are located in the matrix of the mitochondrion, except succinate dehydrogenase, which is the membrane-bound complex II of the ETC (Figure 7.7).

In the absence of oxygen the fermentative pathway occurs (e.g. roots growing in waterlogged soils), pyruvate does not enter the TCA cycle, and most commonly CO_2 and ethanol are produced associated with the formation of NAD^+ from NADH. ATP is only produced in glycolysis when O_2 is absent.

Table 7.3 The enzyme reactions of glycolysis.

Enzyme	EC number	Enzyme reaction	Reaction product
Hexokinase	2.7.1.1	Transfer of phosphate group from **ATP** to glucose yielding glucose-6-phosphate and ADP (1)	Glucose-6-phosphate
Glucose-6-phosphate isomerase	5.3.1.9	Converts glucose-6 phosphate to fructose-6-phosphate (2)	Fructose 6-phosphate
6-Phosphofructokinase	2.7.1.11	Transfer of a phosphate group from **ATP** to fructose 6-phosphate yielding fructose-1,6-bisphosphate and ADP (3)	Fructose 1,6-bisphosphate
Fructose-bisphosphate aldolase	4.1.2.13	Converts the carbon–carbon cleavage of fructose 1,6-bisphosphate to yield glyceraldehyde-3-phosphate and dihydroxy-acetone phosphate (4)	Dihydroxy-acetone phosphate + glyceraldehyde-3-phosphate
Triose-phosphate isomerase	5.3.1.1	Interconversion of glyceraldehyde-3-phosphate and dihydroxy-acetone phosphate (5)	Glyceraldehyde-3-phosphate
Glyceraldehyde-3-phosphate dehydrogenase	1.2.1.12	Phosphorylation of glyceraldehyde 3-phosphate producing 1,3-bisphosphoglycerate and reducing NAD$^+$ to **NADH** (6)	1,3-Bisphosphoglycerate + **NADH**
Phosphoglycerate kinase	2.7.2.3	Interconverts 1,3-bisphosphoglycerate and ADP to 3-phospho-D-glycerate and **ATP** (7)	3-Phosphoglycerate + **ATP**
Phosphoglycerate mutase	5.4.2.1	Catalyses the intramolecular phosphate group transfer between positions 2 and 3 of phosphoglycerate (8)	2-Phosphoglycerate
Enolase	4.2.1.11	Catalyses the dehydration of 2-phospho-D-glycerate to yield phosphoenolpyruvate and H_2O (9)	Phosphoenolpyruvate + H_2O
Pyruvate kinase	2.1.7.40	This enzyme transfers the phosphate group from phosphoenolpyruvate to ADP to form **ATP** under aerobic conditions (10)	**Pyruvate + ATP**

A thorough discussion of the properties of the enzymes can be found in Bowlby (2006).

Phosphorylated hexose sugars are converted to 3-carbon sugars which ultimately produce ATP and pyruvate. The pyruvate can be transported into the mitochondrion where it is oxidized by the tricarboxylic acid (or Krebs or citric acid) cycle. The numbers 1–10 refer to the numbers in the glycolysis diagram in Figure 7.6a.

Bold text highlights ATP utilization and production, NADH production is used for ATP synthesis in the mitochondrion in the presence of oxygen, and pyruvate, the end product of glycolysis.

EC, enzyme commission.

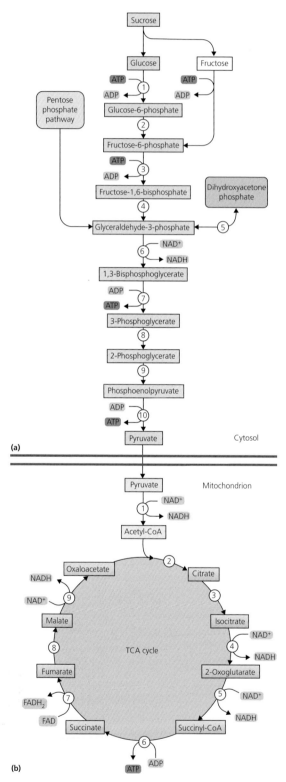

Figure 7.6 Glycolysis and the trichloroacetic acid (TCA) cycle. The numbers in the glycolysis diagram (a) refer to the enzyme reactions shown in Table 7.3. The numbers in the TCA cycle diagram (b) refer to the enzyme reactions shown in Table 7.4.

Table 7.4 The enzyme reactions of the tricarboxylic acid (or Krebs) cycle.

Enzyme	EC number	Reactions catalysed	Reaction product
Pyruvate dehydrogenase complex	2.7.11.2	Links glycolysis to the TCA cycle. Oxidative decarboxylation of pyruvate to acetyl CoA, CO_2 and **NADH** (1)	Acetyl CoA+ CO_2 + **NADH**
Citrate synthase	2.3.3.1	Irreversible aldo condensation of acetyl CoA and **OXALOACETATE** to yield citrate and CoA (2)	Citrate+CoA
Aconitase	4.2.1.3	Isocitrate produced from citrate in dehydration reaction (3)	Isocitrate
Isocitrate dehydrogenase	1.1.1.41	Irreversible NAD^+-dependent oxidative decarboxylation of isocitrate to yield 2-oxoglutarate, CO_2 and **NADH** (4)	2-Oxoglutarate+ CO_2 + **NADH**
2-Oxoglutarate dehydrogenase complex	1.2.4.2	Oxidative decarboxylation of 2-oxoglutarate to yield succinyl CoA, CO_2 and **NADH** (5)	Succinyl CoA+ CO_2 + **NADH**
Succinyl-CoA synthetase	6.2.1.4	Substrate-level phosphorylation of ADP yielding succinate, **ATP** and CoA (6)	Succinate+ **ATP** +CoA
Succinate dehydrogenase	1.3.5.1	Associated with the inner mitochondrial membrane succinate and FAD converted to fumarate and **FADH2** (7)	Fumarate+ **FADH₂**
Fumarase	4.2.1.2	L-Malate is formed by the hydration of fumarate (8)	L-Malate
Malate dehydrogenase	1.1.1.37	L-Malate and NAD^+ converted to oxaloacetate and **NADH** (9)	**OXALOACETATE** + **NADH**
Malic enzyme	1.1.1.39	Oxidative decarboxylation of malate using NAD^+ to produce pyruvate, CO_2 and **NADH**	Pyruvate+ CO_2 + **NADH**

A thorough discussion of the properties of the enzymes can be found in Bowlby (2006).

Pyruvate (and malate) can enter the mitochondrion and be oxidized by the tricarboxylic acid cycle (TCA) and oxaloacetate (bold upper case) is regenerated. A feature of the plant TCA cycle is the presence of malic enzyme in the mitochondrial matrix. NADH and $FADH_2$ are utilized in oxidative phosphorylation in the cytochrome electron transport chain to produce ATP.

The numbers 1–9 refer to the numbers in the TCA cycle diagram in Figure 7.6b.

EC, enzyme commission.

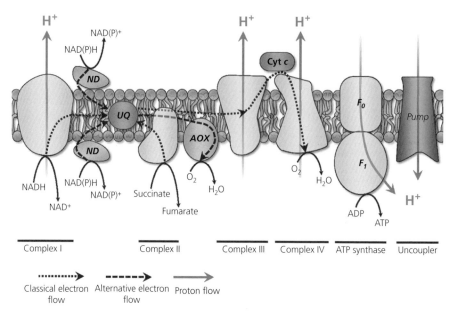

Figure 7.7 The cytochrome electron transport chain. The diagram shows classical electron flow, alternative electron flow involving the alternate oxidase and proton flow. Refer to the text for components.

The electron transport chain and oxidative phosphorylation

Plant mitochondria, as in most other eukaryotes, have an ETC in the inner mitochondrial membrane where the reduced electron carriers are oxidized, with O_2 as the terminal electron acceptor. The mechanism of mitochondrial ATP synthesis is based on the chemiosmotic hypothesis proposed by Nobel Laureate Peter Mitchell. The oxidation of NADH and $FADH_2$ enables the generation of an electrochemical proton gradient as protons flow into the intermembrane and intercristal spaces associated with the flow of electrons to O_2 in the ETC (Figure 7.7). Proton translocation occurs at complexes I, III and IV and generates a pH and electrical gradient. The flow of protons from the intermembrane and intercristal spaces back to the mitochondrial matrix via the H^+-ATP synthase (also known as Complex V) generates ATP. The number of subunits of complexes I, III, IV and V are shown in Table 7.1. Complex II (succinate – ubiquinone oxidoreductase also called 'succinate dehydrogenase') is the smallest complex and has four core subunits and four subunits of unknown function (Millar *et al.*, 2011). The H^+-ATP synthase consists of an F_1 'knob' that protrudes into the matrix and an F_0 'stalk' embedded in the inner membrane. The dimeric H^+-ATP synthase may be responsible for the folding of the inner membrane. The individual complexes of the ETC may be organized into supercomplexes to optimize the process of phosphorylation, rather than free-moving individual respiratory complexes (Eubel *et al.*, 2003).

The alternative electron transfer chain in plant mitochondria

In addition to the classical ETC, plants possess a non-phosphorylating alternative ETC. This is due to the presence of a cyanide-insensitive alternative oxidase (AOX) on the matrix face of the inner mitochondrial membrane and rotenone-insensitive alternative NAD(P)H dehydrogenases (NDs) (Møller, 2001). Essentially the alternative ETC is able to decouple electron transport from ATP formation. This is because proton translocation steps can be by-passed as can be seen in Figure 7.7, where some alternative electron flow pathways can occur whereby no ATP production occurs. The AOX transfers electrons from ubiquinone to oxygen and NDs transfer electrons to ubiquinone. Plant mitochondria also possess plant uncoupling mitochondrial proteins (PUMPs) that transport protons from the intermembrane space to the matrix, with the net effect of also uncoupling ATP formation from electron transport. The uncoupling of respiration from oxidative phosphorylation produces heat and can be seen most dramatically in thermogenesis in the *Araceae* family where inflorescences can reach 40°C to volatilize compounds to attract pollinators. How precisely AOX and PUMPs interact in this latter case is not clear at this time. A more general role for the alternative ETC is to modulate reactive oxygen species (ROS) production, together with antioxidants such as the ascorbate–glutathione cycle. Aerobic metabolism (as opposed to anaerobic metabolism where CO_2 and ethanol are produced and NADH utilized) has the advantage of generating substantially more ATP, but in the process produces ROS because of electron leakage in the form of superoxide ions (O_2^-) and conversion to H_2O_2 by superoxide dismutase. ROS in excess can cause oxidative stress and subsequent DNA, protein and lipid damage. Under stress conditions, ROS can be produced in excess and the alternative ETC can facilitate homeostasis. However, what is now clear is that ROS production can also serve as a signal to initiate pathways to actually ameliorate stress (Foyer *et al.*, 2009).

Plant mitochondria, stress responses and programmed cell death

It has been argued that AOX, likely in conjunction with alternative NAD(P)H dehydrogenases, plays a central role in programming the stress response (Van Aken *et al.*, 2009). This is because the alternative ETC minimizes ROS production to help maintain ROS homeostasis. Knockdown of *AOX1* expression typically causes severe effects only in stressed plants. Changes in AOX expression also alters the level of antioxidants with overexpression of AOX1, causing elevated ascorbate levels in *Arabidopsis*. In addition, the alternative ETC can help maintain carbon flux through the TCA cycle. For example, proline is ultimately synthesized from TCA cycle carbon skeletons and accumulates under salt stress (Mackenzie and McIntosh, 1999).

Extreme perturbations in ROS homeostasis can lead to apoptotic-like programmed cell death (PCD). PCD occurs in the hypersensitive response in defence, and in development as illustrated by the breakdown of the suspensor. Mitochondria likely integrate stress and development signals via ROS/antioxidant levels that determine whether the cell activates its PCD pathway (Reape and McCabe, 2010). The sequence of events in PCD is an early Ca^{2+} flux, mitochondrial membrane permeabilization via opening of permeability transition pores (PTPs) formed at contact points between the inner and the outer mitochondrial membranes, release of apoptogenic proteins (cytochrome c and nucleases also release) and activation of the metacaspase proteases which cause cell destruction (Reape and McCabe, 2010). It is the apoptogenic proteins that are thought to activate the metacaspases, and cytochrome c release increases ROS output and activates PTPs reinforcing the PCD pathway. Although mitochondria are the key regulators of apoptosis, chloroplasts also appear to have a role, presumably related to their electron transfer and ROS activities.

One of the questions in relation to the alternate ETC is as follows: How is this activated in stress responses? It is possible that it is an initial ROS signal that activates the mitochondrial stress response. This could come from plasma membrane-bound NADPH oxidase or, as more recently reported, from a direct effect on succinate dehydrogenase, also known as complex II.

Other functions of plant mitochondria

Photorespiration

This process does not involve the mitochondrial ETC and O_2 consumption. It involves three organelles: the chloroplast, peroxisome and mitochondrion. Oxygen is consumed by the chloroplast and peroxisome. The enzymatic reactions characteristic of photorespiration are presented in Table 7.5 and Figure 7.8. The oxygenase activity of RuBisCO in the chloroplast leads to the formation of 2-phosphoglycolate as well as 3-phosphoglyceric acid. The 2-phosphoglycolate is converted to glycolate, which moves to the peroxisome, and O_2 is used to form glyoxylate and H_2O_2. The enzyme catalase, present in the peroxisome, produces $\frac{1}{2}O_2$ and H_2O from H_2O_2. Glyoxylate is converted to glycine, and glycine is passed to the mitochondrion where serine is produced together with $CO_2 + NADH + NH_4^+$. Serine moves out of the mitochondria into the peroxisome where hydroxypyruvate and then glycerate is formed. Glycerate is transported to the chloroplast where 3-phosphoglycerate is formed which can then be metabolized via the Calvin–Benson cycle of photosynthesis. So the photorespiratory C_2 cycle serves as a carbon recovery system that would otherwise be lost as 2-phosphoglycolate. There is also the interesting argument that under stress conditions (e.g. high

Table 7.5 The main enzyme reactions of photorespiration.

Enzyme	EC number	Reactions catalysed	Reaction product
Ribulose-1,5-bisphosphate carboxylase-oxygenase (in chloroplast)	4.1.3.9	Oxygenation of ribulose-1,5-bisphosphate generates 3-phosphoglycerate and 2-phosphoglycolate (1)	2-Phosphoglycolate participates in photorespiration
Phosphoglycolate phosphatase (in chloroplast)	3.1.3.18	2-Phosphoglycolate hydrolyzed to glycolate (2)	Glycolate produced in the chloroplast passes to the peroxisome
Glycolate oxidase (in peroxisome)	1.1.3.15	Glycolate is oxidized to produce glyoxylate and hydrogen peroxide (3)	$Glyoxylate + H_2O_2$
Glutamate: glyoxylate aminotransferase (in peroxisome)	2.6.1.4	Transamination reactions in which glycine is produced (4)	Glycine is one of the products and exits the peroxisome and enters the mitochondrion
Glycine decarboxylase and Glycine hydroxymethyl transferase complex (in mitochondrion)	1.4.4.2 and 2.1.2.1	$Serine + CO_2 + NADH + NH_4^+$ (5)	Serine is produced and passes from the mitochondrion to the peroxisome

Photorespiration requires the involvement of chloroplasts, peroxisomes and mitochondria. Serine after passing to the peroxisome is utilized to ultimately produce glycerate which passes to the chloroplast to produce 3-phosphoglycerate.
The numbers 1–5 refer to the numbers in the photorespiratory cycle diagram in Figure 7.8.
EC, enzyme commission.

light and water stress), when electron acceptors become limiting, there is surplus electron flow in the chloroplast and photorespiration which can alleviate the damage caused by photoinhibition. Photorespiration with the C_2 cycle can dissipate excess reducing equivalents.

Cytoplasmic male sterility

Cytoplasmic male sterility (CMS) results in the inability to form viable pollen, through either disrupted microsporogenesis or conversion of stamens into other floral organs, frequently carpelloid stamens (Rose and Sheahan, 2012). CMS results from disrupted mitochondrial function. How is mitochondrial function disrupted? The general explanation is that there are disrupted mtDNA genes or new open reading frames that produce novel proteins that have toxic effects (Hanson and Bentolila, 2004). The net effect of the mitochondrial changes on ATP production can also be quite severe, particularly affecting the high energy requirements for pollen production. Homeotic floral morphologies can result from recombination from fusion of mitochondria of different genotypes derived

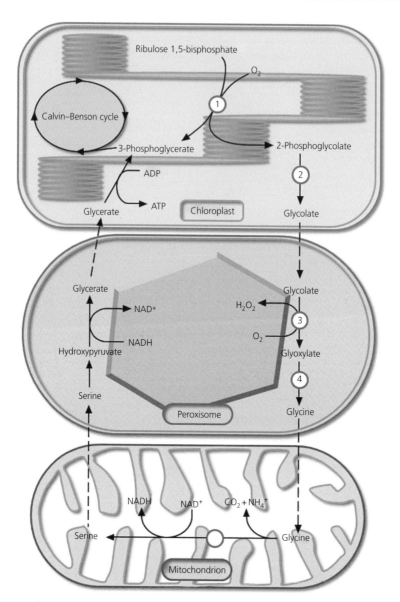

Figure 7.8 Photorespiratory cycle involving chloroplasts, peroxisomes and mitochondria. The numbers refer to the enzyme reactions of photorespiration shown in Table 7.5.

from hybridization including somatic hybridization. How floral development is targeted by mitochondrial genome changes is not understood.

Mitochondria as a biosynthesis centre

In addition to being classically known as the 'power house of the cell' because of its central role in aerobic respiration in eukaryotes, the mitochondria are involved in anabolic functions such as the synthesis of amino acids and lipids

(Mackenzie and McIntosh, 1999). Mitochondria are also intimately involved in the biosynthesis of many vitamins in plants. The entire folate biosynthetic pathway is present exclusively in mitochondria (vitamin B_9). Also located in mitochondria are the first enzyme of pantothenate biosynthesis (vitamin B_5), the terminal enzyme of biotin biosynthesis (vitamin B_7), the last enzyme of ascorbate biosynthesis (vitamin C) and the first enzyme of thiamine biosynthesis (vitamin B_1).

The relationship of mitochondria to other organelles

Photorespiration involves shuttling of metabolites between chloroplasts, mitochondria and peroxisomes; in electron micrographs, these organelles are frequently found in close association. It has also been suggested that the relative position of these three organelles could influence ROS signalling. However, given that most mitochondrial proteins are encoded by nuclear DNA and they need to interact with the fewer, but important, proteins encoded by mitochondrial DNA (Table 7.1), there needs to be a close two-way interaction to assemble some of the key complexes, particularly those of the cytochrome electron transport chain. Further, the nucleus ultimately has to regulate the activities of the mitochondrion in the context of the whole cell and organism. Conversely, and perhaps less obviously, there may be a need to signal the nuclear genome to respond to mitochondrial requirements.

The two-way communication between mitochondria and nucleus is referred to as anterograde (nucleus to mitochondria) and retrograde signalling (mitochondria to nucleus). The nucleus controls the transcription, translation and import of proteins required for mitochondrial gene expression, and it is the pentatricopeptide (PPR) proteins that are important in these control mechanisms. PPR proteins are involved in RNA editing, RNA processing and RNA translation. Nuclear genes (called nuclear restorer genes) that reverse CMS are often PPR genes. These PPR proteins are able to resolve aberrant mitochondrial DNA expression.

Retrograde signalling reflects the need of mitochondrial status to be signalled to the nucleus. Adaptive measures related to oxygen availability, rate of ATP synthesis or ROS production may be required. ROS signalling from mitochondria can result from stress and cause adaptive responses or lead to PCD. The signal transduction pathway from mitochondria to the nucleus is not well understood, but it may involve Ca^{2+} signalling, protein kinases and nuclear transcription factors (Rhoads and Subbiah, 2007). One example of retrograde signalling is the induction of the nuclear gene *AOX1* after inhibition of Complex III by Antimycin A. Another example involves the CMS phenotype which can result from defective mitochondrial retrograde signalling. A nuclear-encoded gene, called *RETROGRADE-REGULATED MALE STERILITY* that is not a PPR gene, when inhibited prevents retrograde signalling and restores fertility in rice.

References

Bowlby, N. (2006) Plant metabolism – respiration. In: *Plant Cell Biology*, (Dashek, W.V. & Harrison, M., eds), Science Publisher, Enfield, NH, pp. 359–398.

Carrie, C., Murcha, M.W. and Whelan, J. (2010) An *in silico* analysis of the mitochondrial import apparatus of plants. *BMC Plant Biology*, **10**, 249.

Chacinska, A., Koehler, C.M., Milenkovic, D., Lithgow, T. and Pfanner, N. (2009) Importing mitochondrial proteins: machineries and mechanisms. *Cell*, **138**, 628–644.

Clifton, S.W., Minx, P., Fauron, C.M.-R. *et al.* (2004) Sequence and comparative analysis of the maize NB mitochondrial genome. *Plant Physiology*, **136**, 3486–3503.

Day, D.A., Millar, A.H. and Whelan, J. (eds) (2004) *Plant Mitochondria: From Genome to Function*. Advance in Photosynthesis and Photorespiration, vol. **17**. Kluwer Academic Publishers, Dordrecht.

Eubel, H., Jänsch, L. and Braun, H.-P. (2003) New insights into the respiratory chain of plant mitochondria. Supercomplexes and a unique composition of complex II. *Plant Physiology*, **133**, 274–286

Foyer, C.H., Bloom, A., Queval, G. and Noctor, G. (2009) Photorespiratory metabolism: genes, mutants, energetics, and redox signaling. *Annual Review of Plant Biology*, **60**, 455–484.

Gillham, M.W. (1994) *Organelle Genes and Genomes*. Oxford University Press, Oxford.

Gray, M.W. (2012) Mitochondrial evolution. *Cold Spring Harbor Perspectives in Biology*, **4**, a011403

Hanson, M.R. and Bentolila, S. (2004) Interactions of mitochondrial and nuclear genes that affect male gametophyte development. *The Plant Cell*, **16** (Supplement 2004), S154–S169.

Kempken, F. (ed) (2010) *Plant Mitochondria*. Advances in Plant Biology, vol. **1**. Springer, New York, NY.

Klodman, J., Senkler, M., Rode, C. and Braun, H.-P. (2011) Defining the protein complex proteome of plant mitochondria. *Plant Physiology*, **157**, 587–598.

Kuroiwa, T., Kuroiwa, H., Sakai, A., Takahashi, H., Toda, K. and Itoh R (1998) The division apparatus of plastids and mitochondria. *International Review of Cytology*, **181**, 1–41.

Logan, D.C. (2006) The mitochondrial compartment. *Journal of Experimental Botany*, **57**, 1225–1243.

Logan, D.C. (2007) *Plant Mitochondria*. Annual Plant Reviews, vol. **31**. Blackwell Publishing Ltd, Oxford.

Logan, D.C. (2010) Mitochondrial fusion, division and positioning in plants. *Biochemical Society Transactions*, **38**, 789–795.

Mackenzie, S. and McIntosh, L. (1999) Higher plant mitochondria. *The Plant Cell*, **11**, 571–585.

Martin, W.F. and Mentel, M. (2010) The origin of mitochondria. *Nature Education*, **3**, 58.

Millar, A.H., Heazelwood, J.L., Kristensen, B.K., Braun, H.-P. and Møller, I.M. (2005) The plant mitochondrial proteome. *Trends in Plant Science*, **10**, 36–43.

Millar, A.H., Small, I.D., Day, D.A and Whelan, J. (2008) Mitochondrial biogenesis and function in *Arabidopsis*. *The Arabidopsis Book*, **6**, e0111. 10.1199/tab.0111, The American Society of Plant Biologists.

Millar, A.H., Whelan, J., Soule, K.L. and Day, D.A. (2011) Organization and regulation of mitochondrial respiration in plants. *Annual Review of Plant Biology*, **62**, 79–104.

Møller, I.M. (2001) Plant mitochondria and oxidative stress: electron transport, NADPH turnover, and metabolism of reactive oxygen species. *Annual Review of Plant Physiology and Plant Molecular Biology*, **52**, 561–591.

Preuten, T., Cincu, E., Fuchs, J., Zoschke, R., Liere, K. and Börner, T. (2010) Fewer genes than organelles: extremely low and variable gene copy numbers in mitochondria of somatic plant cells. *The Plant Journal*, **64**, 948–959.

Reape, T.J. and McCabe, P.F. (2010) Apoptotic-like regulation of programmed cell death in plants. *Apoptosis*, **15**, 249–256.

Rhoads, D.M. and Subbiah, C.C. (2007) Mitochondrial retrograde regulation in plants. *Mitochondrion*, **7**, 177–194.

Rose, R.J. and Sheahan, M.B. (2012) *Plant Mitochondria*. In: Encyclopedia of Life Sciences. eLS. John Wiley& Sons Ltd, Chichester.

Rose, R.J., Thomas, M.R. and Fitter, J.T. (1990) The transfer of cytoplasmic and nuclear genomes by somatic hybridization. *Australian Journal of Plant Physiology*, **17**, 303–321.

Scheffler, I.E. (2011) *Mitochondria*, 2 Edn. John Wiley & Sons, Inc., Hoboken, NJ.

Schuster, W. and Brennicke, A. (1994) The plant mitochondrial genome: physical structure, information content, RNA editing, and gene migration to the nucleus. *Annual Review of Plant Physiology and Plant Molecular Biology*, **45**, 61–78.

Van Aken, O., Giraud, E., Clifton, R. and Whelan, J. (2009) Alternative oxidase: a target and regulator of stress responses. *Physiologia Plantarum*, **137**, 354–361.

Further reading

Hammani, K. and Giege, P. (2014) RNA metabolism in plant mitochondria. *Trends in Plant Science*, **19**, 380–389.

Johnston, I. and B. Williams. (2016) The shrinking mitochondrion. *The Scientist*, **2**, 101–111.

Korobova, E., Ramabhadran, V. and Higgs, H.N. (2013) An actin-dependent step in Mitochondrial Fission Mediated by the ER-Associated Formin INF2. *Science*, **339**, 6118, 464–467.

Lapuente-Burn, E., Moreno-Loshuertos, R., Acín-Pérez, R. *et al.* (2013) Supercomplex assembly determines electron flux in the mitochondrial electron transport chain. *Science*, **340**, 1567–1570.

Smith, A.C. and Robinson, A.J. (2011) A metabolic model of the mitochondrion and its use in modelling diseases of the tricarboxylic acid cycle. *BMC Systems Biology*, **5**, 102, DOI: 10.1186/1752-0509-5-102.

Wallace, D.C. (2010) The epigenome and the mitochondrion: bioenergetics and the environment. *Genes & Development*, **24**, 1571–1573.

Whelan, J. and Murcha, M.W. (2015) *Plant Mitochondria. Methods and Protocols*, Springer, New York, NY.

Youle, R.J. and Van der Bleik, A.M. (2012) Mitochindrial fission, fusion and stress. *Science*, **337**, 1062–1065.

CHAPTER 8

Nucleus

Yogesh Vikal and Dasmeet Kaur

School of Agricultural Biotechnology, Punjab Agricultural University, Ludhiana, India

Cell nuclei, the most prominent compartment of eukaryotic cells, came first into view around 1830, when Robert Brown described areola within plant cells, which he called nuclei. Descriptions of nuclei in cells of different origin, of nucleoli, and of other subnuclear structures like Cajal bodies (CBs) followed; soon, nuclei were recognized as the structure containing the genetic material. The shape of the nucleus may be related to that of the cells or may be completely irregular. In spheroidal, cuboidal, or polyhedral cells, the nucleus is generally a spheroid. In cylindrical, prismatic, or fusiform cells, it tends to be an ellipsoid. The nucleus occupies about 10% of a eukaryotic cell's volume. In general, each somatic nucleus has a specific size that depends partly on its DNA content and mainly on its protein content, and so its size is related to functional activity during the period of nondivision.

Most of the cells are mononucleate, but binucleate cells (some liver and cartilage cells) and polynucleate cells also exist. In the syncytia, which are large protoplasmic masses not divided into cellular territories, the nuclei may be extremely numerous. The same is the case with striated muscle fibers and certain algae, which may contain several hundred nuclei. The contents of the nucleus are present as a viscous, amorphous mass of material enclosed by a complex nuclear envelope (NE) that forms a boundary between the nucleus and cytoplasm and perforated at intervals by the nuclear pores (Figure 8.1). The nucleoplasm or nuclear sap (the fluid substance in which the solutes of the nucleus are dissolved) in a typical interphase cell contains the uncondensed chromatin and the parts of the chromosomes that remain condensed (chromocenters). The nucleus also contains one or more nucleoli, organelles that synthesize protein-producing macromolecular assemblies called ribosomes, and a variety of other smaller components, such as CBs, Gemini of coiled bodies (GEMS), and interchromatin granule clusters. The nuclear matrix consists of a fibrillar network.

Plant Cells and their Organelles, First Edition. Edited by William V. Dashek and Gurbachan S. Miglani.
© 2017 John Wiley & Sons, Ltd. Published 2017 by John Wiley & Sons, Ltd.

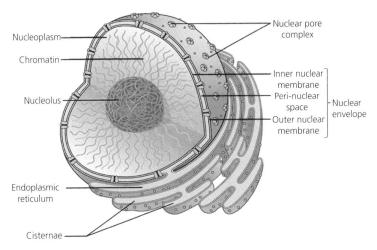

Figure 8.1 Structure of nucleus.

Structural organization of the NE

The NE is organized as a double membrane consisting of the inner nuclear membrane (INM) and the outer nuclear membrane (ONM). The two membranes are separated by a perinuclear space (lumen) of 10–50 nm in width. The membranes of the NE serve as a barrier between the nucleus and the cytoplasm that prevents free passage of ions, solutes, and macromolecules. The ONM is a lipid bilayer, continuous with the membrane of the rough endoplasmic reticulum (RER) and is generally studded with ribosomes (Figure 8.1). The space between the membranes is continuous with the ER lumen and acts as a repository of calcium (Ca), and ion transporters involved in signal transduction, in both the ONM and the INM (Bootman *et al.*, 2009). The INM has a protein lining called the nuclear lamina, which binds to chromatin and other nuclear components. The INM and ONM are fused at specific sites to form aqueous pores (Figure 8.2). Despite the lipid continuity between the NE and the ER, both the ONM and the INM comprise diverse groups of proteins that are typically not enriched in the ER. In higher organisms, the NE plays a vital role in the dissociation and reformation of the nucleus during cell division. It has dynamic interactions with the nucleoplasm and nucleoskeleton and with the cytoplasm and cytoskeleton. Thus, different NE proteins are involved in interactions for transport, chromatin organization, and gene regulation.

Proteins associated with ONM

The proteins of the ONM serve as a connecting link between the nucleus and the cytoskeleton, for nuclear positioning, maintaining the shape of the nucleus, signaling, and more importantly in cell division. The bridge between the INM

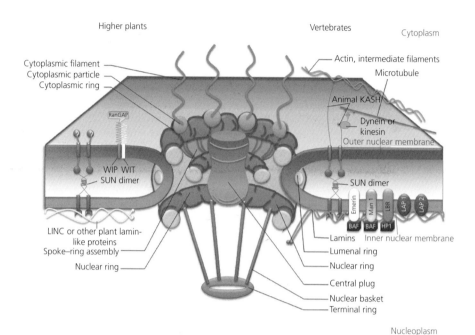

Figure 8.2 Model illustrating structural components of nuclear envelope and nuclear pore complex with comparison between higher plants and vertebrates.

and the ONM described in animal systems involves proteins of the linker of nucleoskeleton and cytoskeleton (LINC) complex (Crisp *et al.*, 2006). This diverse group of integral membrane proteins shares a small Klarsicht, ANC-1, Syne Homology (KASH) domain, which has been shown to interact with spindle Architecture Defective 1/UNC84 (SUN)-domain proteins of the INM within the periplasmic space of the NE (Figure 8.2) (Wilhelmsen *et al.*, 2006). The KASH domain proteins are huge molecules with the largest, being just over 1 MDa. *Schizosaccharomyces pombe* Kms1 is an example of KASH domain proteins that associate with components of the microtubule (MT) cytoskeleton, particularly, dynein or kinesin (Starr and Fischer, 2005; Wilhelmsen *et al.*, 2006). No KASH proteins are known in plants; thus, it is of great interest to identify plant interaction partners of SUN proteins. Both ONM and INM proteins form "bridges" across the perinuclear space that might be involved in separating the two NE membrane leaflets at an even distance of approximately 50 nm. In addition to ONM intrinsic proteins, there are also a number of peripheral and soluble proteins that associate with the cytoplasmic face of the ONM. Ran-GAP is one of the proteins that decorate the plant ONM. The entire surface of the plant NE has MT nucleating activity, while animal cells have one MT-organizing center (MTOC) that nucleates MTs. Intrinsic to eukaryotic MT nucleation sites is the gamma-tubulin ring complex (γ-TuRC), which consists of five gamma-tubulin complex proteins (GCPs) and gamma-tubulin itself. The plant proteins AtGCP2 and AtGCP3 are homologs of yeast GCP2 and GCP3, which form a soluble complex with gamma-tubulin that associates with the plant ONM (Seltzer *et al.*, 2007).

The physical connection between cytoskeleton and chromatin via KASH and SUN proteins is necessary for clustering of telomeres and the formation and anchorage of the meiotic chromosome bouquet at the NE. Components of the LINC complex are also part of signaling pathways during apoptosis. The disruption of SUN–KASH interactions leads to expansion of the NE lumen, explaining its importance in maintaining nuclear structure (Crisp *et al.*, 2006). The LINC complex of animal and yeast cells also associates centrosomes and spindle pole bodies (SPBs) to the NE (Starr and Fischer, 2005) and is involved in centrosome duplication. The proteins that reside in the NE lumen and structural elements within the nucleus are involved in maintaining the dynamics of nuclear structure. Thus, coupling the nucleus to the cytoskeleton allows for controlled movement, positioning, and anchorage of the nucleus inside the cell.

Proteins associated with INM

A number of proteins of the INM have been characterized in animal cells. Lamin B receptor (LBR), lamina-associated polypeptide (LAP) 1, LAP2, emerin, and Man1 as well as the SUN domain proteins are associated with INM. Three INM proteins, namely, LAP, emerin, and Man1 (LEM) share a domain of approximately 40 amino acids. LEM domain proteins function as a promiscuous protein–protein interaction platform, interact with barrier-to-autointegration factor (BAF), and are involved in gene silencing (Gruenbaum *et al.*, 2005). In addition, emerin associates nuclear actin to the INM and Man plays an antagonizing role in the transforming growth factor β (TGF-β) signaling cascade. The emerin is encoded by the *EMD* gene, which, when mutated, produces the X-linked form of Emery–Dreifuss muscular dystrophy (EDMD). Amino acid sensor independent (ASI1-3) are integral INM proteins that act as scavengers of transcription factors that have inadvertently entered the nucleus. Another well-studied mammalian INM protein is the LBR. It has its own multimeric protein complex and apart from a sterol reductase activity, it binds to lamin B, chromatin, and is involved in ribonucleic acid (RNA) splicing. Both emerin and LAP2β associate with several transcriptional regulators, and this association invariably coincides with the repression of the transcription factor target genes. Other INM proteins connect with the nucleoskeleton, in particular the lamina, and SUN proteins even link the nucleoskeleton with the cytoskeleton via the LINC complex (Figure 8.2).

A few INM proteins have homologs in plants. The first bona fide plant INM proteins have been reported in *Arabidopsis* and *Zea mays* (Murphy, Simmons and Bass, 2010; Oda and Fukuda, 2011). Five different SUN domain proteins are present in maize, and only two of them are found in the *Arabidopsis* genome. The two domains of *Arabidopsis*, AtSUN1 and AtSUN2, have homology with animal and yeast INM proteins suggesting a conserved SUN domain. The second family of plant INM proteins is in fact not membrane intrinsic but is strongly associated with the NE very much like the earlier described components of the Ran cycle. The Aurora kinase family members regulate mitotic processes, and two of the three *Arabidopsis* homologs have been found to localize to the NE. Like serine–threonine kinases, they phosphorylate histone H3; and in view of their cellular locations, they have been implicated in chromosome segregation and cytokinesis. Their only currently known role in plants is an involvement in root hair nuclear shape. Nuclei in mature root hairs, which are normally elongated, appear round in the mutant, suggesting an involvement of plant SUN proteins in nuclear morphology. Hence, INM proteins play vital and diverse roles in nuclear function such as chromatin organization, gene expression, and DNA metabolism.

The Nuclear Lamina

A mesh of intermediate filament proteins intrinsic to INM constitute the nuclear lamina which is composed of A/C- and B-type lamins, where lamin A and C are products of the variants of the same gene. Lamin filaments are seen as a square array with a 52-nm repeat or a approximately 20-nm repeat. The filaments lie close to the membrane although they are not integral membrane proteins. Most mammalian cells, for example, contain four different lamins, designated A, B_1, B_2, and C and are 60–80 kilodalton (kDa) fibrous proteins. Like other intermediate filament proteins, the lamins associate with each other to form filaments. The first stage of this association is the interaction of two lamins to form a dimer in which the α-helical regions of two polypeptide chains are wound around each other in a structure called a coiled-coil (Figure 8.3). These lamin dimers then associate with each other to form the filaments that make up the nuclear lamina.

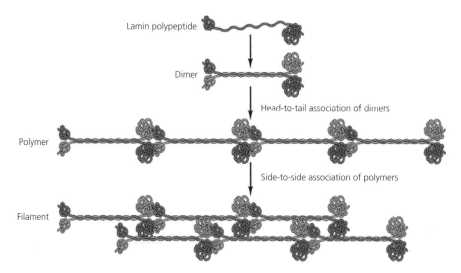

Figure 8.3 Organization of nuclear lamina.

The association of lamins with the INM is facilitated by the posttranslational addition of lipid, in particular, prenylation of C-terminal cysteine residues. B-type lamins have a lipid modification that inserts into the membrane. In metazoa, the nuclear lamina provides one important structural scaffold to which SUN domain proteins are coupled. New ultrastructural studies suggest that a lamina-like structure does exist in plants.

The nuclear lamina in tobacco (*Nicotiana tabacum*) *BY-2* cells closely resembles the animal nuclear lamina both in terms of organization and filament thickness (Fiserova, Kiseleva and Goldberg, 2009). The plant lamin-like proteins belong to a family of coiled-coil proteins which is about twice the size of lamins but with similar overall structure. Nuclear matrix constituent protein 1 (NMCP1) was first identified in *Daucus carota*; NMCP1-like proteins have been found in many plant species, and some localize exclusively to the nuclear periphery (Moriguchi *et al.*, 2005). The mutants, *LITTLE NUCLEI1* (*LINC1*) and *LINC2*, in *Arabidopsis* coding for two NMCP1-related proteins, have reduced nuclear size and different nuclear morphology, suggesting their involvement in plant nuclear organization (Dittmer *et al.*, 2007). Similar to INM proteins, mutations in lamins are linked to aging and a large number of diverse human diseases, which are collectively termed "laminopathies". Lamins act as transcription factor magnets as c-Fos by lamin A/C. c-Fos, together with c-Jun, forms the transcription factor activating protein 1 (AP1), which participates in several crucial cellular processes including cell proliferation and differentiation. The unprocessed lamin A or pre-lamin A is associated with the sterol regulatory element-binding protein 1 (SREBP1) which is synthesized as a precursor protein embedded in the ER membrane. The depletion of cholesterol causes intramembrane proteolysis and the release of the DNA-binding amino-terminal section of SREBP1, which translocates to the nucleus and induces the expression of cholesterol metabolism genes such as peroxisome proliferator-activated receptor-γ (*PPAR-γ*). Pre-lamin A and SREBP1 co-localize to the nuclear rim, which coincides with the downregulation of *PPAR-γ* expression. Another transcription factor that might be sequestered by a lamin is octamer-binding transcription factor 1 (OCT1). OCT1 localizes to the nuclear periphery and correlates with repression of the collagenase gene, a transcriptional target of OCT1. Lamins mediate the attachment of chromatin to the NE during interphase and chromatin detachment during mitosis. They are involved in nuclear stability, particularly in tissues that are exposed to mechanical forces such as muscle fibers. Lamins also play major roles in chromatin function and gene expression and also participate in epigenetic control pathways.

NE Lumen/Perinuclear Space

The NE lumen is the aqueous domain enclosed by the NE membrane that is 20–40 nm wide. The lumen of the NE is the site of Ca^{2+} release, and it is continuous with the lumen of the ER so that Ca^{2+} accumulated anywhere in the ER can diffuse freely to the NE. Its functions overlap with the lumen of the RER, but may carry out unique functions. It provides the environment for one side of the ONM, INM, and pore membrane so that integral proteins of these membranes have lumenal domains. These proteins may simply anchor structures such as the nuclear pore complex (NPC) and lamina, or some of them may provide a functional connection between the lumen and the nucleus, the NPC, or the cytoplasm. These interactions regulate the correct positioning of the nucleus within the cell, but might also have a role in signal transduction. Ca^{2+} signals in the nucleus influence DNA repair and gene transcription and the NE and nucleoplasmic reticulum as a Ca-responsive pool. Several Ca-responsive proteins have been identified in the nucleus including annexin, transcription factors, calmodulin, and Ca-dependent protein kinases and phosphatases.

The mitotic NE vesicles possess inositol triphosphate receptors and transmembrane Ca^{2+} channels that are responsible for localized, transient increases in Ca^{2+} concentration by releasing membrane-contained Ca^{2+} stores in response to an inositol triphosphate (IP3) signal. IP3, produced from membrane phosphoinositides on hormonal stimulation, binds to the IP3 receptor (IP3R) and releases Ca^{2+} from intracellular stores. The IP3R family includes the products of three genes, named *IP3R1*, *IP3R2*, and *IP3R3*, which display specific tissue distribution and sometimes different subcellular localizations. In animals, inositol 1,4,5-trisphosphate receptors initiate localized nuclear IP3-mediated Ca^{2+} signals, and the Ca-signal generated results in movement of nuclear protein kinase C to the NE. In plants, the nuclear Ca^{2+} responds to temperature and to mechanical stimulation. Release of Ca^{2+} from NE vesicles is necessary for vesicle fusion during NE assembly. The regulatory role of Ca^{2+} stored in the lumen of the NE is associated with a switch in the conformation of NPC which may control the diffusion of intermediate-sized molecules. Regulation of the Ca^{2+} level in the nucleus is an example of a complex form of local Ca^{2+} signaling. Increases in Ca^{2+} concentration in the nucleus can have specific effects different from those observed in the cytoplasm. Ca^{2+} is essential in signaling pathways that regulate gene expression by stimulating the translocation of transcription factors from the cytosol to the nucleus; Ca^{2+} also translocates or activates enzymes that regulate nuclear transcription factor activity and modifies the structure of chromatin. The nucleoplasmic reticulum allows the delivery of Ca^{2+} locally deep inside the nucleus. It also forms separate compartments of varying sizes within the nucleoplasm and could therefore determine whether the nucleus functions as an integrator or a detector of oscillating Ca^{2+} signals (Queisser, Wieger and Bading, 2011).

Nuclear pores

The NE is perforated with holes called nuclear pores. These pores regulate the passage of molecules between the nucleus and cytoplasm, permitting some to pass through the membrane, but not others. Building blocks for making DNA and RNA are allowed into the nucleus as well as molecules that provide the energy for constructing genetic material. NPCs are the sole gateways that facilitate this macromolecular exchange across the NE with the help of soluble transport receptors (Figure 8.2). The number of NPCs per cell varies and is correlated with the level of nuclear transport required by the cell. Yeast cells contain 150–250 NPCs (~12 pores/μm^2). Plant NPCs are densely spaced (~50 NPCs/μm^2). It clearly indicates that density of NPCs depends on the activity of the cells, requiring transport of many macromolecules into and out of the nucleus. The NPC is an extremely large structure with a diameter of about 120 nm and an estimated molecular mass of approximately 125 MDa, about 30 times the size of a ribosome composed of 50–100 different kinds of nucleoporins (Nups). In case of yeast, the molecular mass of NPC is 66 MDa. NPC comprises 30 different types of Nups. Electron microscopy (EM) studies have dissected the NPC structure and showed that the general morphology of the NPC is conserved among eukaryotes. NPCs are aqueous channels that show eightfold rotational symmetry with an outer diameter of 100 nm and a central transport channel measuring 40 nm in diameter, referred to as "central plug," through which bidirectional exchange of proteins, RNA, and ribonucleoprotein complexes between the nucleoplasm and cytoplasm occurs. A subset of Nups is stably embedded in the NE, forming a scaffold structure or NPC core, which is thought to stabilize the highly curved and energetically unfavorable pore membrane.

In longitudinal sections, NPC core appears as a tripartite structure consisting of a cytoplasmic and a nucleoplasmic ring that are connected to one another via two sets of eight spokes referred to as the spoke ring complex (SRC) (see Figure 8.2). The SRC forms a central channel with a functional diameter of up to 26 nm and eight smaller channels, each with a diameter of about 9 nm. These channels may facilitate passive diffusion of ions, small soluble proteins, and membrane proteins between the nucleus and cytoplasm. The terminal structures emanating from the cytoplasmic face of the NPC are eight relatively short fibrils that extend approximately 100 nm into the cytoplasm. From the outer periphery of the nucleoplasmic ring, eight filaments extend into the nucleoplasm and terminate in a distal ring to form the nuclear basket or "fish trap." The NPCs appear to attach to the lamina via the SRC, and adjacent NPCs seem to be interconnected via small radial arms within the NE lumen.

Proteomic analysis and X-ray crystallography revealed that 38% of all Nup amino acid residues contain α-solenoid fold, 29% contain phenylalanine–glycine (FG) repeats, and 16% contain β-propeller folds. Other individual fold types accounted for less than 5% of the total Nup pool. The small number of predicted fold types within the Nup proteins and their similar internal symmetries suggest that the bulk of the NPC structures evolved through a series of gene duplications and divergences from a simple precursor set of only a few proteins. The α-solenoid fold comprises a two- or three-α-helix unit, which is repeatedly stacked to form an elongated domain with the N- and C-termini at opposite ends of the molecule. The outer and inner rings (scaffold Nups) are dominated by an evenly distributed meshwork of α-solenoid domains, which is expected to facilitate the formation of a flexible fold. This allows large conformational changes without breaking protein–protein interactions, accommodates nucleo-cytoplasmic transport of different sizes cargoes, and promotes malleability of the NE. Another characteristic of many Nups is the presence of multiple repeats of short sequences. These repeats are thought to be sites where cargo molecules bind to the NPC during transport whose sequences are Gly-Leu–Phe–Gly, X–Phe–X–Phe–Gly, or X–X–Phe–Gly, where X is any amino acid. Because they contain a phenylalanine–glycine (Phe–Gly) pair, these are often referred to an FG repeats (F and G are one-letter abbreviations for phenylalanine and glycine, respectively). Spacers of 3–15 amino acids separate the repeats. These FG repeat regions are unstructured and fill the central channel of the NPC, where they serve as docking sites for karyopherins. Other FG-repeat-containing Nups are components of the cytoplasmic filaments and the nuclear basket. It is therefore reasonable to suggest that Nups rich in FG repeats (FG Nups) coat the central pore surface, thus providing interaction domains for transport receptors within the central pore.

Another feature of many Nups is the presence of a type of alpha helical domain that is able to form a protein structure motif called the coiled-coil by interacting with structurally similar regions of other Nups. Coiled-coils mediate protein–protein interactions, implying that the nuclear basket serves as a recruit-ment platform that brings various factors together within the nucleus. The nuclear basket within the NPC is generally constructed from large coiled-coil proteins: nuclear pore anchor (NUA) in plants, myosin-like protein 1/2 (Mlp1/2) in yeast. The yeast FG Nup complex, Nsp1–Nup82–Nup49 complex (Nup62–Nup58–Nup54 complex in plants), is held together by coiled-coil interactions.

In yeast, only secretory 13 (SEC13) and sec13 homologue 1 (SEH1) contain the signature WD-40 repeat (also known as the WD or β-transducin repeat) and are among the very first β-propellers to be identified. Since then, other Nups containing WD-40 repeats have been identified as β-propeller Nups. β-propellers are ubiquitous disk-shaped domains with an overall diameter of approximately 70 Å and a thickness of approximately 40 Å. The canonical β-propeller core is generated by 4–8 blades that are circularly arranged, which provide a molecular platform that mediates multiple interactions with other proteins. In general, each blade consists of 4–10 repeats of a four-stranded antiparallel β-sheet motif. The central channel of the propeller fold is usually funnel-shaped, with a wider bottom opening, which serves as the entry point to the active site. Both the α-solenoid and the β-propeller folds provide extensive solvent-accessible sur-faces, which appear well suited for binding other proteins. Most of the nuclear pore proteins contain one of two additional structural motifs, the α-solenoid and the β-propeller, that are also present in the proteins that form the coats of coated vesicles. In both coated vesicles and nuclear pores, these proteins are thought to

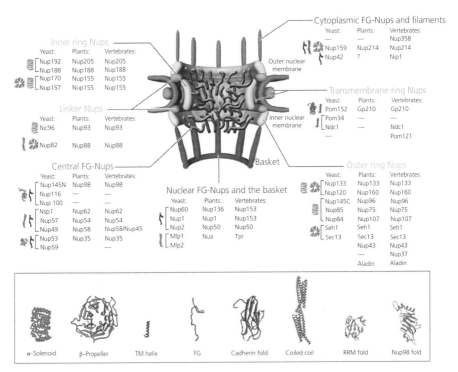

Inner ring Nups

Yeast:	Plants:	Vertebrates:
Nup192	Nup205	Nup205
Nup188	Nup188	Nup188
Nup170	Nup155	Nup155
Nup157	Nup155	Nup155

Linker Nups

Yeast:	Plants:	Vertebrates:
Nic96	Nup93	Nup93
Nup82	Nup88	Nup88

Central FG-Nups

Yeast:	Plants:	Vertebrates:
Nup145N	Nup98	Nup98
Nup116	—	—
Nup100	—	—
Nsp1	Nup62	Nup62
Nup57	Nup54	Nup54
Nup49	Nup58	Nup58/Nup45
Nup53	Nup35	Nup35
Nup59	—	

Outer nuclear membrane

Inner nuclear membrane

Cytoplasmic FG-Nups and filaments

Yeast:	Plants:	Vertebrates:
—	—	Nup358
Nup159	Nup214	Nup214
Nup42	?	Nip1

Transmembrane ring Nups

Yeast:	Plants:	Vertebrates:
Pom152	Gp210	Gp210
Pom34	—	—
Ndc1	—	Ndc1
—		Pom121

Basket

Nuclear FG-Nups and the basket

Yeast:	Plants:	Vertebrates:
Nup60	Nup136	Nup153
Nup1	Nup1	Nup153
Nup2	Nup50	Nup50
Mlp1	Nua	Tpr
Mlp2		

Outer ring Nups

Yeast:	Plants:	Vertebrates:
Nup133	Nup133	Nup133
Nup120	Nup160	Nup160
Nup145C	Nup96	Nup96
Nup85	Nup75	Nup75
Nup84	Nup107	Nup107
Seh1	Seh1	Seh1
Sec13	Sec13	Sec13
	Nup43	Nup43
	—	Nup37
	Aladin	Aladin

α–Solenoid	β–Propeller	TM helix	FG	Cadherin fold	Coiled coil	RRM fold	Nup98 fold

Figure 8.4 Molecular architecture of nuclear pore complex. Source: Reproduced with permission of Grossman, Medalia and Zwerger (2012).

play an important role in membrane curvature. This suggests that some components of coated vesicles and NPCs share a common ancestry.

Although the primary sequence homology between Nups from different model organisms is low, the overall shape and predicted fold types are highly conserved. Most Nups are denoted as "Nup" followed by a number that refers to their molecular mass. Because of the molecular mass differences in various species, a uniform nomenclature for Nups does not exist. However, based on their approximate localization within the NPC, the Nups can be grouped into five classes: transmembrane ring, core scaffold (inner ring, outer ring, and linker), cytoplasmic filaments, nuclear basket, and central FG (Grossman, Medalia and Zwerger, 2012) (Figure 8.4).

Transmembrane ring Nups

Membrane-spanning Nups anchor the NPC to the pore membrane and bind the assembled complex to the NE. In yeast, three proteins, namely, Pom34, Ndc1, and Pom152, are transmembrane Nups. Pom152 is predicted to have a cadherin fold. The cadherin fold predicted in Pom152 may help stabilize the curvature of the pore membrane by stabilizing the interaction between the INM and the

ONM at the lumenal connections. In plants, two transmembrane Nups, glycoprotein of 210kDa (gp210), homolog to Pom152 in yeast, and nuclear division cycle 1 (NDC1), constitute an outer transmembrane ring. In addition to these proteins, vertebrates possess their own unique membrane protein, Pom121. The glycoprotein gp210 has 95% of its mass in the lumen, a single transmembrane domain, and a short carboxyl terminus, which may be part of the NPC core within the pore itself. It is N-glycosylated by high mannose sugar groups on the lumenal domain, a characteristic of the lumenal or extracellular domains of many membrane proteins. The radial arms, which may contain gp210, may simply be there to hold the spokes in place within the pore domain like a rubber band. The gp210 anchors the NPC to the pore membrane and contributes to the complex scaffold. The phosphorylation of gp210 in NE breakdown (NEBD) is required for lamin disassembly. A single serine residue at position 1880 is specifically phosphorylated during mitosis and is important for dissociation of gp210 from the structural elements of the NPC.

Cytoplasmic filaments and the nuclear basket

Two filamentous structures, the cytoplasmic filaments and the nuclear basket, are localized asymmetrically at the cytoplasmic and nuclear faces of the NPC, respectively. The cytoplasmic filaments are primarily composed of two FG Nups: Nup214 and CG1. Although these Nups are not well characterized in plants, those in yeast provide sites at which messenger RNA (mRNA) export factors can maturate messenger ribonucleoprotein particles. Nup82 probably recruits central channel FG-Nups that are directly involved in nuclear transport. Although it does not contain FG-repeats, Nup82 is also classified as a cytoplasmic filament component. Plants use the WIP–WIT complex to anchor Ran-GAP on the NE, whereas yeast Ran-GAP is localized to the cytosol.

The nuclear basket is composed of eight elongated filaments, which protrude approximately 60–80 nm from the nuclear face of the NPC into the nucleoplasm and converge on a distal ring structure. Nups comprising the nuclear basket typically host FG repeats, α/β-regions, and α-helical domains. Each nuclear basket consists of one Nup, NUA in plants, which has long coiled-coil domains, and two FG Nups (Nup136/Nup1 and Nup50 in plants). NUA in plants is thought to constitute the central architectural element that forms the scaffold of the nuclear basket, whereas Nup153 (Nup136/Nup1 in plants) binds to the nuclear coaxial ring linking the NPC core structures to Tpr. The nuclear basket interacts with many nuclear proteins and with the nuclear lamina.

Central FG Nups

Central FG Nups, also known as barrier Nups, show a transverse orientation with respect to the central channel of the NPC and form a selective barrier that mediates nucleocytoplasmic transport. FG-Nups represent one-third of the total pore proteins and transverse the central channel of the NPC, extending into the cytoplasmic and nucleoplasmic sides. It is estimated that there are approximately 160 transport factor binding sites within each NPC, which allow the simultaneous binding of multiple transport factors within a single NPC. Five central FG Nups—Nup98, Nup62, Nup58, Nup54, and Nup35—have been identified in plants and are well conserved in other organisms. Of these, only Nup62 has been characterized in *Arabidopsis*. *Arabidopsis* Nup62 interacts with nuclear transport factor 2 (NTF2), as previously observed in yeast (Nsp1). Knockdown of Nup62 transcripts results in a severe dwarf phenotype and early flowering, indicating an important function for Nup62 at different stages of plant development. Nups that contain FG-repeats usually possess additional domains, such as the autoproteolytic Nup98 folds, coiled-coil domains, and RNA recognition motif (RRM) domains. FG-Nups with coiled-coil domains, such as Nup58/45, are assumed to mediate protein–protein interactions that anchor Nups to the scaffold of the NPC structure. Additionally, these coiled-coil domains may presumably be involved in adjusting the diameter of the central transport channel via a mechanism of intermolecular sliding.

Outer ring, inner ring, and linker Nups (scaffold Nups)

Many of the scaffold proteins are predicted to fold into an α-solenoid fold, a β-propeller fold, or a mixture of the two. Scaffold Nups are composed of three groups of Nup subcomplexes: the outer ring, the inner ring, and the linker. They connect membrane Nups to central FG Nups, thereby bridging the anchoring transmembrane layer and the barrier layer of the NPC. Scaffold Nups are thought to play a key role in maintaining the stability of the NE by ensuring coplanarity of the outer and inner surfaces. The largest and most evolutionarily conserved subcomplex, comprising several Nups, is located at the outer rings and is known as the Nup107/160 subcomplex in plants and vertebrates and the Nup84 subcomplex in yeast. The Nup107/160 subcomplex consists of Nup133, Nup96, Sec13, Nup107, Nup85, Seh1, Nup160, Nup37, and Nup43, and the Nup84, Nup120, Nup85, Nup145C, Sec13, Seh1, and Nup133 are homologs to yeast. In plants, the inner ring consists of Nup205, Nup188, Nup155, and Nup35 (Figure 8.4). Similar to the outer ring Nups, they contain α-solenoid and β-propeller folds and exhibit the typical structural scaffolding motif. The linker Nups are attached between the outer and the inner rings and include Nup93 and Nup88. They act as a bridge between the core scaffold and the FG Nups.

The nucleolus

The nucleolus is a membraneless organelle within the nucleus that manufactures ribosomes, the cell's protein-producing structures. Through the microscope, the nucleolus looks like a large, dark spot within the nucleus. A nucleus may contain up to four nucleoli, but within each species the number of nucleoli is fixed.

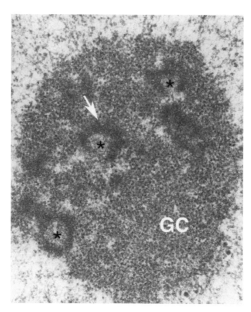

Figure 8.5 An electron micrograph of nucleolus illustrating the fibrillar centers (asterisks), the dense fibrillar component (white arrow), and granular component (GC). Source: Reproduced with permission of Hernandez-Verdun *et al.* (2010).

After a cell divides, a nucleolus is formed when chromosomes are brought together into nucleolar organizing regions (NORs). During cell division, the nucleolus disappears. Nucleolus may be involved with cellular aging and, therefore, may affect the senescence of an organism. The organization and structure of the nucleolus have been well characterized by EM since the 1970s. Nucleoli exhibit (with some exceptions) a strikingly similar organization from protists to humans. The presence of a prominent nucleolus indicates that the cell is actively synthesizing proteins. A prominent nucleolus is present in protein-secreting cells such as those of the pancreas, plasma cells, developing blood cell precursors, and others. The nucleoli in many animal cells have been described in terms of a tripartite structure, with small, lightly staining regions called fibrillar centers (FCs), surrounded by densely stained material called dense fibrillar component (DFC), and granular component (GC). The FCs are clear fibrillar areas of different sizes, ranging from 0.1 to 1 μm, containing fibrils. The FCs and DFC are embedded in the GC that mainly consists of granules 15–20 nm in diameter in a loosely organized distribution (Figure 8.5).

Nucleoli are generally surrounded by a shell of condensed nucleolus-associated chromatin that at particular regions penetrates deeply into the nucleolar body through nucleolar interstices and reaches the FCs. In most plant cells, but not in budding yeast, a heterochromatin layer is observed at the nucleolar periphery by EM and smaller foci of rDNA labeling corresponding to FCs. This heterochromatin is visible with DNA Dapi-staining (a positive ring surrounding a black hole) demonstrating the high DNA content at the periphery compared to within the nucleolus. In plant cells, the transcription sites consist of many foci within the DFC, with the smallest foci representing individual rDNA gene repeats (Shaw and Brown, 2012). The number of rDNA units per cell varies from 150 to 26 000 in plants and correlates positively with genome size. Additional labeling with probes to transcribed spacer regions in the pre-rRNA and with various proteins and other small RNAs known to be involved in ribosome maturation led to a radial model, where the newly formed transcripts moved away from the dispersed genes within the DFC and then subsequently moving into the GC for later processing stages. Different gene families and certain satellite repeats have been identified as being the major blocks of nucleolus-associated chromatin domains (NADs), altogether they correspond to not less than 4% of the total genome sequences. In addition, at the periphery of the nucleolus, a specific domain was designated as the perinucleolar compartment (PNC) which is associated with a specific DNA locus and is highly enriched in RNA-binding proteins and RNA polymerase III transcripts.

Saccharomyces cerevisiae has only two morphologically distinct nucleolar components: fibrillar strands (F) and granules (G). Several features distinguish yeast from human nucleoli. A major difference lies in the internal organization of the organelle (number of components) and occurrence of intranucleolar bodies. In addition, yeast nucleoli lack condensed perinucleolar chromatin and show extensive nuclear membrane attachment. Also, yeast is characterized by closed mitosis implying that its nucleolus does not disassemble during mitosis. In budding yeast, there is a single nucleolus that occupies one-third of the nuclear volume. There is about a 10-fold range difference in size between yeast and human nucleoli ($\approx 0.5\,\mu m$ and from ≈ 0.5 to $9\,\mu m$, respectively), and a human nucleolus is about the size of a yeast nucleus. Nucleoli of different cell types exhibit a variable number of FCs of variable sizes, with an inverse proportion between size and number. It has been proposed that three distinct nucleolar components emerged during evolution from a two-component (fibrillar and granular) nucleolus. This suggestion is based on the size of the intergenic spacers of the rDNA. In addition to nucleolar fibrillar strands and granules, several specialized subnucleolar domains have been described in budding yeast. These include the nucleolar body (NB) and the "no-body," involved in small nucleolar RNA (snoRNA) biogenesis and ribosome surveillance, respectively, as well as a nucleolar domain enriched in poly(A) RNAs. Subnucleolar compartmentalization might facilitate specific reactions such as RNA modification, RNA processing, and RNA degradation. There is one NB per nucleolus. The NB has primarily been implicated in snoRNA maturation. snoRNAs have role in RNA processing, modification, and folding. The transient accumulation of snoRNAs, such as U3, in the NB is involved in 5′-cap trimethylation by the trimethyl guanosine synthase, *Tgs1* and 3′-end processing. One of the best characterized nuclear surveillance pathways is the Trf4/5p-Air1/2p-Mtr4p polyadenylation complex

(TRAMP)–exosome pathway where defective pre-RNPs are targeted for degradation following the addition of short poly(A) tails at the 3′-end of their RNAs by TRAMP. Nuclear bodies such as CBs and speckles have also been identified in plant nuclei. The CB is highly enriched in small nuclear ribonucleoproteins (snRNPs) and small nucleolar ribonucleoproteins (snoRNPs) required for the maturation of pre-mRNAs and pre-rRNAs and shares with the nucleolus the proteins fibrillarin, Nopp 140, and NAP57 and subunits of the RNaseP. Thus nuclear bodies are sites where different factors accumulate, allowing efficient RNA–protein and protein–protein interactions or RNP assembly, to influence or regulate a particular process. The bodies can form or dissemble or increase and decrease in size and number depending on different conditions or activities of the cell, again reflecting the dynamic interplay between nuclear bodies and the surrounding nucleoplasm.

The "no-body" is a nucleolar focus, distinct from the NB, enriched in pre-rRNAs and RNA surveillance components. Small and large ribosomal subunits, TRAMP components, core exosome, and nuclear-specific exosome subunits components have been localized to the "no-body." Other putative nucleolar "surveillance centers," distinct from the NB and "no-body," are composed of foci enriched for polyadenylated small nuclear RNAs (snRNAs) and snoRNAs and a focus detected upon Rnt1 (yeast RNase III) mild overexpression that juxtaposed with primary ribosomal RNA (rRNA) transcripts.

The chromosomal nucleolus organizer

That the nucleoli assembled around the nucleolar organizers (NORs) was first proposed by Barbra McClintock in *Zea mays* in 1934. The NORs are chromosomal regions where multiple rDNA copies cluster in arrays. It has been demonstrated that nucleolar proteins rapidly associate and dissociate with nucleolar components in a continuous exchange with the nucleoplasm. When cells of higher eukaryotes enter mitosis, the nucleoli disintegrate in various stages. The first component to disappear is the DFC which is followed by the disappearance of the GC. Throughout mitosis, the chromosomal NOR, which harbors the tandemly repeated rRNA genes and reveals a striking morphological similarity to the FCs, persists. The reformation of nucleoli at telophase follows the reverse order, that is, the DFC forms around the NORs before the appearance of the GC. At the end of mitosis, when rDNA transcription by the RNA polymerase I resumes, active NORs are directly involved in nucleolar reassembly. Within each active NOR, only a subset of rDNA units are transcribed.

Ribosome biogenesis

Proteomic studies of human cells identified 450 nucleolar proteins including ribosomal proteins (RPs) and those known to be involved in ribosome biogenesis. In addition to these proteins, spliceosomal proteins, and splicing factors, translation factors have also been identified. In case of *Arabidopsis*, 217 proteins have been associated with nucleoli. The organization and size of the nucleoli are directly related to ribosome production. There is correlation between the ribosome biogenesis and the nucleolar organization. In cycling cells, the volume of the nucleoli increases between the G1 and G2 phases and the number of FC doubles in G2. In quiescent cells, at the terminal stage of differentiation when ribosome biogenesis is stopped, small ring-shaped nucleoli or nucleolar remnants (diameter = 0.3 μm) are typically observed in lymphocytes or erythrocytes. These nucleoli are formed by one clear area containing chromatin and dense fibrils at the periphery. The localization of specific proteins, rRNA precursors at various processing stages, and snoRNAs has been analyzed by transcription site mapping, immunocytochemistry, and *in situ* hybridization techniques. It was established that the sites of active RNA polymerase I transcription are localized at the interface between the FCs and the DFC, where early processing of the pre-rRNAs occurs in the DFC and progresses to late processing in the GC. The non-transcribed part of the rDNAs as well as the RNA polymerase I complexes and the transcription machinery such as the upstream binding factor (UBF) and topoisomerase I are localized in the FCs. The FCs appear to be pivotal elements to understand how RNA polymerase I transcription organizes the nucleoli. The FCs are the interphasic counterparts of the mitotic NORs because the nucleoli are reformed around the FCs at the end of mitosis. In the nucleoli, the vectorial distribution of the machineries successively involved correlates with the different processing steps. For example, fibrillarin and nucleolin that participate in the early stages of rRNA processing localize in the DFC along with the U3 snoRNA, whereas proteins B23 and Nop52 involved in intermediate or later stages of ribosome biogenesis have been localized to the GC. Pre-rRNA initially accumulates in DFC, and its first processing steps take place in this region.

These processing modifications encompass methylations and pseudouridylations before the three mature rRNAs (18S, 5.8S, and 28S rRNA) are produced by nuclease cleavage. Nucleolin is one of the major nucleolar phosphoproteins involved in the initial cleavage of pre-rRNA. B23 is another highly abundant nucleolar phosphoprotein that has been implicated in the maturation of pre-ribosomal particles and nucleocytoplasmic transport processes. It serves as a chaperone function in ribosome assembly. The Nopp140 guides the nucleolar proteins fibrillarin and NAP57 to nucleolus and also to the coiled bodies (CBs). Three proteins, Nop60B, Cbf5p, and NAP57, are related to members of the family of prokaryotic transfer RNA (tRNA) pseudouridine 55 synthases (TruB/P35) which pseudouridylate most tRNA at position 55. Each modification is made at a specific position in the pre-rRNA. These positions are specified by "guide RNAs" called snoRNAs. The U3 snoRNA is essential for rRNA processing, leading to the formation of 18S rRNA. An essential step in ribosomal biogenesis is the association of processed rRNAs with 5S rRNA that is transcribed by RNA polymerase III in the nucleoplasm outside the nucleolus, and with up to 80 RPs to form the RNP precursors, which mature into the 40S and 60S subunits of the ribosome (Hernandez-Verdun *et al.*, 2010).

The genes encoding RPs are transcribed by RNA polymerase II, and newly translated RPs move into the nucleus and nucleolus. Once ribosomal subunits are assembled into nearly mature 60S and 40S pre-ribosomal particles, they are exported to the cytoplasm independently of one another. Nuclear export to the cytoplasm of the vast majority of both subunits, via the nuclear pores, requires Crm1, the major export karyopherin, as well as Ran-GTPase system and subset of Nups. Free RPs in excess of the requirement for ribosome subunit production will continue to shuttle into the nucleoplasm, where they are probably ubiquitinylated and exposed to proteasome-mediated degradation.

The rRNA gene is transcribed exclusively by RNA polymerase I in the nucleolus. While in yeast this event generates a 35S rRNA precursor which is processed into the mature 18S, 5.8S, and 26S rRNAs, in mammals a 47S rRNA precursor is generated and processed to give 18S, 5.8S, and 28S rRNAs. There are also numerous similarities in the other component of the ribosome, RPs, which in both cases are transcribed by RNA polymerase II in the nucleoplasm. In growing yeast, it has been established that approximately 40 nascent ribosomes leave the nucleolus every second, 80% of the total RNA is rRNA, and approximately 50% of total protein consists of RPs.

The identification of mRNA-associated proteins in the plant nucleolus suggests their function in mRNA biogenesis. In animals, only a very few mRNAs have ever been identified in the nucleolus. The aberrantly spliced mRNAs are enriched in the nucleolus and the vast majority contained premature termination codons, and are therefore likely to be turned over by the nonsense-mediated decay (NMD) pathway. The localization of two proteins (UPF2 and UPF3) involved in NMD to the nucleolus suggests a novel function for the plant nucleolus in mRNA surveillance/NMD and thereby in mRNA biogenesis. However, the mRNA transcripts with retained introns (or containing unspliced introns) are not turned over by NMD and appear to avoid the NMD pathway. Dicing bodies (D-bodies) are distinct from other nuclear bodies and have been implicated in processing primary microRNA (miRNA) precursors into miRNA. The localization of pre-miRNAs and Dicer-like 1 (DCL1) to D-bodies also suggests a role for the nucleolus in the maturation of miRNAs. Some precursor and mature miRNAs are enriched in the nucleolus of mammalian cells possibly for modification, assembly, or regulation of snoRNA activity. The complexity of RNA/RNP processes involving the nucleolus suggests that it is a center of RNA activity. It has also been speculated that many animal and plant viruses exploit the nucleolus in the production and transport of viral RNPs.

Nucleolar disassembly and assembly

The nucleolus disassembles at the end of G2 as most transcription ceases and the NE breaks down, and then reassembles with the onset of rDNA transcription at the beginning of the following G1 (Hernandez-Verdun, 2011). At the end of mitosis, when rDNA transcription by RNA polymerase I resumes, active NORs are directly involved in nucleolar reassembly. Within each active NOR, only a subset of rDNA units are transcribed. The NOR-related protein becomes visible in nucleus by a silver-staining technique under a light microscope, and it has been named argyrophilic protein of NOR (Ag-NOR). In contrast, inactive NORs are not bound by argyrophilic proteins, they are not associated with the RNA polymerase I machinery, and they are not involved in nucleolar formation. The nucleolus is either organized around a single NOR or alternatively several active NORs coalesce in a single nucleolus once rRNA synthesis has initiated.

During mitosis, the rDNA transcription machinery remains associated with or close to the rDNA throughout the cell cycle. The rDNA transcription is repressed at the entrance to mitosis and maintained repressed during mitosis by phosphorylation of components of the rDNA transcription machinery directed by the cyclin-dependent kinase (CDK) 1-cyclin B. During disassembly, the GC components are lost first, followed by the loss of DFC components (Figure 8.6). Certain proteins, such as RNA polymerase I subunits and UBF,

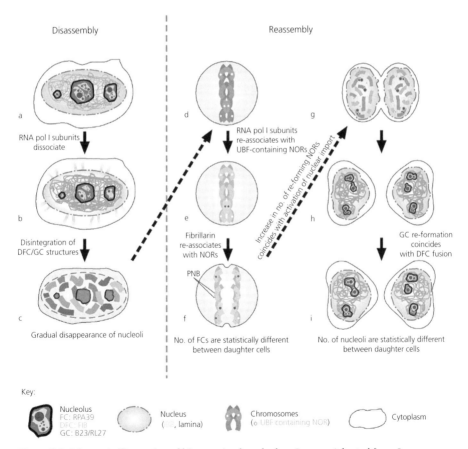

Figure 8.6 Schematic illustration of biogenesis of nucleolus. Source: Adapted from Leung *et al.* (2004).

which modulates DNA conformation, remain with the rDNA arrays; the presence of UBF alone is sufficient to produce a secondary constriction in the mitotic chromosome. The protein B23 associates with the periphery of the mitotic chromosomes as chromosomal "passengers," whereas some nucleolar components diffuse throughout the mitotic cytoplasm. The unprocessed pre-rRNA transcripts persist through the mitotic cell when rDNA transcription is halted. The co-localization of these different factors in the perichromosomal compartment suggests that early and late rRNA processing complexes are at least partly maintained during mitosis. In mitosis, rRNA synthesis ceases as a result of phosphorylation of the relevant nuclear/nucleolar factors and the nucleoli disassemble.

At the end of mitosis, phosphorylation of the relevant factors reverses, RNA synthesis resumes, and nucleoli reform. Nucleolar assembly is an early event starting at telophase, and a complex process that occurs during a relatively long period of the cell cycle (Figure 8.6). Nucleolar reformation begins with the formation of numerous prenucleolar bodies (PNBs) that contain several proteins characteristic of the DFC in functional nucleoli. Nucleolar assembly depends on the coordination between the activation of rDNA transcription and the recruitment and activation of the RNA processing complexes. In addition, translocation of the rRNA processing complexes into the sites of rDNA transcription is linked to the formation of PNBs. Consequently, inactivation of the CDK1-cyclin B occurring normally in telophase is sufficient to release mitotic repression of rDNA transcription. At metaphase and anaphase, fibrillarin is uniformly associated with the surface of all chromosomes. At telophase, however, it appears in numerous PNBs. In a subsequent step, these PNBs then coalesce around the chromosomal NOR(s) into the developing nucleolar body. This site-specific fusion of PNBs requires transcriptional activity of the rRNA genes. Processing proteins (fibrillarin, NPM/B23, nucleolin, Nop52, etc.) from DFC and GC are localized in PNBs as well as the box C+D snoRNA U384 and 45S pre-rRNAs. Thus PNBs are transitory structures that gather the building blocks of the nucleolus machineries. It has been proposed that PNBs move to the sites of RNA polymerase I transcription to deliver the pre-rRNA processing complexes. PNB dynamics in living cells does not reveal such directed movement of PNBs toward the NORs. An analysis by time-lapse fluorescence resonance energy transfer (FRET) demonstrates that proteins of the same pre-rRNA processing machinery interact with each other within PNBs, but not when they are localized at the chromosome periphery. The timing of these interactions suggests that PNBs could be preassembly platforms for pre-rRNA processing complexes.

Functions of nucleolus

Nucleolus is considered a multifunctional domain. The presence of the RNA of the signal recognition particle (SRP) and three SRP proteins found in the nucleolus and SRP-RNA localization differs from the classical sites of ribosome biogenesis. There are also indications that at least some of the tRNAs are processed in the nucleoli, and that the U6 spliceosomal RNA cycles through the nucleolus to undergo methylation and pseudouridylation. The nucleolus is also a domain of sequestration or retention for molecules related to cell cycle, life span, and apoptosis. There is a broad correlation between the genome size and the number of rDNA copies; this has led to a hypothesis that the rDNA may be acting as a sensor for DNA damage, protecting the rest of the genome by inducing DNA repair mechanisms or apoptosis. The nucleolus (and associated bodies, particularly CBs) is involved in the maturation, assembly, and export of RNP particles such as the SRP, telomerase RNP, and processing of pre-tRNAs and U6snRNA. The transcriptional state of rDNA is mediated by an epigenetic network with a close relationship to nucleolar architecture. The interplay of DNA methylation, histone posttranslational modification and chromatin remodeling activities establish silencing at the rDNA locus in higher eukaryotes. In addition, the nucleolus has roles in cellular functions such as regulation of the cell cycle, stress responses, telomerase activity, and aging. Sequestration of specific proteins in the nucleolus or their release is one mechanism by which processes such as the cell cycle or cell death are regulated. Many plant viruses exploit the nucleolus in production and transport of viral RNPs. For example, the coat protein and coat protein read-through proteins of potato leaf roll virus (PLRV) are targeted to the nucleolus, and systemic infection of PLRV is inhibited in fibrillarin-silenced plants suggesting that fibrillarin is also involved in long-distance movement of PLRV.

Plants contain a novel RNA polymerase (RNA polymerase IV) involved in the production of transcripts that give rise to double-stranded RNAs and short interfering RNAs (siRNAs), which guide DNA methylation and results in transcriptional gene silencing. The multifunctional nature of the nucleolus is therefore reflected in the complexity of the protein and RNA composition of the nucleolus and in the dynamic composition changes in response to cellular conditions. Nucleolar dysfunction has severe consequences for human health. Defective ribosome surveillance recently emerged as a possible causal effect for several human diseases with the suggestion that the accumulation of chemically modified ribosomes, for example, oxidized particles, might contribute to the progression of neurodegenerative diseases such as Alzheimer and Parkinson diseases. Ribosome oxidation might alter ribosome function and might result from intracellular exposure to reactive oxygen species (ROS) or environmental exposure to ultraviolet or other debilitating treatments. Finally, several non-ribosomal functions of the nucleolus, for instance, in cell cycle regulation or telomerase trafficking, are directly required for cellular homeostasis.

Chromatin and chromosomes

Within the nucleus, DNA is packaged into the proteinaceous "superstructure," chromatin, by association with histones, RNA molecules, and other proteins. Chromatin is of two types: euchromatin, which undergoes the normal process of condensation and decondensation in the cell cycle, and heterochromatin, which remains in a highly condensed state throughout the cell cycle, even during interphase. Euchromatin is present along the chromosomes where genes are actively transcribed, whereas heterochromatin is mainly present in the telomeric and centromeric regions.

Histone and nonhistone proteins

Both histones and nonhistones are associated with chromatin. Histones are of five types: H1, H2A, H2B, H3, and H4. They are positively charged proteins due to the presence of high percentage of arginine and lysine so that histone proteins interact with the negatively charged DNA molecules. The primary amino acid sequences taken together with physical studies indicate that the H2A, H2B, H3, and H4 histones have two domains. A striking feature of histones, and particularly of their tails, is the large number and type of modified residues they possess. There are over 60 different residues on histones where modifications have been detected either by specific antibodies or by mass spectrometry. There are at least eight distinct types of modifications found on histones, such as acetylation, methylation, phosphorylation, ubiquitination, SUMOylation, ADP ribosylation, deimination, and proline isomerization.

Some groups of nonhistone proteins that are highly heterogeneous in nature are also associated with chromatin. One class of nonhistone proteins that are reproducible component of the nuclear chromatin is the so-called high-mobility group (HMG). HMG proteins act to disrupt the compaction of DNA induced by the histones and may therefore be involved in the derepression of chromatin for transcription *in vivo*. The abundance of the HMG proteins has suggested that they may function to facilitate the formation of nucleosomes by bending DNA in preparation for folding into the compact nucleosome structure. Two HMG proteins, HMG14 and HMG17, have been purified from calf thymus showed high binding affinity for nucleosome in biochemical experiments. Another group of nonhistone proteins are the DNA topoisomerases. Their enzymatic activities are characterized by their ability to alter the overall shape, or topology of DNA. The eukaryotic type I topoisomerase cleaves one of the DNA strands and binds to the 3′-phosphate via a tyrosine residue. The complex then rotates to release torsional stress and reforms the sugar-phosphate backbone upon removal of the topoisomerase I protein. Topoisomerase II is involved in untangling newly replicated DNA molecules and is a major component of the protein "scaffold," to which defined sections of the DNA molecule attach *in vivo*. Topoisomerase II cleaves both strands in a staggered manner with both 5′-ends becoming covalently bound to the protein.

Beads-on-a-string model

When chromatin is isolated from the nucleus of a cell and viewed with an EM, it frequently looks like beads-on-a-sting. If a small amount of nuclease is added to this structure, the enzyme cleaves the string between the beads

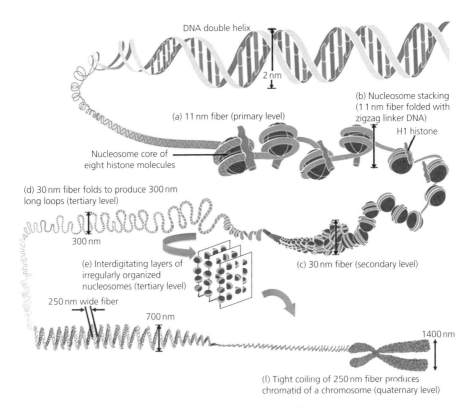

DNA double helix

2 nm

(a) 11 nm fiber (primary level)

(b) Nucleosome stacking
(11 nm fiber folded with
zigzag linker DNA)

H1 histone

Nucleosome core of
eight histone molecules

(d) 30 nm fiber folds to produce 300 nm
long loops (tertiary level)

300 nm

(e) Interdigitating layers of
irregularly organized
nucleosomes (tertiary level)

(c) 30 nm fiber (secondary level)

250 nm wide fiber

700 nm

1400 nm

(f) Tight coiling of 250 nm fiber produces
chromatid of a chromosome (quaternary level)

Figure 8.7 Various levels of organization of chromatin assembly.

leaving individual beads attached to about 200 base pairs (bp) of DNA. If more
nuclease is added, the enzyme chews up the entire DNA between the beads
and leaves a core of proteins attached to a fragment of DNA. Such experiments
demonstrate that chromatin is not a random association of proteins and DNA
but has a fundamental repeating structure. The nucleosome is the fundamental
unit of chromatin, and it is composed of an octamer of the four core histones
(H3, H4, H2A, and H2B) around which 147 bp of DNA are wrapped about two
times, approximately 90 bp per turn. The core histones are predominantly
globular except for their N-terminal "tails," which are unstructured. Within the
nucleosome, the four histone molecules form two pairs, H3–H4 and H2A–H2B,
which in turn, make up a complex $(H3–H4)_2$ plus $(H2A–H2B)_2$ forming the
protein core of the nucleosome. The $(H3–H4)_2$ tetramer is more stable than
the $(H2A–H2B)_2$ tetramer. A linker region of DNA between core particles is
more susceptible to nuclease degradation than the core particle DNA and
is associated with histone H1 and provides partial nuclease protection for 20 bp of
linker DNA. The core particle is shaped like a squat cylinder, with a diameter
of approximately 11 nm and a height of approximately 5.5 nm (Figure 8.7a).

Chromatosome

H1 contains a conserved globular region and extended amino- and carboxy-termini, the latter being rich in lysine and able to interact strongly with DNA. H1-containing chromatin shows a distinct structural motif in which the entering and exiting linker DNA segments are brought together, perhaps promoting an overall zigzag arrangement. Modest depletion of H1 from cells results in reductions in nucleosome repeat length in some tissues, whereas severe depletion is fatal (Lu *et al.*, 2009). H1 depletion also inhibits the proper folding of chromosomes at mitosis. There is a strong linear relationship between H1 content and nucleosome repeat length *in vivo*, a feature that will act to maintain the electrostatic balance between DNA and histones. Together, the core particle and its associated H1 histone are called the chromatosome, the next level of chromatin organization. Chromatosomes are located at regular intervals along the DNA molecule and are separated from one another by linker DNA, which varies in size among different cell types from 30 to 50 bp (Figure 8.7b).

Solenoid model

By folding up into a nucleosome, the DNA condenses by a factor of 6. Nucleosomes are packed one on top of another to form a more compact structure, a 30-nm fiber. It is this packing of 6–8 nucleosomes into a coiled fiber that forms chromatin fibers (Figure 8.7c). Each of these fibers undergoes further folding and coiling to form the metaphase chromosome. More than 30 years ago, in 1976, Finch and Klug proposed that the nucleosome, with linker histone H1 or Mg^{2+} ions, is folded into "30-nm chromatin fibers." In their model, known as "solenoid," consecutive nucleosomes are located next to each other in the fiber, folding into a simple one-start helix. The studies on X-ray crystallography of nucleosome proposed a second model of the "two-start helix" that the 30-nm fiber is arranged in a zigzag manner, such that two "strands" of nucleosomes are stacked on top of each other like coins. The two strands of stacked nucleosomes are then wound into a double helix similarly to the two strands in a DNA double helix, except that the helix is left-handed, rather than right-handed as it is in DNA. In this particular arrangement, which is dependent on the ionic strength of the environment, odd-numbered nucleosomes make contact with other odd-numbered nucleosomes and even-numbered nucleosomes make contact with other even-numbered nucleosomes (Sajan and Hawkins, 2012). During mitosis, this dynamic irregular folding of nucleosome fibers could be a driving force of chromosome condensation and segregation. The folding of the nucleosome fiber determines DNA accessibility for transcriptional regulation and DNA replication in interphase nuclei. The dynamic disordered folding may imply a novel mechanism of genome silencing, which could be quite different from the static local blocking of DNA accessibility through 30-nm chromatin folding (Figure 8.7c). The silencing is established through a dynamic capturing of transcriptional regions inside compact chromatin melt domains. These domains can be considered as drops of viscous liquid, which could be formed by the nucleosome–nucleosome interaction and macromolecular crowding effect. The transcriptional competency of genes may be regulated by changing their "buoyancy" toward the drop surfaces. It has been further speculated that the abundant noncoding, repetitive sequences in higher eukaryotic genomes contribute to the formation of the "chromatin liquid drop" segregated from nucleoplasm because identical DNA sequences tend to be self-assembled (Maeshima and Eltsov, 2008). Histone modifications or the binding of large transcriptional complexes may facilitate the maintenance of the regions on surfaces of the chromatin drops.

Metaphase chromosomes

The 30-nm fiber is arranged in loops that constitute the tertiary structure of chromatin, each anchored at its base by proteins in the nuclear scaffold. On an average, each loop encompasses some 20 000–100 000 bp of DNA and is about 300-nm in length, but the individual loops vary considerably (Figure 8.7d). A consequence of loop formation is that genomic regions that are far apart on the linear DNA molecule are brought in close proximity to one another. This in turn can have profound effects on gene transcription because gene-distal enhancers can now directly interact with gene-proximal promoters. Enhancers can be thousands of kilobases away from their target genes in any direction (or even on a separate chromosome). Such enhancer–promoter interactions are brought about by sequence-specific factors that bind to DNA. The 300-nm loops are packed and folded to produce a 250-nm-wide fiber (Figure 8.7e). Tight helical coiling of this 250-nm fiber in turn produces the structure that appears in metaphase chromosomes and other regions of interphase chromosomes. When chromosomes are depleted of histone proteins, a halo consisting of many loops of DNA (30–90 kb long) anchored along the length of a core/scaffold can be seen under an EM.

The quaternary structure of chromatin refers to the actual positioning of the chromosomes with respect to one another in the nucleus and with respect to the lamina of the INM (Figure 8.7f). Expression of a gene is affected by its three-dimensional (3D) position within the nucleus, with the general consensus being that transcriptionally active genomic regions are further away from the nuclear periphery than those that are silent. Repression near the nuclear periphery appears to be mediated by the interaction of chromatin with lamin proteins of the INM. Epigenetic modifications also impact higher-order chromatin structure by altering the chemical properties of histones and certain DNA bases. A recently developed method known as 3C (chromosome conformation capture) and several other related techniques that employ genomics tools have enabled the identification of these interactions, thereby augmenting our knowledge of higher-order chromatin organization (Sajan and Hawkins, 2012). Because different genes are expressed at different times in different cells, chromatin structure is clearly highly dynamic.

DNA structure

DNA is the prime genetic molecule, carrying all the hereditary information within chromosomes. The most important feature of DNA is that it is usually composed of two polynucleotide chains twisted around each other in the form of a double helix. The primary structure of DNA consists of a string of nucleotides joined together by phosphodiester linkages. The nucleotide consists of a phosphate joined to a sugar, known as 2′-deoxyribose, to which a base is attached. The sugar and base alone are called a nucleoside. To each sugar is attached a nitrogenous base, only four different kinds of which are commonly found in DNA: adenine (A), guanine (G), thymine (T), and cytosine (C). Adenine and guanine are purines, and thymine and cytosine are pyrimidines. A fifth base, 5-methyl cytosine (5mc), occurs in smaller amounts in certain organisms, and a sixth, 5-hydroxy-methyl-cytosine (5hmc), is found instead of cytosine in the T-even phages. The bases are attached to the deoxyribose by glycosidic linkages at N1 of the pyrimidines or at N9 of the purines. Each of the bases exists in two alternative tautomeric states, which are in equilibrium with each other. The nitrogen atoms attached to the purine and pyrimidine rings are in the amino form in the predominant state and only rarely assume the imino configuration. Likewise, the oxygen atoms attached to the guanine and thymine normally have the keto form and only rarely take on the enol configuration.

Native DNA is double helix

The double-helix DNA structure was proposed by James D. Watson and Francis H.C. Crick in 1953. Their proposal was based on the analysis of X-ray diffraction patterns of DNA fibers generated by Rosalind Franklin and Maurice Wilkins, which showed that the structure was helical, and analyses of the base composition of DNA from multiple organisms by Erwin Chargaff and colleagues. The double helix is composed of two polynucleotide chains that are held together by weak, non-covalent bonds between pairs of bases. Adenine on one chain is always paired with thymine on the other chain and, likewise, guanine is always paired with cytosine. This complementary base pairing enables the base pairs to be packed in the energetically most favorable arrangement in the interior of the double helix. In this arrangement, each base pair is of similar width, thus holding the sugar-phosphate backbones an equal distance apart along the DNA molecule. To maximize the efficiency of base-pair packing, the two sugar-phosphate backbones wind around each other to form a double helix, with one complete turn approximately every 10 base pairs. The two strands have the same helical geometry, but base pairing holds them together with the opposite polarity. That is, the base at the 5′-end of one strand is paired with the base at the 3′-end of the other strand. The strands are said to have an antiparallel orientation. This polarity in a DNA chain is indicated by referring to one end as the 3′- and the other as the 5′-end.

DNA types

The 3D structure of DNA described in 1953 by Watson and Crick is termed the "B-DNA structure." The X-ray diffraction pattern of DNA indicates that the stacked bases are regularly spaced 0.34 nm apart along the helix axis. The helix makes a complete turn every 3.4–3.6 nm, depending on the sequence; thus there is about 10–10.5 bp per turn. The B-form of DNA is the most stable configuration for a random sequence of nucleotides under physiological conditions. Under conditions of applied force or twists in the DNA, or under low hydration conditions, it can adopt several helical conformations, referred to as the A-DNA, Z-DNA, and S-DNA. The A-form crystallizes under low hydration conditions and is not normally found for DNA in the cell. It is, however, the structure adopted by double-stranded regions in RNA as well as the transient double-helix between DNA and RNA during transcription. Both A- and B-DNA are right-handed helices, whereas Z-DNA is a left-handed helix and is commonly found in regions of DNA that have an alternating purine–pyrimidine (e.g., 5'-CGCGCGCG-3' or 5'-CGCGCATGC-3') sequences. Parts of some active genes form Z-DNA, suggesting that Z-DNA may play a role in regulating gene transcription.

As a result of the double-helical structure of the two chains, the DNA molecule is a long, extended polymer with two grooves that are not equal in size to each other. The angle at which the two sugars protrude from the base pairs (i.e., the angle between the glycosidic bonds) is about 120° (for the narrow angle) or 240° (for the wide angle). As a result, as more and more base pairs stack on top of each other, the narrow angle between the sugars on one edge of the base pairs generates a minor groove and the large angle on the other edge generates a major groove. If the sugars pointed away from each other in a straight line, that is, at an angle of 180°, then two grooves would be of equal dimensions and there would be no minor and major grooves.

DNA topology

The linear, double-stranded DNA exists in a topologically relaxed state. The active DNA inside a cell is not relaxed, but is supercoiled. This topological state of supercoiling is the next highest order level of DNA organization after the linear relaxed state. Topological aspects of DNA structure arise primarily from the fact that the two DNA strands are repeatedly intertwined. Because their ends are free, linear DNA molecules can freely rotate to accommodate changes in the number of times the two chains of the double helix twist about each other. But if the two ends are covalently linked to form a circular DNA molecule and if there are no interruptions in the sugar-phosphate backbones of the two strands, then the absolute number of times the chains can twist about each other cannot change. Such a covalently closed circular DNA (cccDNA) is said to be topologically constrained. Even the linear DNA molecules of eukaryotic chromosomes are subject to topological constraints due to their entrainment in chromatin and interaction with other cellular components. There are two principal configurations of supercoiled DNA: solenoidal and plectonemic. The plectonemic (or interwound) supercoil is characteristic of DNA in prokaryotes. Solenoidal supercoiling is typical for eukaryotes, where DNA is wrapped around nucleosomal particles. Despite these constraints, DNA participates in numerous dynamic processes in the cell. For example, the two strands of the double helix, which are twisted around each other, must rapidly separate in order for DNA to be duplicated and to be transcribed into RNA. The fundamental topological parameter of a cccDNA is called the linking number (Lk). This quantity is defined as the number of times one strand of DNA winds around the other in the right-handed direction. Another characteristic of a circular DNA is called twist (Tw). Tw is the total number of helical turns in circular DNA under given conditions. Writhing (Wr) is the third important characteristic of circular DNA, describing the spatial pass of the double helix axis, that is, the shape of the DNA molecule as a whole. Wr can be of any sign, and usually its absolute value is much smaller than that of Tw. The unwound DNA and supercoiled DNA are topologically identical but geometrically different. They have the same Lk but differ in Tw and Wr. Although the rigorous definitions of twist and writhe are complex, twist is a measure of the helical winding of the DNA strands around each other, whereas writhe is a measure of the coiling of the axis of the double helix, which is called supercoiling. A right-handed coil is assigned a negative number (negative supercoiling) and a left-handed coil is assigned a positive number (positive supercoiling).

A molecule of cccDNA can readily undergo distortions that convert some of its twist to writhe or some of its writhe to twist without the breakage of any covalent bonds. The only constraint is that the sum of the Tw and the Wr must remain equal to Lk. Topoisomerases regulate the supercoiling of DNA in a cell. These enzymes introduce transient single- or double-stranded breaks into DNA to release torsional tension accumulating during strand separation in a topological domain. Topoisomerases are of two broad types. Type I topoisomerases change the Lk of DNA in steps of one. They make transient single-stranded breaks in the DNA, allowing one strand to pass through the break in the other before resealing the nick. Type I topoisomerases relax DNA by removing super-coils (dissipating writhe). They can be compared to the protocol of introducing nicks into cccDNA with DNase and then repairing the nicks, which as we saw can be used to relax cccDNA, except that type I topoisomerases relax DNA in a controlled and concerted manner. Type II topoisomerases change the Lk in steps of two. They make transient double-stranded breaks in the DNA, through which they pass a region of uncut duplex DNA before resealing the break. Type II topoisomerases require energy from adenosine triphosphate (ATP) hydrolysis for their action. In contrast to type II topoisomerases, type I topoisomerases do not require ATP. Both prokaryotes and eukaryotes have type I and type II topoisomerases, which are capable of removing supercoils from DNA. In addition, however, prokaryotes have a special type II topoisomerase, known as DNA gyrase that is responsible for the negative supercoiling of chromosomes, which facilitates unwinding of the DNA duplex during transcription and DNA replication.

DNA replication

All organisms must duplicate their DNA with extraordinary accuracy before each cell division. Cell division in eukaryotes is carried out in the context of the cell cycle. Unlike prokaryotes, which can double under optimal conditions in as little as 20 min, the eukaryotic cell cycle takes some 18–24 h to complete. The G1/S checkpoint (or restriction checkpoint) regulates entry of eukaryotic cells into the process of DNA replication and subsequent division. The cyclin/CDK complexes, formed during G1 and S phases to determine the phosphorylation/dephosphorylation events controlling the progression of cells through the cell cycle, also have a major control over DNA replication. Replication begins with the ordered assembly of a multiprotein complex called the pre-replicative complex (pre-RC). Pre-RC formation occurs during late M and early G1 phases of the cell cycle and licenses the DNA for replication during S phase. Cell cycle regulation by protein phosphorylation ensures that pre-RC assembly can only occur in G1 phase, whereas helicase activation and loading can only occur in S phase. Checkpoint regulation maintains high fidelity by stabilizing replication forks and preventing cell cycle progression during replication stress or damage.

The complementary nature of the two nucleotide strands in a DNA molecule laid the foundation of DNA replication that each strand can serve as a template for the synthesis of new strand. During the 1950s, the three models, namely, conservative, dispersive, and semiconservative of DNA replication were proposed. These three models make different predictions about the behavior of the two strands of the parental DNA during replication. Following replication, one daughter molecule contains both of the parental strands and the other daughter molecule contains two newly synthesized DNA strands. This model of replication is called conservative. In the second model, the each strand of the daughter DNA molecules would be a combination of old and new DNA, that is, both strands of each daughter molecule contain nucleotides derived from the parental molecule. This type of replication is referred to as random (or dispersive). The third model, in which one strand of the parental DNA serves as a template directing the order of nucleotides on the new DNA strand, is a semiconservative mode of replication. In 1958, M.S. Meselson and F.W. Stahl proved that in bacteria DNA replication was semiconservative. J.H. Taylor, P.S. Woods, and W.L. Hughes demonstrated in 1957 that a eukaryotic cell replicated by semiconservative replication.

Mechanism of replication

The region of replicating DNA at which the two strands of the parental DNA are separated and two new daughter DNA molecules are made, each with one parental strand and one newly synthesized strand, is called a replication fork. Once DNA synthesis has initiated, the elongation of the growing, new DNA strand proceeds via the apparent movement of one or two replication forks. The replication fork(s) are at one or both ends of a distinct replicative structure called a replication eye or bubble. The replication bubble can result from either bidirectional or unidirectional replication. In bidirectional replication, two replication forks move in opposite directions from the origin, and hence each end of the bubble is a replication fork. In unidirectional replication, one replication fork moves in one direction from the origin. In this case, one end of a replication bubble is a replication fork and the other end is the origin of replication. In the common "eye-form" replication structure, or replication bubble, both daughter DNA molecules are synthesized at a replication fork. Initiator proteins are sequence-specific DNA-binding proteins involved in binding at replication fork and initiates replication. Because the two strands of DNA are antiparallel, one new strand must be synthesized in a 5′ to 3′ direction in the same direction as the fork moves, whereas the other strand must be synthesized in an overall 3′ to 5′ direction relative to fork movement.

One of the template DNA strands is oriented 3′ to 5′ at the replication fork, and hence it can be copied continuously by a DNA polymerase extending the new DNA chain in a 5′ to 3′ direction. This new DNA chain is called the leading strand; its orientation is 5′ to 3′ in the same direction as the fork movement (Figure 8.8). It is extending from the replication origin. The other template strand is oriented 5′ to 3′ at the replication fork, and hence copying it will result in synthesis in a 3′ to 5′ direction relative to the direction of fork movement. This new DNA chain, called the lagging strand, is synthesized discontinuously, as a series of short DNA fragments. Each of these short DNA chains is synthesized

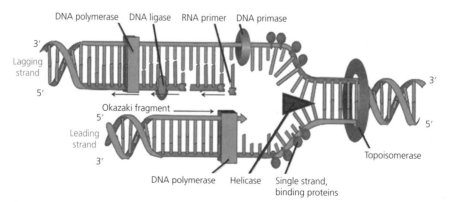

Figure 8.8 Replication fork showing DNA replication of leading and lagging strand.

in a 5' to 3' direction (i.e., opposite to the direction of the replication fork). These short DNA fragments are subsequently joined together by DNA ligase to generate an uninterrupted strand of DNA. Because the leading strand is synthesized continuously and the lagging strand is synthesized discontinuously, the overall process is described as semi-discontinuous. R. Okazaki and colleagues obtained evidence for discontinuous DNA synthesis during replication in 1968. These small DNA segments are intermediates in discontinuous synthesis of the lagging strand, and they are called Okazaki fragments.

Many enzymes used in replication have been isolated by biochemical fractionation. These include not only the DNA polymerases, but also helicases, which unwind the parental DNA duplex to make two new templates. Single-strand DNA-binding (SSB) proteins, also called helix-destabilizing proteins, bind tightly and cooperatively to exposed single-stranded DNA strands. These proteins are unable to open a long DNA helix directly, but they aid helicases by stabilizing the unwound, single-stranded conformation. In addition, their cooperative binding coats and straightens out the regions of single-stranded DNA on the lagging-strand template, thereby preventing the formation of the short hairpin helices that readily form in single-strand DNA. A specialized RNA polymerase called primase forms a short RNA primer complementary to the unwound template strands, which catalyzes the initial joining of nucleotides to start a DNA chain and is then elongated by a DNA polymerase, thereby forming a new daughter strand. DNA ligase joins the Okazaki fragments, and exonucleases can be used to remove incorrectly incorporated nucleotides.

The large linear chromosomes in eukaryotic cells contain many units of replication origin. As with prokaryotes, each origin of replication begins with the binding of a large protein complex; in the case of eukaryotes, it is named the origin recognition complex (ORC). This complex remains on the DNA throughout replication. Then to join the ORC is the minichromosome maintenance (MCM) complex. This complex contains a helicase amongst other enzymes and has a role in both the initiation and the elongation phases of eukaryotic DNA replication, specifically the formation and elongation of the replication fork. MCM is a component of the pre-replication complex, which is a component of the licensing factor. MCM is a hexamer of six related polypeptides (mcm2-7) that form a ring structure. This is required to unwind the DNA. Thus a number of protein complexes bind in a coordinated fashion and is termed "pre-replication complex." These accessory proteins or licensing factors accumulate during G1 of the cell cycle. The initiation complex is licensed to begin replication. It appears at 20–80 initiation sites of DNA. Forks extend from both sides of each initiation site and move in opposite directions until they fuse with an approaching fork from an adjacent site.

Five DNA polymerases— α, δ, β, ϵ, and γ—participate in DNA replication. DNA polymerase α acts as a catalyst of primer synthesis and extends a DNA strand in a 5' to 3' direction. This enzyme contains four polypeptide subunits, one with a polymerase activity (170 kDa), two that comprise a primase activity (50 and 60 kDa), and another subunit of undetermined function (70 kDa). This enzyme, unlike its prokaryotic equivalent, lacks exonuclease activity. It is always associated with a primase. DNA polymerase δ is also a nuclear enzyme with 5′ to 3′ polymerase activity and a 3′ exonuclease activity for proofreading. It does not, however, have a primase partner. It does have greater processivity (the ability to stick to the template and keep copying) than DNA polymerase α. It has two subunits: a polymerase (125 kDa) and another subunit (48 kDa) which is required for effective stimulation of polymerase by proliferating cell nuclear antigen (PCNA). This trimeric protein has a ring structure similar to that of the β2 sliding clamp of *Escherichia coli* DNA polymerase III. The template–primer junctions are recognized by the multi-subunit replication factor C (RFC). Like the γ complex in *E. coli*, this enzyme is an ATPase, and it helps to load on the processivity factor PCNA. Thus RFC is carrying out a similar function to the bacterial γ-complex. DNA polymerases β and ε are involved in repair of nuclear DNA. DNA polymerase β is a single polypeptide of 36 kDa, and has no 3′ to 5′ exonuclease activity. DNA polymerase γ replicates mitochondrial DNA. Not only do we have different enzymes, but eukaryotic cells have more copies of these enzymes than do prokaryotes.

The ends of the DNA are known as telomeres. The telomeres have an unusual sequence; several thousand copies of a tandem repeat which is species-specific, but is G-rich. The vertebrate version is TTAGGG. This repeating sequence is added onto the 3′-end by telomerases and is tucked into the end forming a cap. Capping proteins associate to prevent nuclease erosion of the ends. The telomerases contain an integral RNA component that has a sequence complementary to the repeat sequence, and these acts as a template for a reverse transcriptase reaction. The telomerase then moves to the new 3′-end once the repeat has been added and does it all again. The telomerase protein component is a type of reverse transcriptase. The other strand then copies the sequence by normal lagging strand mechanisms.

RNA structure, function, and synthesis

RNA consists of a long chain of nucleotide units. Each nucleotide consists of a nitrogenous base, a ribose sugar, and a phosphate. RNA is very similar to DNA, but it differs in a few important structural differences: each of the ribose rings contains a 2′-hydroxyl, and RNA uses uracil in place of thymine. The presence of 2′-hydroxyl group causes the helix to adopt the A-form geometry rather than the B-form most commonly observed in DNA. Thus, RNA exhibits a very deep and narrow major groove and a shallow and wide minor groove. In some viruses, RNA can also acts as a genetic material known as genetic RNA. All RNA species that are produced by transcription are termed nongenetic RNA. Most cellular

RNA is single-stranded, although some viruses have double-stranded RNA. The single RNA strand is folded upon itself, either entirely or in certain regions. This helps in the stability of the molecule. In the unfolded region, the bases have no complements. Because of this, RNA does not have the purine–pyrimidine equality that is found in DNA. Unlike DNA, RNA can form 3D structures. As a result, RNA can also exhibit catalytic activity. There are three types of commonly found cellular RNA: rRNA, mRNA, and tRNA. Other types of RNA, namely, heterogeneous nuclear RNA (hnRNA), snoRNA, and interfering RNA also exist in the cell. mRNA is the only coding RNA; all other RNAs are noncoding. Structure, function and synthesis of various diverse types of RNAs are discussed here.

Ribosomal RNA

rRNA is found in the ribosomes (the machinery responsible for protein synthesis). It comprises about 60–70% of the total RNA of the cell. rRNA contains the four major RNA bases with a slight degree of methylation, and shows differences in the relative proportions of the bases between species. Its molecules appear to be single polynucleotide strands that are unbranched and flexible. At low ionic strength, rRNA behaves as a random coil, but with increasing ionic strength the molecule shows helical regions produced by base pairing between adenine and uracil and guanine and cytosine. The rRNA strands unfold upon heating and refold upon cooling. Ribosomal RNA is stable for at least two generations. The ribosome consists of proteins and RNA. The eukaryotic 80S ribosome consists of two subunits: 40S and 60S. The 40S ribosome contains one rRNA (18S rRNA = 1900 bases) and about 35 different proteins. The 60S ribosome contains three rRNA (5S = 120 bases, 5.8S = 160 bases, and 28S = 4700 bases) and about 50 proteins. The 5S rRNA has its own gene; the others are synthesized as a single transcript which is then cleaved to release the mature RNA molecules that become part of the ribosome. The prokaryotic 70S ribosome consists of two subunits: 30S and 50S. The 30S subunit contains 16S rRNA, while the 50S subunit contains 23S and 5S rRNA. rRNA is formed from only a small section of the DNA molecule, and hence there is no definite base relationship between rRNA and DNA as a whole.

Transfer RNA

tRNA is a relatively small (~75-base) molecule having a molecular weight of about 25 000–30 000, and the sedimentation coefficient of mature eukaryote tRNA is 3.8S. It constitutes about 10–20% of the total RNA of the cell. tRNAs are L-shaped molecules. The amino acid is attached to one end, and the other end consists of three anticodon nucleotides. The anticodon pairs with a codon in mRNA ensuring that the correct amino acid is incorporated into the growing polypeptide chain. The L-shaped tRNA is formed from a small single-stranded RNA molecule that folds into the proper conformation. Four different regions of double-stranded RNA (dsRNA) are formed during the folding process to form a cloverleaf secondary structure. The 3'-end always terminates in a cytosine–cytosine–adenine (–CCA) sequence. The 5'-end terminates in guanine or cytosine. Many of the bases are bonded to each other, but there are also unpaired bases. In addition to the usual bases A, U, G, and C, tRNA contains a number of unusual bases, and in this respect differs from mRNA and rRNA. Most of the unusual bases are formed by methylation. Only 0.025% of DNA codes for tRNA. Synthesis of tRNA occurs near the end of cleavage stages. tRNA is an exception to other cellular RNAs in that a part of its ribonucleotide sequence (–CCA) is added after it comes off the DNA template. Precursor tRNA molecules transcribed on the DNA template contains the usual bases. These are then modified to unusual bases. The unusual bases are important because they protect the tRNA molecule against degradation by RNase. This protection is necessary because RNA is found floating freely in the cell. Electron density maps have revealed that tRNA has a tertiary structure. This structure is due to hydrogen bonds (i) between bases, (ii) between bases and ribose-phosphate backbone, and (iii) between the backbone residues.

tRNA carries the amino acids and transfers them to the growing protein. Since 20 amino acids are coded to form proteins, it follows that there must be at least 20 types of tRNA. It has, however, been shown that in several cases there are at least two types of tRNA for each amino acid. Thus, there are many more tRNA molecules than amino acid types. Each tRNA molecule is coded by one gene, with some exceptions. The starting amino acid in eukaryote protein synthesis is methionine, while in prokaryotes it is *N*-formyl methionine. The tRNA molecule specific for these two amino acids are methionyl tRNA (tRNA^{Met}) and *N*-formyl-methionyl IRNA (tRNA$_f^{Met}$), respectively. These tRNAs are called initiator tRNAs, because they initiate protein synthesis.

Messenger RNA

mRNA is always single-stranded and consists of only 3–5% of the total cellular RNA. The molecular weight of an average sized mRNA molecule is about 500 kDa and its sedimentation coefficient is 8S. The cell does not contain large quantities of mRNA. This is because mRNA, unlike other RNAs, is constantly undergoing breakdown to its constituent ribonucleotides by ribonucleases. There are few unusual substituted bases. Although there is a certain amount of random coiling in extracted mRNA, there is no base pairing. In fact, base pairing in the mRNA strand destroys its biological activity. The mRNA molecule has characteristic structural features. At the 5'-end of the mRNA molecule in most eukaryote cells and animal virus molecules a "cap" is present. This is blocked methylated structure, m7Gpp Nmp Np or m7Gpp Nmp Nmp Np. where: N = any of the four nucleotides and Nmp = 20 methyl ribose. The rate of protein synthesis depends on the presence of the cap. Without the cap, mRNA molecules bind very poorly to the ribosomes. The coding region of mRNA consists of about 1500 nucleotides on the average and translates protein. It is made up of 73–93 nucleotides.

The initiation codon is AUG in both prokaryotes and eukaryotes. Eukaryotic mRNAs usually have a series of 100–200 adenosine residues added to the 3'-end. These are not included in the DNA sequence; instead, these are added by a template-independent enzyme, poly(A) polymerase. The presence of cap and poly(A) tail is thought to stabilize the mRNA. Of all common mRNAs, only those that code for histones lack poly(A) tails. Eukaryotic mRNA molecules contain significant sequences upstream and downstream of the actual coding regions known as 5' and 3' untranslated regions. These untranslated regions (UTRs) confer increased stability to the mRNA and also UTRs contain sequences that control translation.

Small nucleolar RNAs

Since the late 1980s, it has become evident that a remarkably large number of small metabolically stable RNA species are localized to the nucleolus; these are called snoRNAs. The snoRNAs are present in the cell in the form of small nucleolar ribonucleoprotein particles (snoRNPs). During rRNA processing and ribosome assembly, each human pre-rRNA is predicted to transiently associate with approximately 150 different snoRNA species. A large group of snoRNAs shares two short-sequence elements, called boxes C and D; these are essential for the stable accumulation of the snoRNAs and for binding of a common snoRNP protein component, fibrillarin. The C and D boxes are positioned near to the 5' and 3' ends of the snoRNA, respectively, and they are frequently brought together by short complementary sequences found in close proximity to these elements. The middle part of the snoRNA is generally loosely structured and may contain additional, usually imperfect, copies of the authentic C and D boxes, referred to as boxes C' and D'. Another major class of snoRNAs is defined by an evolutionarily conserved "hairpin–hinge–hairpin–tail" structure. The single-stranded hinge region and the short 3'-end tail contain two conserved sequences, called boxes H and ACA. Like the box C and D elements, boxes H and ACA are required for the synthesis or stability of this class of snoRNAs. Another class of snoRNA is the RNA component of RNase MRP. This is an endoribonuclease that is structurally and functionally related to the ubiquitous endonuclease RNase P.

The complex population of snoRNAs is generated by a range of expression strategies. In vertebrates, a few snoRNAs are transcribed from their own promoter and terminator signals by RNA polymerase II (for U3, U8, and U13) or by RNA polymerase III (for RNase P and MRP RNAs). However, the large majority of vertebrate snoRNAs are excised from the introns of pre-mRNAs which also generate functional mRNAs from their exonic regions. In plants and yeast, snoRNAs can be generated by another pathway. Polycistronic transcripts are synthesized from which the snoRNAs are excised by a mechanism that does not involve RNA splicing. In yeast, both polycistronic snoRNAs and intron-encoded snoRNAs are found, while in plants the majority of characterized snoRNAs are synthesized from polycistronic transcripts and no intron-encoded snoRNAs have yet been identified. In contrast, no polycistronic snoRNA transcripts have yet been detected in metazoans.

The snoRNAs play role in the cleavage of the pre-rRNA, in the formation of the correct pre-rRNA structure, and in designating the sites of pre-rRNA modification by 2′-O-methylation and pseudouridine formation. The snoRNAs function in various aspects of nucleolar ribosome biogenesis; they probably assist in the complex conformational changes that rRNAs undergo during processing and packaging with RPs.

Small nuclear RNAs

snRNAs comprise a small group of highly abundant, non-polyadenylated, noncoding transcripts that function in the nucleoplasm. Six snRNA species (Ul–U6), ranging in size from 90 to 216 nucleotides, were originally identified in mammalian nuclei and found to have several features in common: all (except U6) possess m2′2′7G caps at the 5′-end, all are rich in uridine residues, and all are primarily nuclear molecules. The snRNAs can be divided into two classes on the basis of common sequence features and protein cofactors: Sm-class RNAs are characterized by a 5′-trimethylguanosine cap, a 3′ stem–loop and a binding site for a group of seven Sm proteins (the Sm site) that form a heteroheptameric ring structure. Lsm-class RNAs contain a monomethylphosphate cap and a 3′ stem–loop, terminating in a stretch of uridines that form the binding site for a distinct heteroheptameric ring of Lsm proteins. The Sm class of snRNAs comprises U1, U2, U4, U4$_{atac}$, U5, U7, U11, and U12, whereas the Lsm class is made up of U6 and U6$_{atac}$. snRNAs combine with small nuclear protein subunits to form snRNPs. Some of the proteins in the snRNPs may be involved directly in splicing; others may have structural roles or may be required just for assembly or interactions between the snRNP particles. The uridine-rich snRNPs except U7 form the core of the spliceosome and catalyze the removal of introns from pre-mRNA. Like the ribosome, the spliceosome depends on RNA-RNA interactions as well as protein–RNA and protein–protein interactions. Some of the reactions involving the snRNPs require their RNAs to base pair directly with sequences in the RNA being spliced; other reactions require recognition between snRNPs or between their proteins and other components of the spliceosome.

The Sm-class snRNA genes share several common features with protein-coding genes, including the arrangement of upstream and downstream control elements. Sm-class genes are transcribed by a specialized form of RNA poly-merase II that is functionally similar to the polymerase II used by the mam-malian protein coding genes. The Lsm-class snRNA genes (*U6* and *U6atac*) are transcribed by polymerase III using specialized external promoters. The run of uridines that forms the Lsm-binding site at the 3′-end also doubles as a Pol III transcription terminator. In higher eukaryotes, Lsm-class snRNAs never leave the nucleus. Therefore, there are few parallels between Lsm-class genes and genes that encode proteins.

Heterogeneous nuclear RNA

Eukaryotic genes that code for mRNA are copied by RNA polymerase II into transcripts collectively termed hnRNA. The great majority of these nuclear transcripts undergo subsequent covalent modifications, through which some (but, importantly, not all) are converted into mRNA. The terms hnRNA and pre-mRNA are often used interchangeably, although only a subpopulation of hnRNAs are actually mRNA precursors. hnRNAs in cells do not normally occur as naked polynucleotides, but they are found in complexes with specific proteins, which are termed heterogeneous nuclear ribonucleoprotein (hnRNP) complexes or hnRNP particles. The proteins of the hnRNP particles are as abundant as histones in the nucleus of growing cells and comprise approximately 80% of the mass of hnRNP particles. hnRNP is simply a metabolically inert packaging device, involving a regular array of stable hnRNA-protein contacts analogous to the nucleoprotein organization of chromatin or viral nucleocapsids. The packaging must be done, however, in such a way that the pre-mRNA is folded into a structure that can be spliced. This includes the requirement that pre-mRNA is also accessible for interaction with snRNPs.

MicroRNA

MiRNA genes are transcribed by RNA polymerase II as large primary transcripts (pri-miRNA) that are processed by a protein complex containing the RNase III enzyme Drosha, to form an approximately, 70 nucleotide precursor miRNA (pre-miRNA). This precursor is subsequently transported to the cytoplasm where it is processed by a second RNase III enzyme, DICER, to form a mature miRNA of approximately 22 nucleotides. The mature miRNA is then incorporated into a ribonuclear particle to form the RNA-induced silencing complex, RISC, which mediates gene silencing.

miRNAs are small noncoding RNAs that serve as posttranscriptional regulators of gene expression in plants and animals. They comprise approximately 1% of genes and are often highly conserved across a wide range of species. The miRNAs come from endogenous transcripts that can form local hairpin structures, which ordinarily are processed such that a single miRNA molecule accumulates from one arm of a hairpin precursor molecule. MiRNAs are sometimes encoded by multiple loci, some of which are organized in tandemly co-transcribed clusters. MiRNAs usually induce gene silencing by binding to target sites found within the 3'UTR of the targeted mRNA. This interaction prevents protein production by suppressing protein synthesis and/or by initiating mRNA degradation. Since most target sites on the mRNA have only partial base complementarity with their corresponding miRNA, individual miRNAs may target as many as 100 different mRNAs. Moreover, individual mRNAs may contain multiple binding sites for different miRNAs, resulting in a complex regulatory network. MiRNAs have been shown to be involved in a wide range of biological processes such as cell cycle control, apoptosis, and several developmental and physiological processes. In addition, highly tissue-specific expression and distinct temporal expression patterns during embryogenesis suggest that miRNAs play a key role in the differentiation and maintenance of tissue identity.

Short interfering RNAs

Short interfering RNAs (siRNAs) are derived from long, double-stranded RNAs that are transcribed endogenously or introduced into cells by viral infection or transfection. siRNA duplexes are produced by processing of these long, double-stranded RNAs by the unusual Dicer ribonuclease, and one strand of the duplex is then incorporated into a ribonucleoprotein complex, the RNA-induced silencing complex (RISC). The siRNA component guides RISC to mRNA molecules bearing a homologous antisense sequence, resulting in cleavage and degradation of that mRNA (Cejka, Losert and Wacheck, 2006). This process is termed RNA interference (RNAi). RNAi was first discovered in 1998 by A. Fire and W. Mello in *Caenorhabditis elegans*, when it was noted that introducing a dsRNA that was homologous to a specific gene resulted in the posttranscriptional silencing of that gene. siRNA-mediated gene silencing provides a fast and reliable way to characterize a gene knockdown phenotype *in vitro* with striking potency at relatively low compound costs (Fire *et al.*, 1998). A. Fire and W. Mello were awarded the Nobel Prize in Physiology or Medicine in 2006. Functional genomic approaches delineating a complete signal transduction pathway using multiple siRNAs are feasible as well.

Both miRNA and siRNA are 21–22 nucleotides long, but siRNAs arise from cleavage of mRNAs, RNA transposons, and RNA viruses, whereas miRNAs are cleaved from RNA molecules transcribed from sequences that are distinct from other genes. Each miRNA is cleaved from a single-stranded RNA precursor that forms small hairpins, while multiple siRNAs may be produced from the cleavage of an RNA duplex consisting of two different RNA molecules or from the cleavage of longer hairpins arising within a single RNA molecule. miRNAs silence genes that are distinct from those from which they were transcribed as compared to siRNAs which silence the genes from which they were transcribed.

Synthesis of RNA

There may be 1000–60 000 different species of mRNA in a cell. These mRNA types differ only in the sequence of their bases and in length. When one gene (cistron) codes for a single mRNA strand or when it contains the genetic information to translate only a single protein chain (polypeptide), the mRNA is said to be monocistronic. This is the case for most of the eukaryotic mRNAs. On the other hand, polycistronic mRNA carries several open reading frames, each of which is translated into a polypeptide. These polypeptides usually have a related function (they often are the subunits composing a final complex protein), and their coding sequence is grouped and regulated together in a regulatory region, containing a promoter and an operator. Most of the mRNAs found in bacteria and archea is polycistronic. Initiation of transcription is a complex process in which the following steps are recognized: (i) transcription factor proteins bind to the promoter element upstream of the coding sequence, (ii) the DNA-dependent RNA polymerase binds to the promoter/protein complex, (iii) the RNA polymerase complex separates the DNA strands, (iv) the RNA polymerase begins synthesizing RNA, and (v) termination factors bind to termination codon to terminate RNA synthesis. Unlike DNA polymerases, RNA polymerases do not need primers. Transcription only requires binding of the RNA polymerase to the promoter. *E. coli* has a single RNA polymerase responsible for all types of gene transcription. Eukaryotes have three transcription polymerases: RNA polymerase I transcribes rRNA; RNA polymerase II transcribes hnRNA, which are the source of the mRNA used as templates for protein synthesis; and RNA polymerase III transcribes tRNA, 5S rRNA, and a few other small RNA molecules. The RNA polymerases are large multiprotein complexes with about 10 subunits.

Nucleocytoplasmic transport, nuclear import, and nuclear export

The NPCs play a fundamental role in the physiology of all eukaryotic cells as they are the only passage channels for dispensing the traffic of molecules between the nucleus and cytoplasm. NPC acts as a highly efficient molecular sieve for small polar molecules, ions, metabolites, and macromolecules (proteins and RNAs) that shuttle continuously between the nucleus and the cytoplasm. Small molecules and some proteins (diameter $\geq 9\,nm$) with molecular mass less than approximately $50\,kDa$ pass freely across the NE through open aqueous channels in either direction: cytoplasm to nucleus or nucleus to cytoplasm via passive diffusion. Most proteins and RNAs, however, are unable to pass through these open channels. Such proteins are responsible for all aspects of genome structure and function; they include histones, DNA polymerases, RNA polymerases, transcription factors, splicing factors, and many others. Thus macromolecules are transported by a selective, energy-dependent mechanism by an active process that acts predominantly to import proteins to the nucleus and export RNAs to the cytoplasm. They are transported through regulated channels in the NPC that, in response to appropriate signals, can open to a diameter of more than $25\,nm$, a size sufficient to accommodate large ribonucleoprotein complexes, such as ribosomal subunits. Thus this kind of regulation helps the nuclear proteins to get selectively imported from the cytoplasm to the nucleus, while RNAs are exported from the nucleus to the cytoplasm. This regulated traffic of proteins and RNAs through the NPC thus determines the composition of the nucleus and play a key role in gene expression. The central channel is approximately $40\,nm$ in diameter, which is wide enough to accommodate the largest particles able to cross the NE.

Nucleocytoplasmic transport

Transport of almost all macromolecules into and out of the nucleus achieved through a common active mechanism requires the assistance of soluble NTFs and transport signals, which together form the "soluble phase" of nuclear transport. Most NTFs belong to the karyopherin (Kap) family of proteins and these NTPs specifically bind to transport signals which are found on their cognate substrates and translocate them through the NPC channel. Transport signals found on nuclear protein cargoes consist of short amino acid sequences called nuclear localization sequences/signals (NLSs; for import) or nuclear export sequences/signals (NESs; for export). A consensus three-step mechanism for macromolecular cargoes has provided the framework for all subsequent investigations into the molecular details of nuclear transport. First, nuclear transport substrates are recognized and bound by NTFs. Second, this NTF–cargo complex docks to the NPC by binding to FG Nups and translocates through the NPC. Third, on reaching its target compartment (either the nucleoplasm or cytoplasm), the complex dissociates.

A key player in the translocation process is a small GTP-binding protein called Ran, which is related to the Ras proteins. Enzymes that stimulate GTP binding to Ran are localized to the nuclear side of the NE, whereas enzymes that stimulate GTP hydrolysis are localized to the cytoplasmic side. Ran-GTP is predominantly found in the nucleus and drives the release of import cargo inside the nuclear compartment by binding to importing Kaps. In the case of nuclear export, the presence of Ran-GTP increases the affinity of exporting Kaps for NES containing cargo. On reaching the other side of the NPC, GTP hydrolysis induces the release of cargo into the cytoplasm. The resulting NTF and Ran-GDP are subsequently recycled back into the transport pathway. Consequently, there is a gradient of Ran-GTP across the NE, with a high concentration of Ran-GTP in the nucleus and a high concentration of Ran-GDP in the cytoplasm.

Nucleocytoplasmic transport through an NPC is highly efficient, with a rate of approximately 100 to 500 translocation events per second. Several models have been proposed to explain nuclear transport. First, the Brownian affinity gating model had been proposed where FG-repeat-containing filaments bristling out of the NPC are supposed to move vigorously back and forth and prevent non-FG-domain-binding proteins from reaching the central channel. Then, polymer brush model proposed by Lim *et al.* (2007) extends these considerations. In this model, FG repeat-containing proteins are also assumed to form extended brush-like polymers that reversibly collapse upon binding of transport receptors. In contrast, the selective phase/hydrogel model proposes that FG–FG interactions form a hydrogel within the central channel. NTRs can bind these domains via low-affinity interactions and transiently open the mesh, thus permeating the channel by a solubility-diffusion process. Thus far, no comprehensive study has been performed to clearly distinguish between these and other proposed models.

Transport can also be modulated at the level of the transport factors and their cargos. A common method is to regulate the binding strength or accessibility of an NLS or NES in a cargo for its cognate karyopherin. This often involves phosphorylation in or near the signal sequence of the cargo. Transcription factors, for example, are functional only when they are present in the nucleus; therefore, regulation of their import to the nucleus is a novel means of controlling gene expression. Regulated nuclear import of both transcription factors and protein kinases plays an important role in controlling the behavior of cells in response to changes in the environment, because it provides a mechanism by which signals received at the cell surface can be transmitted to the nucleus. One way of regulation is where transcription factors (or other proteins) associate with cytoplasmic proteins that mask their nuclear localization signals so these proteins remain in the cytoplasm as their signals are no longer recognizable. Alternatively, nuclear import of other transcription factors is regulated directly by their phosphorylation, rather than by association with inhibitory proteins.

However, in a physiological context, NPC events and non-FG Nup-binding sites are required to provide transport efficiency, regulation, and directionality. There are specific factors for mediating transport directionality and cargo release. The Ran-GTPase for karyopherins and the Gle1–IP_6 (export protein Gle1 bound to inositol hexakisphosphate) activated Dbp5 (DEAD-box protein) for mRNA export. These use non-FG NPC docking sites that increase their local concentration at their respective sites of action. Although the FG domains do not form a gradient of docking sites in the translocation channel, there are non-FG high-affinity docking sites located at terminal transport steps that influence efficiency.

The movement of RNAs and ribosomes, from the nucleus to the cytoplasm is a critical component in a cell's ability to make the proteins necessary for essential biological functions. Since proteins are synthesized in the cytoplasm, most of RNAs are exported from the nucleus to the cytoplasm. The export of mRNAs, rRNAs, and tRNAs is a critical step in gene expression in eukaryotic cells. Like protein import, the export of RNAs through NPCs is an active, energy-dependent process that requires the Ran-GTP-binding protein. RNAs are transported across the NE as RNA–protein complexes, which in some cases are large enough to visualize by EM. The substrates for transport are ribonucleoprotein complexes rather than naked RNAs, and RNAs are targeted for transport from the nucleus by nuclear export signals on the proteins bound to them. These proteins are recognized by exportins and transported from the nucleus to the cytoplasm as described earlier. Pre-mRNAs and mRNAs are associated with a set of at least 20 proteins (forming hnRNPs) throughout their processing in the nucleus and eventual transport to the cytoplasm. At least two of these hnRNP proteins contain nuclear export signals and are thought to function as carriers of mRNAs during their export to the cytoplasm. Ribosomal RNAs are assembled with RPs in the nucleolus, and intact ribosomal subunits are then transported to the cytoplasm. Nuclear export signals present on RPs mediate their export from the nucleus. For tRNAs, the specific proteins that mediate nuclear export remain to be identified. In contrast to mRNAs, tRNAs, and rRNAs, which function in the cytoplasm, the snRNAs function within the nucleus as components of the RNA processing machinery. Perhaps surprisingly, these RNAs are initially transported from the nucleus to the cytoplasm, where they associate with proteins to form functional snRNPs and then return to the nucleus. Proteins that bind to the 5′ caps of snRNAs appear to be involved in the export of the snRNAs to the cytoplasm, whereas sequences present on the snRNP proteins are responsible for the transport of snRNPs from the cytoplasm to the nucleus.

Nuclear import

Several proteins remain within the nucleus after their import from the cytoplasm, but many others shuttle back and forth between the nucleus and the cytoplasm. Some of these proteins act as carriers in the transport of other molecules, such as RNAs; others coordinate nuclear and cytoplasmic functions (e.g., by regulating the activities of transcription factors). Protein import through the NPC can be operationally divided into two steps, distinguished by whether they require energy or not. In the initial step, a transport receptor recognizes a signal-bearing target protein (cargo) and forms a receptor–target protein complex that does not require energy. In first step, nuclear localization signals are recognized by a cytosolic receptor protein, only then the receptor–target protein complex formation takes place which further docks on the near side of the NPC but do not pass through the nuclear pore. The prototype receptor, called importin, consists of two subunits. One subunit (importin α) binds to the basic amino acid-rich nuclear localization signals of proteins such as T antigen and nucleoplasmin. The second subunit importin β (Kap β) binds to the cytoplasmic filaments of the NPC, bringing the target protein to the nuclear pore. Other types of NLSs, such as those of RPs, are recognized by distinct receptors that are related to importin β, and they function similar to importin β during the transport of their target proteins into the nucleus. The second step in nuclear import, translocation through the NPC, is an energy-dependent process that requires GTP hydrolysis. Following import through the NPC, complex disassembly is initiated by nuclear binding of Ran-GTP to the receptor. Ran-GTP binds to importin β, releasing importin α and the target protein in the nucleus. The Ran-GTP importin β complex is then transported back to the cytoplasm, where the Ran-GTPase-activating protein (Ran-GAP) stimulates hydrolysis of the bound GTP to form Ran-GDP. This conversion of Ran-GTP to Ran-GDP is accompanied by the release of importin β (Figure 8.9). In the case of Kap-mediated transport, directionality is enforced by the distribution of the GTP- and GDP-bound states of the small GTPase Ran, which is essential for the nuclear transport of RNA and proteins.

Nuclear export

Proteins are targeted for export from the nucleus by specific amino acid sequences, called NESs. Like NLSs, NESs are recognized by receptors within the nucleus that direct protein transport through the NPC to the cytoplasm.

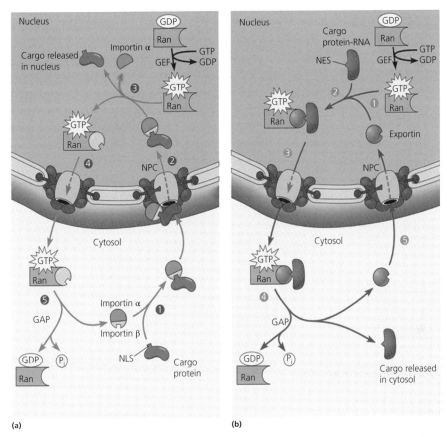

(a)   (b)

Figure 8.9 Nucleocytoplasmic transport of macromolecular cargoes: (a) nuclear import cycle (b) and nuclear export cycle.

Interestingly, the NESs (called exportins) are related to importin β. Like importin β, the exportins (Kapβ) bind to Ran, which is required for nuclear export as well as for nuclear import. Strikingly, however, Ran-GTP promotes the formation of stable complexes between exportins and their target proteins, whereas it dissociates the complexes between importins and their targets. This effect of Ran-GTP binding on exportins orders the movement of proteins containing nuclear export signals from the nucleus to the cytoplasm. Thus for nuclear export, Ran-GTP in the export receptor–target protein complex is hydrolyzed to Ran-GDP by the Ran-GTPase-activating protein (Ran-GAP), resulting in complex disassembly inside the cytoplasm. The Ran guanine nucleotide exchange factor (Ran-GEF) is localized to the nucleus and generates high Ran-GTP concentrations, and Ran-GDP is imported into the nucleus. Exportin is then transported back to the nucleus.

Until recently, it was believed that the NPC was the sole pathway between the cell nucleus and cytoplasm for these materials. In addition to the canonical pathway of mRNA export going through the NPC, it has been revealed now that large RNA transport granules can be assembled in the cell nucleus and exported via a budding mechanism previously thought to be used only by the herpes virus (Sean *et al.*, 2012).

The dynamics of NE biogenesis during mitosis

The NE is a highly specialized membrane architecture composed of the INMs and ONMs, NPCs, and the lamina (in metazoans), which delineates the eukaryotic cell nucleus as a compartment from rest of the cell components. Sophisticated collaboration of these molecular machineries is necessary for the structure and functions of NE. The NE not only provides a unique environment to regulate the trafficking of macromolecules between nucleoplasm and cytoplasm but also provides anchoring sites for chromatin and the cytoskeleton. Through these interactions, the NE helps to hold the nucleus intact within the cell and chromosomes within the nucleus, thereby regulating the expression of certain genes. NE is not stationary; rather, it is continuously reconstructed during cell division. At the beginning of mitosis, the chromosomes condense, the nucleolus disappears, and the NE breaks down, resulting in the release of most of the contents of the nucleus into the cytoplasm. At the end of mitosis, the process is reversed: The chromosomes decondense, and NEs re-form around the separated sets of daughter chromosomes. NE reorganization in metazoans during mitosis presents the most unique feature as it disassembles and re-forms each time most cells divide. The process is controlled largely by reversible phosphorylation and dephosphorylation of nuclear proteins resulting from the action of the Cdc2 protein kinase, which is a critical regulator of mitosis in all eukaryotic cells. Mutations in certain NE proteins are associated with a diversity of human diseases, including muscular dystrophy, neuropathy, lipodystrophy, torsion dystonia, and the premature aging condition progeria.

Closed and open mitosis

Although a membrane-enclosed nuclear genome can be found in all eukaryotcs, there is a critical difference in the cell-cycle-dependent dynamics of the NE between "lower" eukaryotes (e.g., yeast and filamentous fungi) and metazoans (i.e., "higher" eukaryotes). The NE of metazoan cells completely disintegrates during cell division to allow the mitotic spindle to access chromosomes. As a consequence, every dividing cell has to reform the NE and reestablish the identity of the nuclear compartment. Although open mitosis is often cited as characteristic of higher eukaryotes, the degree to which mitosis is open or closed varies markedly among plants and metazoans, However, in some higher eukaryotes, semi-closed mitosis is also accomplished by certain cell types. Here, the NE only partially opens up near to centrosomes to allow cytoplasmic spindle MTs to reach the nuclear interior without the need for major rearrangements of NE components. In syncytial cells, this ensures that spindle MTs capture the correct chromosomes in the common cytoplasm. The mitotic spindles form through open fenestrae in a degenerate NE, while the NPCs disassemble to various degrees. Otherwise, during mitosis in metazoans, NE breaks down and the separation of the nucleoplasm from the cytosol vanishes, until the NE re-forms after a cell completes division. The NE finally breaks down during anaphase.

The lower eukaryotes undergo closed mitosis, where the NE stays intact and spindle MTs can either form inside the nucleus or are able to penetrate an intact nuclear membrane. Here, MTOCs are either constantly part of the NE (e.g., in budding yeast *S. cerevisiae*) or are inserted into the NE during mitotic entry (e.g., in *S. pombe*), and in both cases MTOCs direct the formation of a nuclear spindle. The establishment of a nuclear spindle requires nuclear uptake of tubulin. The NE and NPCs remain intact throughout the entire cell cycle. A prevalent intermediate between open and closed mitosis is the partial disassembly of the NE is also found in some lower eukaryotes. For instance, the filamentous fungus, *Aspergillus nidulans*, undergoes semi-closed mitosis and the NE remains mostly intact throughout the cell cycle, but there is a partial disassembly of the NPC to achieve the rapid influx of tubulin. These "remnant" NPCs are freely permeable to the assembling spindle components and also provide the framework for the dispersed Nups to reassemble at the end of mitosis.

Interphase

NE remodeling in proliferating cells is a highly dynamic process that involves a vast number of molecular players. By the end of interphase in G2, the nuclei have duplicated their genome, doubled the number of NPCs, and increased the surface area of the NE. The surrounding ER network is continuous with the NE, but not enriched in NE proteins.

NE breakdown

A cell needs to accurately segregate not only the genetic material and all the organelles but also the NE membranes with its specific protein components. Disassembly of the NE, which parallels a similar breakdown of the ER, involves changes in all three of its components: the nuclear membranes are fragmented into vesicles, the NPCs dissociate, and the nuclear lamina depolymerizes (Figure 8.10). Depolymerization of the nuclear lamina through phosphorylation of the lamin filaments catalyzed by the Cdc2 protein kinase results into individual lamin dimers. Cdc2 (as well as other protein kinases activated in mitotic cells) phosphorylates all the different types of lamins, and treatment of isolated nuclei with Cdc2 has been shown to be sufficient to induce depolymerization of the nuclear lamina.

In 2000, P. Collas and J.C. Courvalin proposed ER retention model which suggests that some NE components are retained in the mitotic ER network

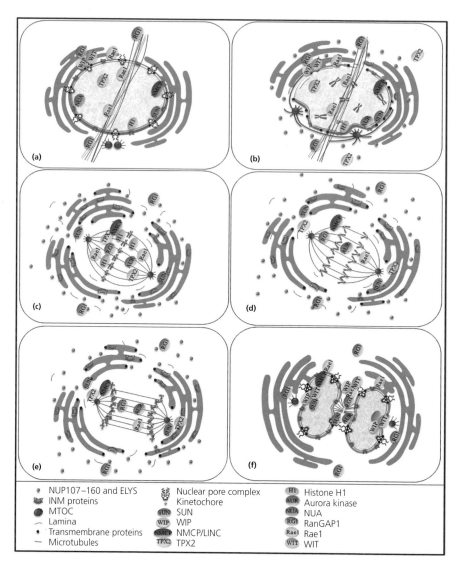

Figure 8.10 Schematic illustration of biogenesis of nuclear envelope showing mitotic locations of associated proteins: (a) interphase, (b) prophase, (c) metaphase, (d) early anaphase, (e) late anaphase, and (f) telophase.

during cell division, but numerous other ones localize to diverse mitotic structures and play crucial role in consecutive stages of the division process. Both the localization patterns and a variety of developmental phenotypes point to these functions. Through the use of fluorescent protein fusions and live cell imaging, it has been shown that some NE proteins relocate to the ER during NE disassembly in cultured animal and plant cells. In the animal cells, it has been shown that the NE marker (LBR-GFP) exhibits a significant change in mobility between interphase and mitosis, with LBR being predominantly immobile within the intact NE in contrast to showing a high diffusion rate within mitotic membranes. The presence of NE proteins in the ER during mitosis has also been shown by immunolabeling of native proteins in mammalian cells. In plant cells, photobleaching data indicate that LBR-GFP shows a high level of mobility within the NE membranes at interphase, and apparently lacks the significant fraction of immobilized protein present in mammalian cells.

The possible mechanisms responsible for NEBD have been well described in animal and to some extent in plant cells. In animal cells, NEBD is the result of MT-dependent stretching of the nuclear lamina, which causes deformation of the NE and changes in nuclear pore distribution. Initial tearing of the NE occurs on the opposite side of the nucleus where the tensile forces are at their greatest; the tearing is possibly linked to the localized disassembly of nuclear pores creating a focal point for the membrane perforation. After the initial tearing of the nuclear membrane, structure of the NE is lost rapidly. In tobacco, *BY-2* cells co-expressing an MT marker (GFP-MBD) and an ER/Golgi marker (Nag-DsRed) to highlight the NE, it is observed that NEBD occurs before the disappearance of the preprophase band (PPB), an MT structure characteristic of plant cell division. After dissociation, nuclear membranes ruffle in area of NE in closest proximity to the PPB. The spatio-temporal link between the loss of the PPB and breakdown of NE, thus suggesting that plant NEBD is initiated by membrane tearing and involves the attachment of MTs to the surface of the NE.

Preprophase/prophase

NE notably itself acts as an MT organizing center (MTOC). Plant cells undergo drastic MT array rearrangements during cell division, forming cortical and radial MTs, the PPB, the spindle, and phragmoplast structures. One of the initial events is the selective loss of Nups from the NPCs. Nup98 (a dynamic NPC component) is released before the rest of the NPC components that disassemble synchronously. At the onset of mitosis, the cortical MTs depolymerize and rearrange into the PPB surrounding the nucleus. This initial cytoskeletal change is crucial for the fate of a dividing cell, since this transient MT array demarcates the future cortical division site, where a cell will separate into two daughter cells. Ran-GAP1 is an NE-associated protein that is delivered to the PPB in an MT-dependent manner, and it remains associated with the cortical division site during mitosis and cytokinesis, constituting a continuous positive marker of the plant division plane. Ran-GAP1 is thus a molecular landmark left behind by the PPB, which later guides the phragmoplast and the forming cell plate. The silencing of Ran-GAP1 in *Arabidopsis* roots leads to mutation with mispositioned cell walls with division plane defects. At this stage, another NE-associated protein, Rae1, is targeted to the PPB. This localization of Rae1 reflects its association with mitotic MTs throughout mitosis as well as at least partial involvement of the PPB in spindle assembly, since the RNAi inhibition of *Nicotiana benthamiana* Rae1 (NbRae1) in *BY-2* cells leads to the formation of disorganized or multipolar spindles and defects in chromosome segregation.

In plants, the PPB indeed marks the plane perpendicular to the axis of symmetry, the spindle, and the central positioning of the nucleus. At this stage, the NE, acting as an MTOC, promotes the nucleation of MTs on its surface.

An essential factor of the MT-nucleating complex is the γ-tubulin ring complex, which is conserved among the kingdoms. In mammals, the minimal complex functioning as an MTOC is composed of γ-tubulin, γ-tubulin complex protein 2 (GCP2), and GCP3, which all have orthologs in the *Arabidopsis*. A nuclear rim-associated fraction of histone H1 has been shown to have MT-organizing activity in *BY-2* cells and to promote MT nucleation through the formation of complexes with tubulin and the elongation of radial MTs. Both processes seem to require phosphorylation events carried out by a CDK known as M-phase-promoting factor (MPF) kinase $p34^{cdc2}$ and its regulatory protein, cyclin B (CYCB). This is believed to abolish the structural protein–protein interactions that are necessary for NE integrity, which leads to its disassembly. Nuclear lamins are phosphorylated by protein kinase C and, subsequently by cdc2, which results in their depolymerization and dispersal in mitotic cells. Nups and several INM proteins are also substrates of cdc2. In case of Nups, this presumably causes their release from, and subsequently the disassembly of, the NPC, which explains the dispersed localization of Nups that is observed in mitotic cells. The effect of phosphorylation is less clear for INM proteins, but it is also assumed to abolish their ability to interact with lamins and/or chromatin, which allows the INM to detach from chromosomes. The CDK/CYCB complex promotes PPB disassembly in plants, the depolymerization of nuclear lamins in vertebrates, *C. elegans*, and yeast, and the disassembly of Nups in animal cells. Among plant nuclear pore proteins with dynamic mitotic relocalization, there are homologs, for instance, NUA (the *Arabidopsis* homolog of Tpr/Mlp1/Mlp2/Megator) and Rae1 (Boruc, Zhou and Meier, 2012).

Metaphase
The Ran gradient

The Ran gradient controls the spindle assembly in animal cells. Plant Ran-GAPs, the proteins that activate Ran-GTPase allowing the conversion of Ran-GTP to Ran-GDP, thus aiding the maintenance of the Ran gradient and directionality of nuclear import/export, are located at the outer side of the NE during interphase. During mitosis, these proteins are found associated with MTs within the mitotic apparatus. A specific N-terminal WPP (tryptophan–proline–proline) domain in the plant Ran-GAP mediates the novel location of this protein during mitosis, which differs from that observed for animal cells. An antibody raised against the *Arabidopsis* Ran2 protein also differs from the location seen for mammalian cells at interphase, the plant protein being found in the perinuclear and NE region but not in the nucleoplasm, whereas the animal Ran proteins are usually present in the nucleus. These subtle differences in protein location between plant and animal Ran and Ran-associated proteins and kingdom-specific targeting domains high-light the fact that high concentrations of Ran-GTP around chromosomes (and high Ran-GDP concentration at the cell periphery) attract importins and release NLS containing cargo proteins. These cargos are, for instance, spindle assembly factors, such as targeting protein for Xklp2 (TPX2), Rae1, and NuMA (for nucleus and mitotic apparatus). *Arabidopsis* TPX2 is nuclear in interphase, but it is actively exported in prophase, enriched around the NE, and then accumulates in the vicinity of the prospindle. After its release from importin-dependent inhibition, TPX2 promotes spindle formation around chromosomes through MT nucleation. Simultaneously, human TPX2 targets Aurora A to the spindle and activates it. The coordination of chromosomal and cytoskeletal events in mitosis is partly mediated by the chromosomal passenger complex.

Aurora kinases (in *Arabidopsis*, Aurora1 and Aurora2) are thought to play this role through mediating the positioning information of the PPB to the formation of the bipolar prophase spindle. At the onset of prophase, AtAurora1 and AtAurora2 are associated with the ONM and then gradually migrate to the poles of the prospindle as mitosis progresses. Tobacco NbRae1, a homolog of Rae1/mrnp41 in metazoans, Gle2p (for GLFG lethal 2p) in *S. cerevisiae*, and Rae1 in *S. pombe*, exhibit a mitotic function besides its role as an mRNA export factor associated with the NPC. NbRae1 associates with the spindle and has been shown to function in the proper spindle organization and chromosome segregation. NbRae1 silencing results in delayed progression of mitosis, which leads to plant growth arrest, reduced cell division activities in the shoot apex and the vascular cambium, and increased ploidy levels in mature leaves. Together, these results suggest a conserved function of the Rae1 protein in spindle organization among eukaryotes, which is distinct from their roles at the interphase NE. In metaphase, while histone H1 relocalizes along the condensed chromosomes, Aurora3 and Aurora1 are associated with centromeric regions of chromosomes and Ran-GAP1 localizes to kinetochores and the spindle. It still remains enigmatic as to how *Arabidopsis* Ran-GAP1, which lacks the SUMOylation domain, is targeted to kinetochores. Ran-GAP1 is found only on the attached sister chromatids. The exact timing of kinetochore association and the function of plant Ran-GAP1 at this cellular location remains to be verified. The cell cycle dynamics of NMCP1 and NMCP2 have been investigated by Kimura and coworkers (2010). Both proteins associate with the NE in interphase, disassemble simultaneously during prometaphase, and reaccumulate around the reforming nuclei (Figure 8.10). However, while NMCP1 is mainly localized to the spindle and accumulated on segregating chromosomes, NMCP2 disperses in the mitotic cytoplasm in vesicular structures that can be distinguished from the bulk endomembrane system. This vesicular signal might represent the NE membranes absorbed into the ER network upon NEBD.

Recruitment of nuclear membrane

The recruitment of NE membranes around chromatin arises from the intact tubular ER network *in vitro*. The ends of ER tubules bind directly to DNA, and it is proposed that by virtue of the highly dynamic and mobile nature of the ER the membranes can quickly spread over the surface of the chromatin, with the membranes expanding and forming flattened sheets (Anderson and Hetzer, 2007). In another study, it has been demonstrated that other membrane binding and fusion events in the cell, NSF (*N*-ethyl-maleimide-sensitive fusion protein) and SNARE (soluble NSF receptors) are necessary for NE formation and successful pore complex assembly. Because membranes are typically delicate structures that are easily disrupted during cell fractionation and fixation, it is easy to see how data obtained from different systems lead to the postulation of sometimes opposing models. There are essentially two different ideas about how the NE is formed: (i) by vesicle fusion and (ii) by reshaping of ER into NE sheets. Based on biochemical data and EM observations, it was initially proposed that the NE fragments into NE vesicles. Two *Arabidopsis* homologs of the spindle pole body protein Sad1 were initially discovered in a survey for cytokinesis-related genes. These *Arabidopsis* SUN domain proteins are NE markers in plants. Oda and Fukuda (2011) carefully followed the localization dynamics of both proteins through the cell cycle using transgenic *Arabidopsis* plants and stably transformed BY-2 cells. Both groups reported the localization of SUNs in mitotic ER membranes and an asymmetric reassociation with the decondensing telophase chromatin, with an envelope-like structure first appearing at the surface next to the spindle poles and a delayed reappearance of the envelope at the surface close to the phragmoplast (Figure 8.11). This might indicate that NE assembly lags behind at the phragmoplast proximal surface of the daughter nuclei, and potentially this area remains open longer to nonrestricted exchange between nucleus and cytoplasm. Alternatively, because SUN1/2 are nuclear proteins, it might indicate that nuclear pores at the phragmoplast proximal surface lag behind in regaining full import capacity. These scenarios can also be distinguished by following ONM and NPC proteins as well as generic markers for active nuclear import.

Anaphase/telophase
Anaphase compaction

During NE formation, chromatin undergoes a series of conformational changes, from a state of maximal chromatin compaction in late anaphase, right before NE formation, to transcription- and replication-competent decondensation in interphase. Anaphase compaction requires the kinesin-like DNA-binding protein (Kid), which ensures the formation of a compact chromosome cluster during anaphase and the proper enclosure of the segregated chromatin mass into a single nucleus. Interestingly, Kid loading onto anaphase chromosomes is dependent on "importin β," adding to the growing number of mitotic processes

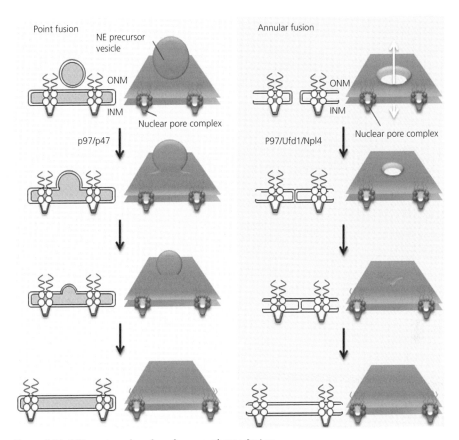

Figure 8.11 Different modes of nuclear membrane fusion.

regulated by this transport receptor. Recently, a compelling case has been made that mechanistically links NE formation and chromatin decondensation. The hexameric ATPase Cdc48/p97, which has been implicated in membrane fusion and ubiquitin-dependent processes, extracts Aurora B from chromatin, which results in its inactivation and subsequent chromatin decondensation. The inhibition of Cdc48/p97 blocked NE formation suggests that chromatin decondensation is required for NE formation, possibly by opening chromatin structure. It has been speculated that this open chromatin structure provides binding sites for INM proteins and, therefore, drives NE formation. This is closely related to the NSF, which regulates membrane fusion events and is involved in many aspects of vesicular transport. Originally identified as the product of the *Cdc48* gene in yeast, p97 and an associated protein, p47, mediate re-formation of the Golgi apparatus at the end of mitosis, and are also involved in the establishment of transitional ER. This is exemplified by the fusion of a vesicle with a target organelle, as occurs in the secretory and endocytic pathways. Point fusion could also represent the mechanism of nuclear membrane expansion during nuclear re-formation *in vitro*. A second type of fusion process, annular fusion, is required for the closure of gaps in the double nuclear membranes. This is fundamentally different from point fusion as it requires the constriction of a membrane ring followed by resolution of the INM and ONM, as the ring is eliminated by fusion at its cytoplasmic surface (Figure 8.11). Annular fusion also probably occurs in the fenestrated membranes of the ER and Golgi apparatus, as well as in the final step of cytokinesis.

Formation of phragmoplast

As chromosomes migrate to opposing spindle poles, a plant-specific MT structure, the phragmoplast, is formed to allow the completion of cell division through the assembly of a new cell wall between the separating sister nuclei. The phragmoplast is a complex assembly of MTs, microfilaments (MFs), and ER elements, which is specifically found in plant cells during late cytokinesis. It carries out the scaffold formation for cell plate assembly succeeding formation of a new cell wall separating the two daughter cells. Besides the proteins involved in vesicular trafficking and fusion, some NE-associated proteins have been found to mark the phragmoplast and/or the cell plate as well. The localization of Rae1 and SUN1/2 at the cell plate and the phragmoplast for Rae1 (Figure 8.10) suggests a tight linkage between the NE components and the cytoskeleton during mitosis. Thus, it is of great interest to identify plant interactors of SUN proteins both at the NE and at the cell plate. Apart from Rae1, other nuclear rim-associated proteins colocalize with SUNs at the cell plate as well. For instance, *Arabidopsis* ONM proteins, WIP1, WIP2, WIT1, and WIT2, are redistributed to the cell plate during cytokinesis. Both WITs and WIPs are required for Ran-GAP1 anchoring to the NE in the root meristem, but only one of the protein families, either WIPs or WITs, is sufficient to target Ran-GAP1 to the NE in differentiated cells. The cell plate localization of Ran-GAP1 (as well as its PPB and cortical division site association), on the other hand, is independent on both WIPs and WITs, suggesting that interphase and mitotic targeting of Ran-GAP1 require different mechanisms.

The dynamics of nuclear pore complex biogenesis

NPCs serve as transport channels across the nuclear membrane, a double lipid bilayer that physically separates the nucleoplasm and cytoplasm of eukaryotic cells. These multiprotein nuclear pores also play a role in chromatin organization and gene expression. Given the importance of NPC function, it is not surprising that a large number of human diseases and developmental defects have been linked to its malfunction. In order to fully understand the functional repertoire of NPCs and their essential role for nuclear organization, it is critical to determine the sequence of events that lead to the formation of nuclear pores.

This is particularly relevant since NPC number and possibly composition are tightly linked to metabolic activity. Most of our knowledge is derived from NPC formation that occurs in dividing cells at the end of mitosis when the NE and NPCs reform from disassembled precursors. NPC biogenesis into an intact NE also occurs during interphase, requiring the insertion of a pore complex into an intact NE in order to increase the number of pore complexes prior to cell division, presumably to meet increasing demands for metabolic activity. Importantly, this process is not restricted to dividing cells, but it also occurs during cell differentiation.

NPC formation in interphase

During interphase, the nuclear volume increases, but the density of the NPC within the NE remains almost constant, indicating that NE growth and NPC number is proportionated. New NPCs assemble *de novo* and are not formed by growth or division of preexisting NPCs because newly formed complexes do not contain components of preexisting NPCs. Moreover, depressions/ dimples within the NE (new NE patches), presumably occurring at sites of membrane fusion between the INM and the ONM, are the most probable sites for insertion of nascent complexes, whereas mature NPCs are embedded in an immobile network. Nascent NPC formation is initiated by fusion of the INM and ONM using Nups Pom121 and Ndc (Figure 8.12). However, overexpression of Pom121 decreases the space between the ONM and INM at numerous sites in the NE. Reticulons are ER-resident proteins that are evolutionarily conserved membrane proteins that stabilize tubular sections of the ER. Thus, reticulons may induce bending of the nuclear membrane and stabilize a highly curved state that may induce pore formation. A SUN (Sad1-UNC-84 homology) domain-containing INM protein, Sun1, localizes to Pom121-containing nascent pores and is suggested to transiently interact with Pom121. This interaction promotes membrane fusion at early stages of NPC formation. Pom121 is recruited in an initial slow process, prior to the faster association of the soluble Y-shaped NPC Nup107/160 subcomplex probably attached at both sides of the NE. Its recruitment depends on the membrane-curvature-sensing domain, the so-called amphipathic lipid-packing sensor (ALPS) domain of Nup133. Owing to the structural similarities of Nup133 to COPII and clathrin vesicle coats, it has been suggested that the Nup107/160 subcomplex stabilizes the highly curved pore membrane by constructing a coat-like structure. The order and mode of subsequent attachment of additional Nups or preformed complexes to the scaffold is still unclear.

NPC formation in mitosis

During mitosis, with the coating of chromatin by the NE, NPCs assemble in the reforming nuclei. This fascinating example of protein self-assembly is coordinated by the stepwise recruitment of a subset of NPC proteins to chromatin,

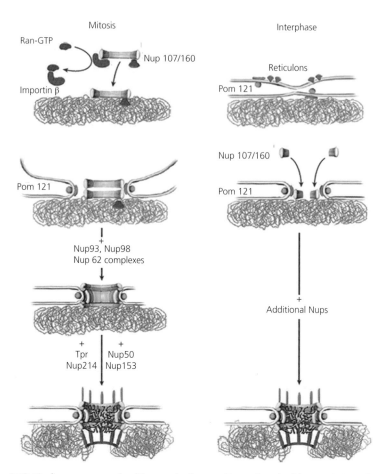

Figure 8.12 Nuclear pore complex biogenesis. Source: Reproduced with permission of Grossman, Medalia and Zwerger (2012).

and some progress has been made in determining how this process occurs. These intermediate states of assembly may serve as important models for understanding the tenets of NPC assembly, particularly as the mechanisms guiding interphase and telophase. NPC reassembly after open mitosis share common elements. In *A. nidulans*, the NPC is partially disassembled at the onset of mitosis, leading to the dispersion of 14 Nups, while 12 other Nups remain attached to the NE as a remnant pore structure. Interestingly, the *Aspergillus* Nup84 subcomplex, though normally part of this remnant framework, has been shown to be dispensable, leaving a minimal framework comprising the three transmembrane Nups (AnPom152, AnPom34, and AnNdc1), AnNup170, and AnGle1. Remarkably, a triple deletion of all the known transmembrane Nups of the fungus is fully viable. However, in eukaryotes undergoing open mitosis, the NE and NPCs disassemble completely at the onset of mitosis (prophase) and reassemble around the segregated chromosomes at the end of mitosis (telophase).

There are two main ways in which this reassembly is accomplished. The simplest way is to assemble the central region of the spoke–ring complex on the chromatin surface, then surround this by flattened membrane cisternae. These form the INM and ONM, as well as the POM followed by fusion to form sealed membranes around each NPC to create an aqueous channel between the nucleus and cytoplasm. The alternative approach is to create a largely continuous double membrane using the appropriate p97-fusion apparatus, then to insert the NPC, or NPC subunits. Although the identity of the fusion apparatus is still unknown, the large NPC membrane protein gp210 is one candidate. This is because of its large lumenal domain, and the presence of hydrophobic sequences that are similar to the fusogenic peptides that are seen in certain viral envelope glycoproteins.

The earliest events involved in NE reassembly begin in mid-anaphase, and includes the association of certain soluble NPC proteins with the chromatid surfaces. These include a substoichiometric amount of Nup153, which is a component of the nuclear basket, Nup 50, which helps to form cytoplasmic filaments and Nup133, which is associated with the NPC spoke–ring complex. Soon after the association of these soluble Nups, membranes begin to associate with the lateral and polar margins of the chromatids. At this time, INM proteins such as LAP2, LBR, and emerin concentrate at the membrane–chromatin interface. This presumably reflects the appearance of specific chromatin-associated binding sites for these proteins. A protein called Mel-28/ELYS is critical for the association of the Nup107/160 complex to chromatin. Because Mel-28/ELYS contains an AT-hook domain (a small DNA-binding motif), it is likely that this step occurs by direct binding to DNA. Interestingly, RanGTP stimulates Mel-28/ELYS recruitment, presumably by releasing importin β from one of the Nup107/160/ELys components. The NE is likely to be reformed from membranes delivered by an intact, continuous ER network. ER tubules are assumed to attach to and sequentially spread across the chromatin surface to form membrane sheets that enclose the pores. It is plausible that membrane Nups residing within the ER during mitosis dock onto the prepores and thus seal the NE. Further, Nup93, Nup98, and Nup62 complexes are sequentially recruited to the newly assembled pores. The Nup93 complex influences the size exclusion property of the NPC and, together with Nup98 and Nup62 complexes, comprises most of the FG-Nups that interact with transport receptors. Therefore, concomitant with the association of these three complexes, selective nuclear import across the NPC is reestablished. In a last maturation step, peripheral Nups, such as Nup214, Tpr, and the major pools of Nup50 and Nup153, associate with the NPC core structure to form the asymmetrical parts of the NPC (Grossman, Medalia and Zwerger, 2012).

Due to the differences in chromatin accessibility, cell cycle stage, mitotic kinase activity, and NE topology and organization, NPC formation mechanisms fundamentally differ in interphase and postmitosis. First, the activity of the CDKs, CDK1 and CDK2, seems to be essential for interphase NPC assembly, but it is not required for postmitotic NPC assembly. Second, the mode and kinetics of recruiting Nup107/160 occur in two distinct mechanisms: Elys/Mel-28 is involved during postmitotic assembly, whereas interphase recruitment of Nup107/160 is mediated by the ALPS domain component, Nup133. Third, the order of initial assembly steps in interphase, that is, the attachment of Pom121 and then the Nup107/160 complex, is reversed during postmitotic NPC formation, in which Pom121 attaches to prepores containing Nup107/160 complexes (Figure 8.12). NPCs most likely assemble from newly translated Nups during interphase, whereas after mitosis, NPCs form from existing components or subcomplexes that disassembled. A group of Nups have been shown to be essential for pore assembly; the largest subcomplex being the Nup107/160 complex, whose depletion results in NPC-free nuclear membranes. The third known transmembrane Nup, gp210, is not expressed in all cell types and thus is unlikely to be essential for pore assembly.

Cell cycle control

Cell cycle regulation is important to ensure faithful segregation of genetic material and thereby allow normal development and maintenance of multicellular organisms. The cell cycle is controlled by numerous mechanisms ensuring correct cell division. Failure to coordinate such processes leads to genome instability leading to birth defects and cancer. G1, S, G2, and M phases are the divisions of the standard cell cycle. The concept of "checkpoints" was introduced in 1989 by L.H. Hartwell and T.A. Weinert, inferring that an uncompleted cell cycle event sends an inhibitory signal to later events. The cells have checkpoint controls ensuring that the correct sequence of events is firmly maintained. The 2001 Nobel Prize in Physiology or Medicine was awarded to L.H. Hartwell, P. Nurse, and T. Hunt for their ground-breaking work on cell cycle regulation. Starting in the late 1960s, L.H. Hartwell used budding yeast to identify mutants that blocked specific stages of cell cycle progression. P Nurse, working in fission yeast in the 1970s, went on to isolate mutants that could also speed up the cell cycle, thus focusing his attention on the original CDK kinase, cdc2. In the 1980s, T. Hunt identified proteins in sea urchin extracts, the levels of which varied through the cell cycle, hence termed as "cyclins." These entire three scientists have made important advances in cell cycle research including the identification of checkpoints, mechanisms coupling cell morphology to the cell cycle, and identification of additional classes of kinases, cyclins, and inhibitors. Three checkpoints have been amply documented: the DNA damage checkpoint, which arrests cells in G1, S phase, G2, or even mitosis in case of DNA lesions; the DNA replication checkpoint, which ensures that mitosis is not initiated until DNA replication is complete and also that no DNA is replicated twice; and, the spindle assembly checkpoint (SAC), which delays anaphase onset until

Table 8.1 Cyclin–CDK complexes are activated at specific points of the cell cycle.

CDK		Cyclin	Cell cycle phase activity
CDK4		Cyclin D1, D2, D3	G_1 phase
CDK6		Cyclin D1, D2, D3	G_1 phase
CDK2		Cyclin E	G_1/S phase transition
CDK2		Cyclin A	S phase
CDK1	(cdc2)	Cyclin A	G_2/M phase transition
CDK1	(cdc2)	Cyclin B	Mitosis
CDK7		Cyclin H	CAK, all cell cycle phases

Source: From Vermeulen *et al.* (2003).
CAK, CDK-activating kinase.

all chromosomes are properly attached to the mitotic spindle. Checkpoints consist of at least three components: a sensor, which detects the error; a signal, generated by the sensor via a signal transduction pathway; and, finally, a response element in the cell cycle machinery to block cell cycle progression.

The core components of the eukaryotic cell cycle control system are CDKs and their regulatory subunits (cyclin). CDKs and cyclins are conserved throughout evolution. CDKs act by catalyzing the covalent attachment of phosphate groups derived from ATP to protein substrates, therefore, inducing changes in the enzymatic activity or capacity of the substrate to interact with other proteins. CDKs are activated by direct binding to its regulatory subunits cyclins. The first cyclins to be identified were cyclin A and B. The original mitotic cyclins were first described in *Xenopus* egg extracts as one of the components of the maturation-promoting factor (MPF) later found to be a heterodimer composed of CDK1 and cyclin B. CDK protein levels remain stable during the cell cycle, in contrast to their activating proteins, the cyclins. Cyclin protein levels rise and fall during the cell cycle and in this way they periodically activate CDK. Different cyclins are required at different phases of the cell cycle and are given in Table 8.1. The major drivers of the plant cell cycle are the A- and B-type CDKs. A-type CDKs (CDKA) are most closely related to mammalian CDK1 and CDK2; they are constitutively present during the cell cycle; and they control the G1-S and G2-M transition points. B-type CDKs (CDKB) are specific to plants and accumulate in a manner that is dependent on the phase of the cell cycle, reaching a maximum level at the G2-M transition, the phase at which their activity is necessary.

The intracellular localization of different cell cycle-regulating proteins also contributes to a correct cell cycle progression. Cyclin B contains a nuclear exclusion signal and is actively exported from the nucleus until the beginning of the prophase. The CDK-inactivating kinases Wee1 and Myt1 are located in the nucleus and Golgi complex, respectively, and protect the cell from premature mitosis. The 14-3-3 group of proteins regulates the intracellular

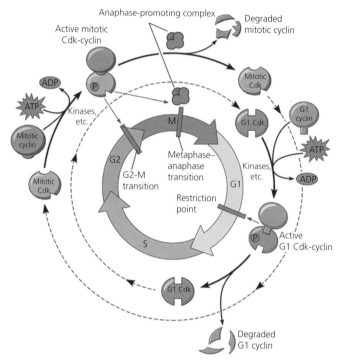

Figure 8.13 Cell cycle check points.

trafficking of different proteins. During interphase, the CDK activating kinase, Cdc25, is kept in the cytoplasm through interaction with 14-3-3 proteins. Sequestration of the CDK1-cyclin B complex in the cytoplasm following DNA damage is also mediated by 14-3-3 proteins. The three D-type cyclins (cyclin D1, cyclin D2, and cyclin D3) bind to CDK4 and CDK6, and CDK-cyclin D complexes are essential for entry in G1. Unlike the other cyclins, cyclin D is not expressed periodically, but is synthesized as long as growth factor stimulation persists. Another G1 cyclin is cyclin E that associates with CDK2 to regulate progression from G1 into S-phase. Cyclin A binds with CDK2 and this complex is required during S phase. Toward the end of the S-phase, M-cyclins (cyclins-A/ B) gene expression is switched on, and their relative concentration within the cell increases, leading to the accumulation of M-cyclins–CDK complexes during G2. In late G2 and early M, cyclin A complexes with CDK1 to promote entry into M phase of cell cycle (Figure 8.13). Mitosis is further regulated by cyclin B in complex with CDK1. Cyclins A and B contain a destruction box and cyclins D and E contain a PEST sequence (segment rich in proline (P), glutamic acid (E), serine (S) and threonine (T) residues): these are protein sequences required for efficient ubiquitin-mediated cyclin proteolysis at the end of a cell cycle phase. *Arabidopsis thaliana* has 12 distinct bona fide CDKs and more than 49 different cyclins. To date, the exact composition of different CDK–cyclin complexes in plants is largely unknown. A critical decision point at which cells can enter an additional round of cell division is the G1 restriction point, which is situated at the G1-S transition where cells become committed to the mitotic cell cycle. The activity of the CDKA–CYCD complexes in plants can be inhibited by their association with CDK inhibitory proteins (ICK/KRP and SIM) that respond to stress stimuli or developmental signals. Selective destruction of CYCA and CYCB by the anaphase-promoting complex (APC) marks the exit from mitosis.

Cyclin–CDK activity is modulated by the three basic mechanisms—cyclin availability, stoichiometric inhibition, and inhibitory phosphorylation. Cyclin level can be changed by transcriptional regulation and/or by specific proteolysis. Cyclins are targeted for ubiquitin-dependent degradation by the proteosome via two ubiquitin–ligase systems: Skp1-Cullin-F box protein complex (SCF) and APC. The SCF recognizes cyclins as substrates when they are phosphorylated, whereas the APC, also called the cyclosome, ubiquitinates cyclins when specificity factors (fizzy and fizzy-related) become activated. Assembled cyclin-CDK complexes can be kept inactive by stoichiometric inhibitors (cyclin kinase inhibitors; CKIs). Cyclin–CDK complexes can also be inactivated by phosphorylation of tyrosine and threonine residues close to the active site of the CDK subunit. This phosphorylation is mediated by Wee1-type protein kinases, and the inhibitory phosphate groups are removed by Cdc25-type phosphatases.

In addition to CDK1, several other mitotic kinases are known to influence cell division (e.g., Polo, Aurora B, Mps1, and BubR1) and are mostly involved in controlling the spatial–temporal order of later mitotic events, like removal of cohesion complexes from chromosomal arms during prophase (Polo), correction of improper interactions between spindle MTs and kinetochores (Aurora B, BubR1), or setting off sister chromatid separation during metaphase, by alleviating APC inhibition (BubR1, Mps1). Some of these proteins are activated by cyclin B–CDK1 complexes, creating a positive feedback loop that rapidly amplifies to the fullest cyclin B–CDK1 activity. M-cyclins–CDKs pave their own way to destruction, since once activated they will also phosphorylate core subunits of the APC (Cdc16, Cdc27, Cdc23) and also Cdc20, main co-activator of the APC during mitosis, thus stimulating its E3 ubiquitin ligase activity, resulting in mitotic cyclins degradation and CDK1 inactivation in a negative feedback loop vital to ensure mitotic exit. The activity of the APC–Cdc20 complex is further regulated by SAC signaling, which monitors chromosome alignment at the metaphase plate by targeting Cdc20, creating an intricate network with different levels of regulation that influence each other, ultimately leading to a controlled mitotic exit. When each sister chromatid of a chromosome binds MTs from opposite spindle poles bi-orientation is achieved, the checkpoint is satisfied, and the APC is activated. At this point, the APC triggers sister chromatid separation and CDK1 inactivation, resulting in anaphase onset and mitotic exit. The checkpoint acts on Cdc20, ultimately causing APC inactivation in the presence of unattached or improperly attached kinetochores, it is suggested that checkpoint effectors act by either blocking the access of substrates to Cdc20 or by preventing the release of ubiquitinated substrates.

Sister chromatids are attached to the mitotic spindle through a multiprotein complex, the kinetochore, directly assembled upon centromeric DNA. The kinetochores persistently assemble over DNA organized around centromeric nucleosomes that contain specialized histone H3-like proteins, referred to as CENP-A (Cse4 in budding yeast). Most chromosome imbalances during development cause embryonic lethality, and chromosome instability has been associated with tumorigenesis. To minimize the production of aneuploid progeny during mitosis, cells have evolved a checkpoint capable of delaying sister chromatid separation in the presence of non-attached or improperly attached kinetochores. The SAC monitors the interaction between spindle MTs and kinetochores; and if chromosomes are not properly bi-oriented, it will prevent activation of the APC–Cdc20 and separation of sister chromatids and exit from mitosis. The mechanical interactions between the mitotic spindle and the kinetochore are monitored by Mad1, Mad2, BubR1/Mad3, Bub1, Bub3, and Mps1, an evolutionary conserved group of proteins first identified in a budding yeast. BubR1 is localized at the outer kinetochore, where it has been proposed to function as a mechanosensor monitoring CENP-E activity, in addition to its role in SAC signaling. CENP-E is a kinetochore motor protein, involved in generating kinetochore-MT attachment by ensuring the transport of a mono-oriented chromosome along the spindle fibers. BubR1 has also been shown to play a significant role in the inhibition of the APC–Cdc20. BubR1 is itself phosphorylated in mitosis; its phosphorylation sites regulate kinetochore attachment, tension, and mitotic exit, suggesting that BubR1 might be an effector of multiple kinases involved in discrete aspects of kinetochore attachments and checkpoint regulation. Mad2 interacts with Mad1, a coiled-coil protein that is stably bound to kinetochores throughout mitosis. Mad1 has an essential role in checkpoint response, since it is required for Mad2 localization at kinetochores and for the *in vivo* formation of Mad2–Cdc20 complex. Several pathways converge to turn off SAC signaling after chromosome bi-orientation is achieved. Dynein is involved and uses its minus-end-directed motor activity to strip Mad1, Mad2, RZZ complex, Mps1, CENP-F, and other components of the checkpoint from kinetochores. As soon as attachment occurs, dynein activity redistributes these proteins to the poles, thus preventing checkpoint continual signaling. Another mechanism of inactivation is dependent on p31comet. This protein stops the positive feedback loop based on closed-to-open Mad2 hetero-dimerization, because it binds specifically to closed-Mad2. Additionally, APC-induced proteolysis is also a relevant mechanism for checkpoint silencing.

Summary

The defining characteristic of a eukaryotic cell is the nucleus. The nucleoplasm is surrounded by a double-membrane system, the NE. The ONM and the INM are separated by the perinuclear space. The lipid bilayer of the ONM is continuous with the ER, thus allowing for direct insertion of NE membrane proteins and translocation of proteins into the perinuclear space. The INM has a distinct protein composition and specialized functions. The NE surrounds and protects the chromatin, forming a barrier that spatially uncouples transcription from translation. Like the ER, the NE lumen acts as a repository of Ca, and ion transporters in both the ONM and INM are involved in signal transduction The NE also provides a central anchor site for numerous nuclear and cytoplasmic structures, such as the nuclear lamina and the spindle organizer. In higher organisms, the NE plays role in the dissociation and reformation of the nucleus during cell division. The NE components have been shown not only to separate the nucleoplasm from the cytosol and to constitute a selective barrier for nucleo-cytoplasmic transport but are also involved in nuclear mobility, signal transduction, chromatin attachment, and transcriptional activation and repression. Trafficking between the nucleoplasm and the cytoplasm occurs through NPCs, large multiprotein complexes that form selective channels perforating the NE double membrane. NPCs have eightfold rotational symmetry perpendicular to the membrane, and are asymmetric with respect to the plane of the NE. The complex can be characterized as having three substructures: the cytoplasmic fibrils, a central core, and the nuclear basket. The NPC is composed of approximately 30 different proteins termed Nups, which are broadly conserved among yeast, vertebrates, and plants. Each Nup is repeated 8-, 16-, 32-, or 48-fold to form this multiprotein complex. NPCs function as gatekeepers of the nucleus, allowing the free diffusion of small molecules, while regulating the transport of macromolecules with high specificity. The precise dynamic localization of a given protein, and the order of disassembly/reassembly of plant NE/NPC components, could be tackled with high-resolution imaging techniques, such as multicolor confocal laser scanning microscopy, in-lens field emission scanning EM, and 3D structured illumination microscopy.

The nucleolus is the most prominent structure in a cell nucleus. It is the site of rRNA transcription, pre-rRNA processing, and ribosome subunit assembly. The nucleolus is a dynamic structure that assembles around the clusters of rRNA gene repeats during late telophase, persists throughout the interphase, and then disassembles as cells. DNA storage in eukaryotic cell's nucleus is achieved through the chromatin fiber. The basic chromatin building block is the nucleosome: a histone core composed of four pairs of protein dimers (histone proteins H2A, H2B, H3, and H4) around which 147 bp of DNA are wound 1.75 turns. Nucleosomes are packed one on top of another to form a more compact structure called a 30-nm fiber. Each of these fibers undergoes further folding and coiling to form the metaphase chromosome. Whereas many proteins are selectively transported from the cytoplasm into the nucleus, most RNAs are exported from the nucleus to the cytoplasm. Since proteins are synthesized in the cytoplasm, the export of mRNAs, rRNAs, and tRNAs is a critical step in gene expression in eukaryotic cells. Like protein import, the export of RNAs through NPCs is an active, energy-dependent process that requires the Ran-GTP-binding protein. The snRNAs function within the nucleus as components of the RNA processing machinery. These RNAs are initially transported from the nucleus to the cytoplasm, where they associate with proteins to form functional snRNPs and then return to the nucleus. miRNAs are endogenously encoded small noncoding RNAs, derived by processing of short RNA hairpins that can inhibit the translation of mRNAs bearing partially complementary target sequences. siRNAs, which are derived by processing of long double-stranded RNAs and are of exogenous origin, degrade mRNAs bearing fully complementary sequences. Both have role in gene silencing. The eukaryotic cell cycle is guarded at three checkpoints: at the G1/S boundary, the G2/M boundary, and the metaphase/anaphase boundary. Progress through the chromosome cycle can be halted at these checkpoints if the conditions for successful cell division are not met. The transition from one cell cycle phase to another occurs in an orderly fashion and is regulated by different cellular proteins. Key regulatory proteins are the CDKs, a family of serine/threonine protein kinases that are activated at specific points of the cell cycle. An integrated understanding on how the cell cycle machinery operates in an entire plant context contributes to growth, cell differentiation, and the formation of new tissues and organs.

References

Anderson, D.J. and Hetzer, M.W. (2007) Nuclear envelope formation by chromatin-mediated reorganization of the endoplasmic reticulum. *Nature Cell Biology*, **9**, 1160–1166.

Bootman, M.D., Fearnley, C., Smyrnias, I., *et al.* (2009) An update on nuclear calcium signalling. *Journal of Cell Science*, **122**, 2337–2350.

Boruc, J., Zhou, X. and Meier, I. (2012) Dynamics of the plant nuclear envelope and nuclear pore. *Plant Physiology*, **158**, 78–86.

Cejka, D., Losert, D. and Wacheck, V. (2006) Short interfering RNA (siRNA): tool or therapeutic? *Clinical Science*, **110**, 47–58.

Crisp, M., Liu, Q., Roux, K. *et al.* (2006) Coupling of the nucleus and cytoplasm: role of the LINC complex. *Journal of Cell Biology*, **172**, 41–53.

Dittmer, T.A., Stacey, N.J., Sugimoto-Shirasu, K. and Richards, E.J. (2007) *LITTLE NUCLEI* genes affecting nuclear morphology in *Arabidopsis thaliana*. *The Plant Cell*, **19**, 2793–2803.

Fire, A., Xu, S., Montgomery, M.K., *et al.* (1998) Potent and specific genetic interference by double-stranded RNA in *Caenorhabditis elegans*. *Nature*, **391**, 806–811.

Fiserova, J., Kiseleva, E. and Goldberg, M.W. (2009) Nuclear envelope and nuclear pore complex structure and organization in tobacco BY-2 cells. *The Plant Journal*, **59**, 243–255.

Grossman, E., Medalia, O. and Zwerger, M. (2012) Functional architecture of the nuclear pore complex. *Annual Review of Biophysics*, **41**, 557–584.

Gruenbaum, Y., Margalit, A., Goldman, R.D. *et al.* (2005) The nuclear lamina comes of age. *Nature Reviews Molecular Cell Biology*, **6**, 21–31.

Hernandez-Verdun, D. (2011) Assembly and disassembly of the nucleolus during the cell cycle. *Nucleus*, **2**, 189–194.

Hernandez-Verdun, D., Roussel, P., Thiry, M. *et al.* (2010) The nucleolus: structure/function relationship in RNA metabolism. *WIREs RNA*, **1**, 415–431.

Kimura, Y., Kuroda, C. and Masuda, K. (2010) Differential nuclear envelope assembly at the end of mitosis in suspension-cultured *Apium graveolens* cells. *Chromosoma*, **119**, 195–204.

Leung, A.K.L., Gerlich, D., Miller, G. *et al.* (2004) Quantitative kinetic analysis of nucleolar breakdown and reassembly during mitosis in live human cells. *The Journal of Cell Biology*, **166** (6), 787–800.

Lim, R.Y.H., Fahrenkrog, B., Köser, J. *et al.* (2007) Nanomechanical basis of selective gating by the nuclear pore complex. *Science*, **318** (5850), 640–643.

Lu, X., Wontakal, S.N., Emelyanov, A.V. *et al.* (2009) Linker histone H1 is essential for *Drosophila* development, the establishment of pericentric heterochromatin, and a normal polytene chromosome structure. *Genes and Development*, **23**, 452–465.

Maeshima, K. and Eltsov, M. (2008) Packaging the genome: the structure of mitotic chromosomes. *Journal of Biochemistry*, **143**, 145–153.

Moriguchi, K., Suzuki, T., Ito, Y. *et al.* (2005) Functional isolation of novel nuclear proteins showing a variety of subnuclear localizations. *The Plant Cell*, **17**, 389–403.

Murphy, S.P., Simmons, C.R. and Bass, H.W. (2010) Structure and expression of the maize (*Zea mays* L.) SUN-domain protein gene family: evidence for the existence of two divergent classes of SUN proteins in plants. *BMC Plant Biology*, **10**, 269.

Oda, Y. and Fukuda, H. (2011) Dynamics of *Arabidopsis* SUN proteins during mitosis and their involvement in nuclear shaping. *The Plant Journal*, **66**, 629–641.

Queisser, G., Wieger, S. and Bading, H. (2011) Structural dynamics of the cell nucleus: basis for morphology modulation of nuclear calcium signaling and gene transcription. *Nucleus*, **2**, 1–7.

Sajan, S.A. and Hawkins, R.D. (2012) Methods for identifying higher-order chromatin structure. *Annual Review of Genomics and Human Genetics*, **13**, 59–82.

Sean, D.S., Ashley, J., Jokhi, V., *et al.* (2012) Nuclear envelope budding enables large ribonucleoprotein particle export during synaptic Wnt signaling. *Cell*, **149**, 832–846.

Seltzer, V., Janski, N., Canaday, J. *et al.* (2007) *Arabidopsis* GCP2 and GCP3 are part of a soluble gamma-tubulin complex and have nuclear envelope targeting domains. *The Plant Journal*, **52**, 322–331.

Shaw, P. and Brown, J. (2012) Nucleoli: composition, function and dynamics. *Plant Physiology Preview*, **158**, 44–51.

Starr, D.A. and Fischer, J.A. (2005) KASH 'n Karry: the KASH domain family of cargo-specific cytoskeletal adaptor proteins. *BioEssays*, **27**, 1136–1146.

Vermeulen, K., Dirk, R., Bockstaele, V. and Berneman, Z.N. (2003) The cell cycle: a review of regulation, deregulation and therapeutic targets in cancer. *Cell Proliferation*, **36**, 131–149.

Wilhelmsen, K., Ketema, M., Truong, H. and Sonnenberg, A. (2006) KASH-domain proteins in nuclear migration, anchorage and other processes. *Journal of Cell Science*, **119**, 5021–5029.

Further reading

Eibauer, M., Pellanda, M., Turgay, Y. *et al.* (2015) Structure and gating of the nuclear pore complex. *Nature Communications*, **6**, 7532, doi:10.1038/ncomms8532.

Gavrilov, A.A. and Razin, S.V. (2015) Compartmentalization of the cell nucleus and spatial organization of the genome. *Molecular Biology*, **49** (1), 21–39.

Guo, T. and Fang, Y. (2014) Functional organization and dynamics of the cell nucleus. *Frontiers in Plant Science*, http://dx.doi.org/10.3389/fpls.2014.00378.

Knockenhauer, K.E. and Schwartz, T.U. (2016) The nuclear pore complex as a flexible and dynamic gate. *Cell*, **164** (6), 1162–1171.

Koh, J. and Blobel, G. (2015) Allosteric regulation in gating the central channel of the nuclear pore complex. *Cell*, **161** (6), 1361–1373.

Petrovská, B., Šebela, M. and Doležel, J. (2015) Inside a plant nucleus: discovering the proteins. *Journal of Experimental Botany*, **66** (6), 1627–1640.

Scofield, S., Jones, A. and Murray, J.A.H. (2014) The plant cell cycle in context. *Journal of Experimental Botany*, **65**, 10, 2557–2562.

CHAPTER 9

Plant cell walls

James E. Bidlack[1] and William V. Dashek[2]

[1] Department of Biology, University of Central Oklahoma, Edmond, OK, USA
[2] Retired Faculty, Adult Degree Program, Mary Baldwin College, Staunton, VA, USA

Introduction

Constituents of the plant cell wall (CW) are among the most abundant biological molecules on earth and the functions of the CW are diverse. The CW provides mechanical strength, maintains cell shape, controls cell expansion, regulates transport, provides protection, functions in signaling processes, and stores food reserves (Brett and Waldron, 1990). The CW has even been referred to as a cell organelle (Mauseth, 1988), and more information is needed to have a complete understanding of how CW structure and function are integrated.

Structure

There are two types of CWs: primary and secondary (Hayashi, 2006; Albersheim *et al.*, 2010). Primary CWs (Figure 9.1a and b) are thin, elastic, and permit cell expansion while providing strength (Rose, 2003). Cells that possess only primary walls can lose their specialized form, divide, and subsequently differentiate into new cells (Evert, 2006). In some instances, cells possessing primary walls can exhibit thickenings and layering, for example, certain collenchyma cells (Bidlack and Jansky, 2014). The middle lamella is an area of union between adjacent cells' primary walls. Secondary walls are thick, often consisting of layers and deposited when cell expansion is complete (Carpita *et al.*, 2001). Examples of cells possessing secondary walls are xylem fibers, tracheids, vessel elements, and sclereids (see Chapter 1). These cell types possess lignified, spiral, scalariform, reticulate, or pitted walls (Figure 9.2).

Plant Cells and their Organelles, First Edition. Edited by William V. Dashek and Gurbachan S. Miglani.
© 2017 John Wiley & Sons, Ltd. Published 2017 by John Wiley & Sons, Ltd.

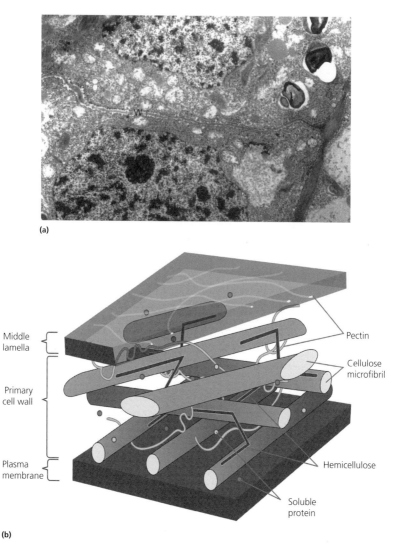

Figure 9.1 (a) Electron micrograph of two adjoining plant cells with an intervening primary wall. From: Courtesy of J. Mayfield. (b) Model of the primary cell wall. From: https://en.wikipedia.org/wiki/Cell_wall.

Primary CW structure

Typical primary plant CWs are composed of cellulose microfibrils (9–25%) and an interpenetrating matrix of hemicelluloses (25–50%), pectins (10–35%), and proteins (10%) (Esau, 1977; Goodwin and Mercer, 1983; Salisbury and Ross, 1992). Albersheim (1975) and coworkers (Bauer *et al.*, 1973; Keegstra *et al.*, 1973; Talmadge *et al.*, 1973) describe the primary CW composition as cellulose fibers bound together by molecules made of many sugar units. Approximately 90% of the CW consists of carbohydrates (mostly pentoses and hexoses), and the

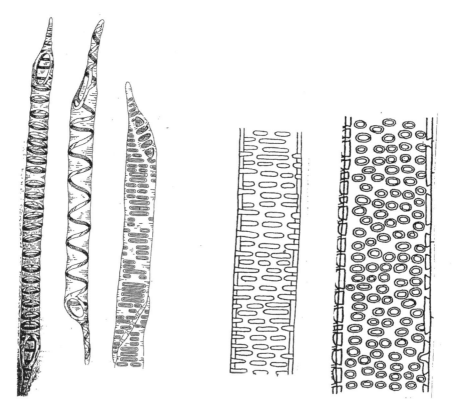

Figure 9.2 Various cell wall thickenings. Left to right: annular, spiral, scalariform, reticulate, pitted. From: Esau (1977).

remaining 10% is protein (Albersheim, 1975). Cellulose forms the framework of the CW (Esau, 1977), while hemicelluloses function as cross-links between noncellulosic and cellulosic polymers (Keegstra *et al.*, 1973). Pectins function to provide cross-links and structural support to the CW (Preston, 1979; Fry, 1986), whereas proteins can be either structural (extensin) or enzymatic (Goodwin and Mercer, 1983).

Cellulose is composed of approximately 8000 D-glucopyranose residues linked by $\beta1\rightarrow4$ glycosidic bonds (Goodwin and Mercer, 1983). Hydrogen bonds hold about 40 of these glucan chains together to form a cellulose microfibril (Albersheim, 1975). Cellulose microfibril arrangement in the primary wall is random (Esau, 1977). Cellulose microfibrils are linked to hemicellulosic polysaccharides composed of xylans, mannans, galactans, or combinations thereof. In one example, xyloglucans attach to cellulose microfibrils through hydrogen bonds and glycosidically bond to arabinogalactans which, in turn, link to rhamnogalacturans that bond back to cellulose via arabinan–galactan junctions (Albersheim, 1975). Cross-links made of pectin (based on $\alpha1\rightarrow4$-linked-polygalacturonan) and extensin (glycoprotein rich in hydroxyproline) enable support and extension, respectively, of the CW (Preston, 1979; Goodwin and Mercer,

1983; Fry, 1986). Proteins other than extensin found in the CW include enzymes such as the cellulase synthases (Giddings *et al.*, 1980; Herth, 1985), hydrolases (Freudenberg and Nesh, 1968), and oxidases (Goodwin and Mercer, 1983) needed for CW thickening, modification, and lignification, respectively, during secondary growth.

Reorganization, *de novo* synthesis, and insertion of new wall polymers lead to rearrangement of the CW during cell growth (Hatfield, 1989). This enables inclusion of lignin into the wall and strengthening of the CW matrix. In forage legumes and especially grasses, the order of maximum CW component deposition is hemicellulose, followed by that of cellulose (1–6 days later), and then lignin (up to 14 days after maximum hemicellulose deposition) (Bidlack and Buxton, 1992). As a definition, secondary walls are derived from primary walls after thickening and inclusion of lignin into the CW matrix (Theander and Aman, 1984). The secondary CW is laid down to the inside of the primary wall.

Secondary CW structure

Secondary CWs of plants contain cellulose (40–80%), hemicellulose (10–40%), and lignin (5–25%) (Salisbury and Ross, 1992). The arrangement of these components allow cellulose microfibrils to be embedded in lignin, much like the steel rods are embedded in concrete to form prestressed concrete (Salisbury and Ross, 1992). In wood, three layers of secondary CW, referred to as the S_1, S_2, and S_3 lamellae, result from different arrangement of the microfibrils (Goodwin and Mercer, 1983). The first-formed (outermost) S_1 lamella has both left- and right-handed microfibril helices, while the S_2 and S_3 (innermost) lamellae each have one helix of microfibrils in opposite directions to each other (Figure 9.3). During secondary CW formation, lignification occurs in the S_1 and S_2 lamella as

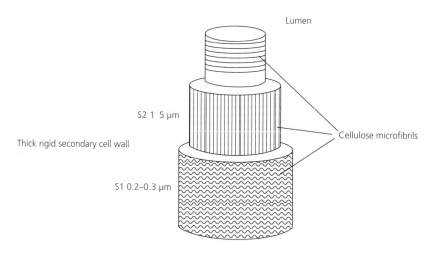

Figure 9.3 Layers of the secondary cell wall.

well as in the primary wall and middle lamella. Lignification rarely occurs in the S_3 lamella (Goodwin and Mercer, 1983).

Cellulose and hemicellulose appear to be more structurally organized in the secondary CW when compared to the primary CW (Hatfield, 1989). Changes that occur during maturation of the plant cell include loss of water from the CW matrix (Selvendran, 1983) and alterations/replacements that decrease polysaccharide branching in primary CWs (Huber and Nevins, 1979; Preston, 1979; Carpita, 1983). Hydrolysis of load-bearing polysaccharides may occur in the CW to enable cell expansion (Labavitch, 1981), sometimes resulting in transient loss of specific polysaccharides (Hatfield, 1989). Enzymes such as the β-glucanases can be secreted to selectively degrade polysaccharides and thereby decrease their concentrations in the CW matrix (Huber and Nevins, 1979). These modifications, which set the stage for secondary CW formation, result in a tighter and more rigid structure. The inclusion of lignin adds further rigidity to the matrix.

Lignin monomers originate from the action of phenylalanine ammonia lyase (PAL—legumes and grasses), tyrosine ammonia lyase (TAL—grasses only), and other phenylpropanoid-related enzymes directing metabolites through, among other things, lignin biosynthesis (Freudenberg and Neish, 1968; Hahlbrock and Grisebach, 1979; Goodwin and Mercer, 1983; Lewis and Yamamoto, 1990). The shikimic acid pathway and phenylpropanoid metabolism lead to synthesis of the following acid lignin monomers: *p*-coumaric (*p*CA), ferulic (FA), diferulic (DFA), sinapic (SA), cinnamic (CA), and benzoic (BA). Enzymes subsequently catalyze the formation of three alcohols—*p*-coumaryl, coniferyl, and sinapyl, all of which interact and polymerize to from lignin for incorporation into the secondary CW.

Control and development of the CW during plant growth is not completely understood. Current knowledge suggests that cellulose is formed by plasmalemma enzymes, matrix polysaccharides and proteins are formed in the cytomembrane system, and lignin is formed within the CW (Delmer, 1987; Bolwell, 1988). Lignification, which accompanies secondary CW formation, arises from generation of free radicals that react spontaneously to form lignin and even some linkages to wall polysaccharides (Brett and Waldron, 1990). Free radical linkages between lignin monomers and polysaccharides may constitute what is referred to as non-core lignin, while polymerization of monomeric free radicals result in highly condensed core lignin.

Cellulose is hydrogen-bonded to hemicellulose (Albersheim, 1975), whereas ester and ether bonds connect hemicellulose to non-core lignin in secondary CWs (Jung, 1989). Diverse and complex nature of lignin monomers and hemicellulosic moieties in ligno-hemicellulosic bonds make stereotypic conceptualizations of secondary wall structures for all plants extremely difficult.

Research, directed mostly toward the understanding of ligno-hemicellulosic linkages in grasses, has been performed on Italian ryegrass (*Lolium multiflorum* Lam.) (Hartley, 1972), perennial ryegrass (*Lolium perenne* L.) (Hartley, 1972;

Morrison, 1974), sugar cane (*Saccharum officinarum* L.) (Atsushi *et al.*, 1984), wheat (*Triticum aestivum* L.) (Scalbert *et al.*, 1985), and barley (*Hordeum vulgare* L.) (Mueller-Harvey and Hartley, 1986). These and other studies (Chesson *et al.*, 1983) have shown that an ester bond connecting arabinose to (non-core) lignin is the major ligno-hemicellulosic linkage in plant secondary CWs. Ferulic and *p*-coumaric acids are major non-core lignin monomers found between hemicellulose and core lignin (Jung, 1989) although diferulic, sinapic, cinnamic, and benzoic acid constituents can also be found. Each of these "cinnamic acids," named as such because they were derived from *trans*-cinnamic acid in phenylpropanoid metabolism, is uniquely bound between hemicellulose and core lignin in the secondary CW matrix.

Theoretical structure of secondary CW structure in grasses

Representative molecular arrangement and bonding among secondary CW components in grasses is demonstrated in Figure 9.4. Relative quantities of secondary CW components in this representative grass are shown as 45–60% cellulose, 20–40% hemicellulose, and 5–10% lignin (Bidlack *et al.*, 1992). Glucan chains of cellulose, shown as long, ribbon-like fibers, are held together by hydrogen bonds. Cellulose microfibrils consist of approximately 40 of these bonded chains. Three or four glucan chains occur for each β-1→4-D-xylopyranose chain of hemicellulose with an occasional L-arabofuranose molecule linked α-1→3 to the xylopyranose chain. The hemicellulosic xylans and arabinoxylans, represented by the actual molecules or as long, triangular rods, are hydrogen-bonded to cellulose and ester- or ether-bonded to non-core lignin.

In Figure 9.4, lignin is shown as an embedding matrix of polymerized lignin monomers. Two types of lignin, namely core and non-core, are encountered. The non-core portion of lignin binds to the hemicellulosic fraction of the secondary CW and the core lignin forms an amorphous matrix. A representative portion of core lignin is presented in the upper right-hand corner of the figure, and individual non-core components are shown in molecular form bound to hemicellulose.

Ester bonds between hemicellulose and non-core lignin shown in Figure 9.4 include linkage between the O-5 position of arabinose in arabinoxylan and *p*-coumaric, ferulic, and diferulic acids (Scalbert *et al.*, 1985; Mueller-Harvey and Hartley, 1986), as well as a hypothetical linkage between the O-3 position of xylose and cinnamic acid. Some of these lignin monomers, such as ferulic acid (Mueller-Harvey and Hartley, 1986; Jung, 1989) may be so intimately associated with the hemicellulosic fraction that they fail to cross-link to lignin.

Linkages between non-core and core lignin are demonstrated by both ester and ether bonds. The predominant ester linkage between non-core and core lignin in secondary CWs is encountered with *p*-coumaric acid, which bonds via its own carboxyl group (Hartley and Jones, 1976; Mueller-Harvey and Hartley,

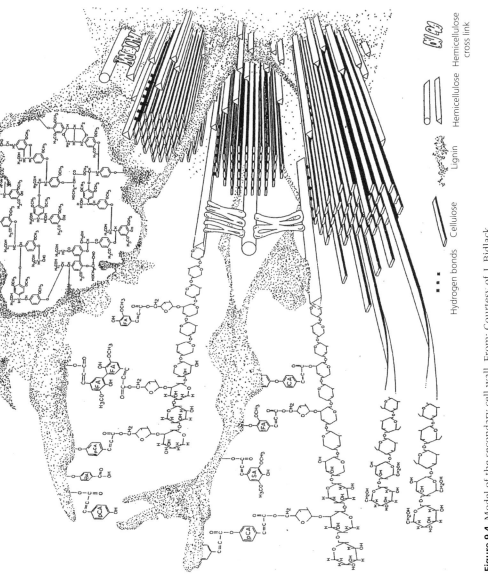

Figure 9.4 Model of the secondary cell wall. From: Courtesy of J. Bidlack.

1986) or requires contribution of a carboxyl group from the core (Jung, 1989). Other ester bonds linking core to none-core lignin include those between the core and diferulic (Scalbert *et al.*, 1985) and sinapic acids (Hartley and Jones, 1976; Newby *et al.*, 1980). Ether linkages between core and non-core lignin are encountered with ferulic, *p*-coumaric, and cinnamic acids (Newby *et al.*, 1980; Scalbert *et al.*, 1985) as well as with benzoic acid (Newby *et al.*, 1980). In the case of *p*-coumaric, sinapic, and benzoic acids, linkages may exclude involvement of the hemicellulosic fraction.

Biosynthesis

Lignin biosynthesis

Lignin biosynthesis starts with the shikimic acid pathway. Steps leading to synthesis of lignin monomers from this pathway have been reviewed (Hahlbrock and Grisebach, 1979; Bolwell, 1988). The shikimic acid pathway begins with the merging of D-erythrose 4-phosphate and phosphoenolpyruvate to form shikimc acid which, in turn, follows a path via tyrosine or phenylalanine (via *trans*-cinnamic acid) to *trans-p*-coumaric acid. *trans-p*-coumaric acid is then converted into the three lignin precursors: *p*-coumaryl, coniferyl, and sinapyl alcohols (Figure 9.5). Polymerization of these precursors forms lignin.

Formation of cinnamic or coumaric acid from phenylalanine or tyrosine is regulated by the enzymes, PAL and TAL. Legumes primarily follow the PAL route and grasses usually follow the TAL route to coumaric acid (Goodwin and Mercer, 1983). Another enzyme that distinguishes legumes from grasses is cinnamate 4-hydroxylase, which catalyzes the conversion of cinnamic acid to coumaric acid. Although PAL, TAL, and cinnamate 4-hydroxylase play key roles in lignin biosynthesis, enzymes further down the pathway are more closely related to the final products. Three enzymes—4-coumarate:CoA ligase, cinnamoyl-CoA:NADPH oxidoreductase, and cinnamyl alcohol:NADPH oxidoreductase—catalyze the conversion of coumaric acid to lignin precursors.

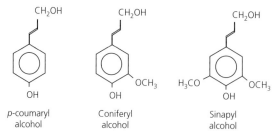

Figure 9.5 Chemistry of monolignols. From: Zhong and Zheng-Hua (2009).

CW peroxidases catalyze the final stages of lignification and are involved in oxidative polymerization of cinnamyl alcohols (Elstner and Heupel, 1976) and peroxide formation (Mader and Fussl, 1982; Pedreno *et al.*, 1989). Peroxidases increase CW cross-links (Lamport and Catt, 1981), which resist CW degradation.

Chemical composition

Comparative compositions of monocot and dicot primary CWs are presented in Tables 9.1 and 9.2. Cellulose is a β-1→4-linked glucan that exists in the wall as 10–23 mm microfibrils (Figure 9.6; Brown and Saxena, 2007; Keegstra, 2010). The cellulose is bonded to certain hemicelluloses (Iiyama *et al.*, 1994). These are polysaccharides that are insoluble in water but extractable by aqueous alkali. Table 9.3 presents a variety of hemicelluloses present in plant CWs as follows: xylan, glucuronoxylan, arabinoxylan, mannan, glucomannan, and galactoglu-comannan. The main hemicellulose in primary walls is xyloglucan (O'Neill and York, 2003; Scheller and Uliviskov, 2010).

Most of pectin (~70%) consists of homopolymeric, partially methylated, and sometimes aceytlated α-(1→4)-D-galacturonic acid (Table 9.4; Mohnen, 2008). However, there are "non-gelling" areas of rhamnogalacturonans I and II (Willats *et al.*, 2006; Bacic, 2006). Whereas the former (~35% of pectin) consists of alternating α-(1→2)-L-rhamnosyl-α-(1→4)-D-galacturonosyl regions containing branches of L-arabinose and D-galactose side chains, the latter (~10% of pectin) contains D-xylose, L-fucose, D-glucuronic acid, D-apiose, and some

Table 9.1 Chemical composition of primary and secondary walls in Monocots (% dry weight).

Chemicals	Primary wall	Secondary wall
Polysaccharides		
Cellulose	20–30	35–45
Hemicellulose (matrix)		
Xylans	20–40	40–50
Mixed linked glycans	10–30	?
Xyloglucan	1–5	?
Mannans and glucomannans	2	?
Pectins	5	0.1
Proteins		
Structural	1	?
Phenolics (ferulic and *p*-coumaric acids)	1–5	0.1–1.5
Lignin	?	20

Source includes work done by Bacic (2006), Vogel (2008), and others
?, Under investigation/unknown at this time.

Table 9.2 Chemical composition of primary and secondary cell walls in Dicots (% dry weight).

Chemicals	Primary wall	Secondary wall
Polysaccharides		
Cellulose	15–30	45–50
Hemicellulose (matrix)		
Xylans	5	20–30
Mixed linked glycans	None	None
Xyloglucan	20–25	?
Mannans and glucomannans	5–10	3–5
Pectins		
Homogalacturonan	20–35	0.1
Rhamnogalacturonans I and II		
Proteins		
Structural	10	?
Enzymes	?	?
Phenolics (ferulic and *p*-coumaric acids)	?	?
Lignin	0	7–10

Source: Data adapted from a variety of sources including Bacic (2006) and Vogel (2008) references therein.
?, Under investigation/unknown at this time.

Figure 9.6 Shadow cast preparation of cell wall depicting cellulose microfibrils. From: R.D. Preston, deceased, University of Leeds.

Table 9.3 Types of plant cell wall hemicelluloses (see Zhong and Zheng-Hua (2009) for structures).

Type	Characteristics
Xyloglucan*	Most abundant hemicellulose in primary walls of non-graminae plants. Backbone consists of 1,4-linked β-D-glucan residues with 1,6-α-xylosyl residues added to the backbone
Arabinoxylan, glucuronoxylans, and glucuronoarabinoxylans	Minor components of primary walls of dicots and non-graminae monocots
Xylans and substituted xylans	Backbone is 1,4-linked β-D-Xyl residues
Arabinoxylans in grasses	Backbone of 1,4-linked β-D-Xyl p residues; side chains of α-L-Araf; may contain ferulic acid
Galactoglucomannans	Mannose-containing polysaccharides; backbone of 1,4-linked β-D-mannose residues

*Sanhu et al. (2009) present the structure of xyloglucan. The structures of glucuronoxylan, glucurono-arabinoxylan, glucomannan, and galactoglucomannan are presented in Zhong and Zheng-Hua (2009).

Table 9.4 Pectin components of the plant cell wall.

Component	Characteristics
Homogalacturonan*	Linear chain of 1,4-linked α-D-galactosyl-uronic acid with methyl esterification of some of the COOH groups pectin=homogalacturonan with a high degree of CH_3 esterification pectic acid=homogalacturonan with a low degree of CH_3 esterification
Rhamnogalacturonan I	Backbone of α-D-Galp A-(1,2)-α-L-rhap-(1); can contain linear and branched chains of L-arabinofuransoyl and/or D-galactopyranosyl residues
Rhamnogalacturonan II	Exists as a dimer that is cross-linked by a 1:2 borate–diol diester; ester is between the apiosyl residue in the side chain of end monomer subunit
Xylogalacturonans	β-D-Xylp residues linked to C3 of the 1,4-linked galacturonan backbone
Apiogalacturonans	β-D-Apif residues attached to the C2 of the 1,4-linked δ-D-galacturonan backbone

*The structures of these compounds and additional information can be found at the Complex Carbohydrate Research Center at the University of Georgia (see https://www.ccrc.uga.edu).

unusual acids. Both rhamnogalacturonans I and II are thought to be attached to homogalacturonan. Certain regions of pectin may be covalently linked to xyloglucan (Park and Cosgrove, 2015).

Lignin renders the wall hydrophobic and water impermeable. This noncarbohydrate polymer (Figure 9.7) is derived from monolignols. Lignin is deposited in secondary walls where it may comprise 15–36% of the wood's weight. Zhong and Zheng-Hua (2009) reported that tracheary and vessel elements as well as fibers

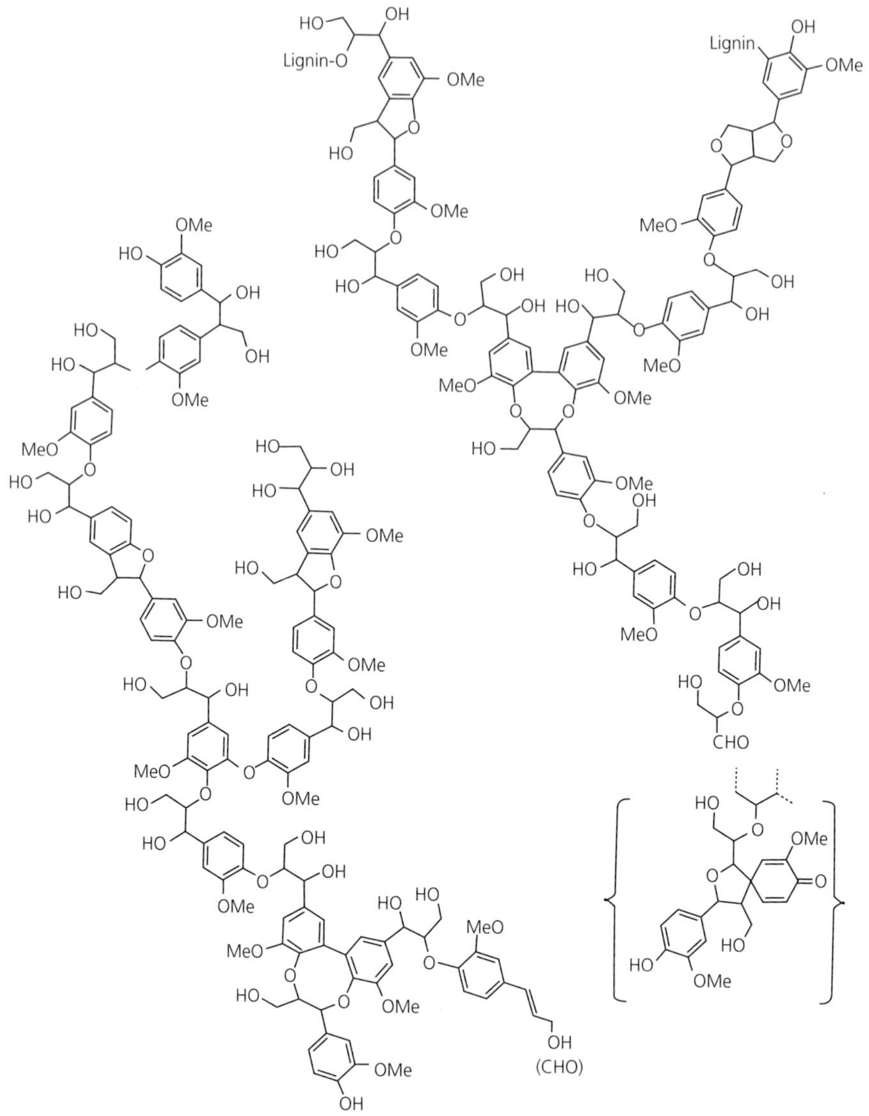

Figure 9.7 Hypothetical structure showing the lignin polymer. From: https://en.wikipedia.org/wiki/Lignin.

can contain abundant lignin. In grasses, lignin can be attached to feruloylated arabinoxylan (Gruppen *et al.*, 1992; Bunzel *et al.*, 2001).

The plant CW can contain a variety of proteins (Cassab, 1998; Jamet *et al.*, 2006). Table 9.5 summarizes both the nonenzymatic and enzymatic proteins. Expansins appear to play a significant role in cell expansion (see discussion later). Models of the primary and secondary CWs are depicted in Figures 9.1b and 9.4, respectively.

Table 9.5 Nonenzymatic and enzymatic proteins in plant cell walls.

Protein	Composition	Function	References
Nonenzymatic			
Arabinogalactan proteins (AGPs)lysine-rich, AGP peptides), chimeric AGPS	Not covalently linked to the cell wall; protein content <10%; carbohydrate content >90% wt/wt; protein rich in Hyp, Ala, Thr, Gly, and Ser; carbohydrate side chains linked by o-glycosylation to the OH groups of Ser and Hyp	Occur in higher plants and liverworts; mediators between the cell wall, the plasmalemma, and the cytoplasm; maybe soluble signals	Sardar and Showalter (2007), Seifert and Roberts (2007), Ellis *et al.* (2010)
Expansins	Multigenetic proteins	Wall-loosening proteins; disrupt hydrogen bonds	Jones and McQueen-Mason (2003), Li *et al.* (2003), Asha *et al.* (2007)
Extensins	Hyp-rich and threonine-rich glycoproteins	Cell elongation	Kieliszewski and Lamport (1994)
Glycine-rich	Glycoprotein but not extensively glycosylated	Defense mechanism	Movsavi and Hotta (2005)
Proline-rich	Contain one specific repetitive motif, variations of (Pro-Hyp-Val-Tyr-Lysln)	Specific functions unknown	Cassab (1998), Minorsky (2002)
Enzymatic			
Peroxidases		Possible role in polymerizing lignin; capable of oxidizing coniferyl alcohol	Showalter (1993), Whetten *et al.* (1998)
Kinases		Transmembrane sensors	Cosgrove (2001), Wagner and Kohorn (2001), Liu *et al.* (2006)
Esterases		Catalyzes cleavage and formation of ester bonds between cell wall polysaccharide and phenolic acid	Fasay and Ju (2007)
Pectinmethylesterases			

Other Enzymes
Phosphatases, Alpha-mannosidases, β-Mannosidases, β-1,3, Glucanases, β-1,4 Glucanases, Arabinosidases, Alpha-galactosidases, β-Galactosidases, β-Glucuronosidases, β-Xylosidases, Ascorbic acid oxidase, Polygalacturonase, Transeliminase, Rhamnogalacturonalyase, Xylosidase

For further discussion of cell wall proteins, see review article by Showalter (1993).

Biogenesis

The new CW begins with the formation of the phragmoplast by the fusion of Golgi-derived vesicles into dumble-shaped vesicle–tube–vesicle structures (Figure 9.8a; Verma, 2001). Nebenfuhr *et al.* (1999) followed the distribution of Golgi stacks during mitosis and cytokinesis by using green fluorescent protein-tagged soybean alpha-1,2-mannosidase. During telophase and cytokinesis, many Golgi stacks redistributed around the phragmoplast. Microtubules are present in the cortical matrix and may play a role in cellulose microfibril orientation by orienting cellulose synthase complexes in the plasmalemma (Wightman and Turner, 2010).

The employment of isolated protoplasts has yielded fruitful information regarding the mechanics of CW synthesis (Liu *et al.*, 1992; Chen and Chiang, 1995; Nishiyama *et al.*, 1995). Wall-less protoplasts can be obtained by heating cells with wall-degrading enzymes (Table 9.6) or mechanical methods (Cocking, 1972; Rao and Prakash, 1995). Then, the protoplasts can be maintained in culture where they can be made to regenerate CWs. Liu *et al.* (1992) reported that in the initial stages of CW regeneration following several hours of protoplast culture endoplasmic reticulum-associated, electron-dense bodies were observed. Prior to microfibrillar orientation, the bodies were found near the plasmalemma. In the investigation of Nishiyama (1995), an interwoven network of microfibrils was seen in *Candida albicans* protoplasts cultivated for 3 h. At 24 h, CW regeneration was complete. For protoplasts of another species, microfibrils were observed following 7 days of culture.

The synthesis of noncellulosic polysaccharides occurs in the Golgi apparatus (Dhugga, 2005; Lerouxel *et al.*, 2006), which contains nucleotide sugars (Figure 9.8b; Baldwin *et al.*, 2001). The number of genes involved in nucleotide sugar formation is approximately 100 (Bar-Peled and O'Neill, 2011). Recent investigations (Seifert and Roberts, 2007; Reiter, 2008) suggest that nucleotide sugar interconversions take place in the Golgi. These nucleotide sugars can be transported to the Golgi via transporters (Hirschberg *et al.*, 1998; Baldwin *et al.*, 2001). The UDP-galactose and UDP-rhamnose involved in pectin synthesis can be derived from UDP-glucose.

Of particular research interests are the Golgi-based, nucleotide sugar derivatives involved in pectin synthesis. A significant derivative is UDP-galacturonic acid (Gu and Bar-Peled, 2004). This compound is generated via the epimerization of UDP-α-D-glucuronic acid to UDP-α-D-galacturonic acid through the action of UDP-D-glucuronic acid 4-epimerase (Figure 9.9).

In addition to the enzymes mediating nucleotide sugar interconversions, pectin synthesis (Table 9.7) requires glycosyltransferases (Bouton *et al.*, 2002) as well as methyltransferases, acetyltransferases, ferulosyltransferases (Scheller *et al.*, 2007), galacturonidase (Sterling *et al.*, 2006), and possibly xylosyltransferases and an arabinosyltransferase (Scheller *et al.*, 2007). Mohnen (2008) stated that pectin synthesis requires at least 67 transferases. Pertinent contemporary reviews of pectin synthesis are those of Scheller *et al.* (2007), Mohnen (2008), and Harholt *et al.* (2010).

Fusion of Golgi-derived vesicles along phragmoplast microtubules

.
.

Formation of a tubular-vesicular network

.
.

Tubular network forms a fenestrated sheet involving a fusion tube and a zone of adhesion

.
.

New cell wall

(a)

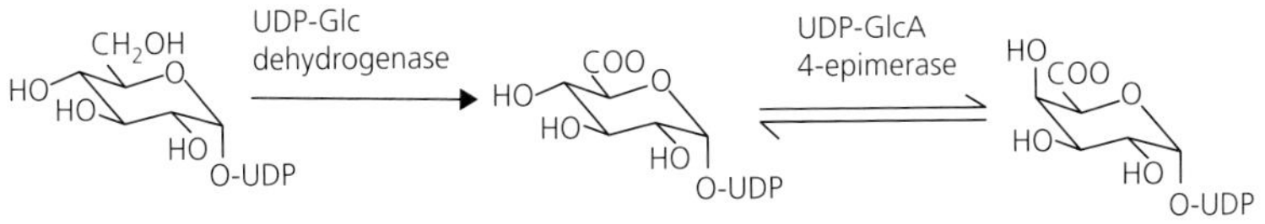

UDP-alpha-glucose

| UDP glucose dehydrogenase

UDP-alpha–ᴅ–glucuronic acid

| UDP-ᴅ–glucuronic acid 4 epimerase

UDP-alpha–ᴅ-galacturonic acid

Precursor to homogalacturonans, rhamnogalacturonans, selected galacturonans

(b)

Figure 9.8 (a) Steps in the formation of the phragmoplast. (b) Structure and biosynthesis of UDP-galacturonic acid.

Table 9.6 Some examples of cell wall degrading enzymes.

Enzyme	Reference
Cellobiohydrolase	Vincken *et al.* (1994)
Cellulosomes	Doi and Kosugi (2004)
Chitinase	Grover (2012)
endo-o-Glycosyl-hydrolases	Fry (1995)
endo-(1→4)-β-ᴅ-glucanase	Takahisa *et al.* (1984)
Pectinases	Showalter (1993)
Xylanases	Phupiewklam *et al.* (2011)
exo-o-Glycosylhydrolases	Fry (1995)
α-β-ᴅ-Glucosidase	
α-ᴅ-Galacturonidases	
α-ʟ-Fucosidase	
endo- and *exo*-polygalacturonidase	Schachta *et al.* (2011)
endo- and *exo*-pectinmethylgalacturonases	Schachta *et al.* (2011)
Pectinmethylesterase	Willats *et al.* (2001b)

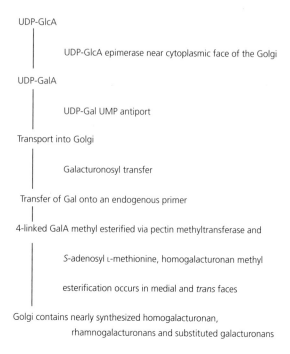

UDP-GlcA

|

        UDP-GlcA epimerase near cytoplasmic face of the Golgi

UDP-GalA

|

        UDP-Gal UMP antiport

Transport into Golgi

|

        Galacturonosyl transfer

Transfer of Gal onto an endogenous primer

|

4-linked GalA methyl esterified via pectin methyltransferase and

|

        S-adenosyl L-methionine, homogalacturonan methyl

        esterification occurs in medial and *trans* faces

Golgi contains nearly synthesized homogalacturonan,
        rhamnogalacturonans and substituted galacturonans

Pectin synthesis requires at least 67 transferases

Addition of side chains occurs in the *trans*-cisternae

Figure 9.9 Pectin biosynthetic pathway.

Table 9.7 Enzymes for cell wall synthesis.

Enzyme	Reference
Pectin synthesis	Scheller *et al.* (2007)
Acetyltransferases	Mohnen (2008)
Ferulosyltransferases	Scheller *et al.* (2007)
Galacturonosyltransferase	Caffall (2008)
UDP-decarboxylase GlcA	Harper and Bar-Peled (2002)
UDP-GlcA4 epimerase	Gu and Bar-Peled (2004)
Glycosyltransferase	Yin *et al.* (2010)
Xylan fucosyltransferase	Perrin *et al.* (1999)
Xylan *endo-trans*-glycosylase	Fry *et al.* (1992)
Cellulose synthase	Paredez *et al.* (2006)

    Cellulose synthesis (Figure 9.10) employs UDP-glucose (Lai-Kee-Hin *et al.*, 2002), cellulose synthase (Suzuki *et al.*, 2006; Diotallevi and Mulder, 2007), and the cytoskeleton (Paradez *et al.*, 2006). Spinning disk confocal microscopy utilizing fluorescent protein fusion to cellulose synthase revealed that the enzyme was aligned with cortical microtubules (Paradez *et al.*, 2006). Various inhibitors have aided investigations of cellulose synthesis. Morlin is an inhibitor of cortical

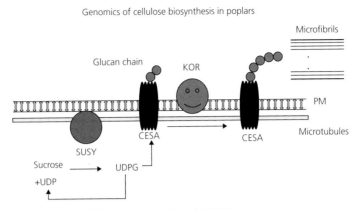

Figure 9.10 Synthesis of cellulose. From: Joshi *et al.* (2004).

microtubule dynamics (DeBolt *et al.*, 2007). The deposition of cellulose can be affected by thaxlomin and synthetic herbicides (Acebes *et al.*, 2010). Rosette structures associated with the plasmalemma transfer glucose from UDP-glucose yielding extracellular glucan chains.

The hemicellulose glucomannan is synthesized in the Golgi and then secreted. This hemicelluose is synthesized via glucomannan synthetase (Zhong and Zheng-Hua, 2009). Zhang and Staehelein (1992) concluded that xyloglucan synthesis is confined to the *trans*-Golgi cisternae and *trans*-Golgi network (TGN) but that rhamnogalacturonan I, a pectin component, is synthesized throughout the Golgi cisternae.

Lignin synthesis occurs via polymerization of monolignols through a free radical mechanism (Zhong and Zheng-Hua, 2009; Wong *et al.*, 2011). This synthesis involves glycosyltransferases that can glycosylate sinapic acid and sinapyl alcohol (Lim *et al.*, 2001). Monolignols are synthesized in the cytoplasm and most likely transported through the plasmalemma via transporters (Zhong and Zheng-Hua, 2009).

Some of the key enzymes involved with lignin biosynthesis are listed in Table 9.8.

Function

One of the very significant functions of plant CWs is cell-to-cell communications. This is accomplished by plasmodesmata (see Figures 9.11 and 9.12).

While the origin of primary and secondary plasmodesmata is not completely understood, there are some theories that suggest primary formation during cytokinesis by entrapment of the endoplasmic reticulum (Figure 9.13).

Perhaps, the most important function of the CW is its role in cell expansion. It has been known for decades that auxin, a plant hormone, can promote cell elongation (Rayle and Cleland, 1992). The mechanism appears to involve CW

Table 9.8 Enzymes involved with lignin biosynthesis.

Enzyme	Reaction catalyzed	Reference
Phenylalanine ammonia lyase	Deanimates phenylalanine to yield cinnamate	Whetten and Sederoff (1995)
Cinnamate-4-hydroxylase, a cytochrome P450-linked monoxygenase	Hydroxylation of cinnamic acid to p-coumaric acid	
Coumarate 3-hydroxylase	Catalyzes the hydroxylation of p-coumarate to from caffeate	
Coumaroyl-coenzyme A3-hydroxylase		
Caffeate-o-methyltransferase	Methylates caffeic acid generates ferulic acid; employs S-adenosyl methioninie	
Caffeoyl-coenzyme A-methyltransferase	Function not yet varied	

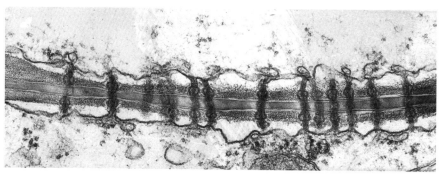

Figure 9.11 Occurrence of plasmodesmata in the cell wall. From: E. Newcomb, University of Wisconsin–Madison.

loosening (Rayle and Cleland, 1992). How is loosening achieved? Various observations suggested that auxin-promoted proton efflux thereby lowering the wall's pH (Acid Growth Theory). This enables expansins, which possess an acidic pH optima, to temporarily affect wall structure. Cosgrove (2000) proposed that expansins weaken non-covalent bonding between wall polysaccharides but does not cause a lasting alteration in wall structure. This weakening would permit stress relaxation and turgor-driven polymer creep (Marga *et al.*, 2005). Suslov *et al.* (2010) reported that heat inactivation of endogenous CW protein inhibited acid-induced cell extension. Keller and Van Volkenburgh (1998) proposed that auxin-induced expansion of tobacco leaf tissues does not require acidification of the CW. Then, Schopfer (2001) concluded that the OH radical may be involved in auxin-induced CW loosening. The OH radical may be generated by an apoplastic peroxidase (Liszkay *et al.*, 2003). Last, wall-associated kinases appear to

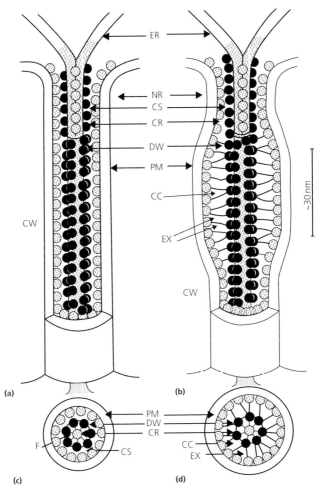

Figure 9.12 Diagram of structure of plasmodesmata. From: Courtesy of B. Ding with permission.

be covalently bound to the CW in *Arabidopsis* and may have receptor activity in the control of cell expansion (Wagner and Kohorn, 2001).

The plant CW can function as an ion exchanger (Meychik and Yermakov, 2001). Ions are thought to move through the water-filled spaces between cellulose microfibrils. Subsequent movement of the ions to the cytosol may involve ion channels in the plasmalemma (Malmstrom *et al.*, 1997; Brett and Waldron, 1990).

A number of investigations have raised the possibility that CWs may serve as defense agents against pathogens (Keegstra, 2010). Humphrey *et al.* (2007) suggested that plant CWs contain sensing agents. Narvaez-Vasquez (2005) reported the possibility of signal peptides in the cell wall. As early as 1985, Albensheim and Darvill noted that xyloglucan can be degraded to oligosaccharins, short chain oligosaccharides. The oligosaccharins may serve as the signaling agents (Fry *et al.*, 1993).

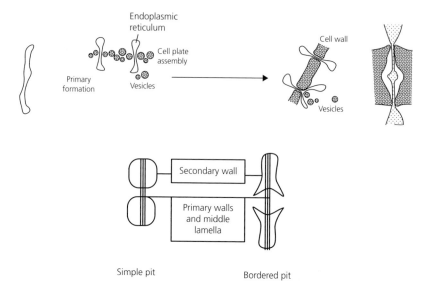

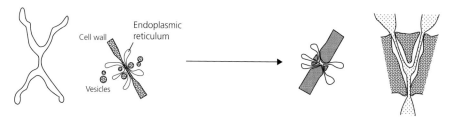

Figure 9.13 Origin of the primary and secondary plasmodesmata. From: Hypothetical model derived from discussions with Bill Dashek.

Because of the alleged role of fossil fuel usage in increasing global warming (see literature in Dashek and McMillin, 2009), scientists have searched for other sources of biofuels. Certain plant CW polymers, for example, cellulose, have received considerable attention as possible biofuels (Pauly and Keegstra, 2010).

Plant CW structure and function are fascinating topics with many applications and implications in all aspects of plant growth and development. Virtually, all aspects of life on earth are affected by the CW and its components, be it food, shelter, clothing, or the integral contribution of the CW to the biosphere itself. Some of the more exciting and innovative approaches to studying CW structure and function include genes involved in CW synthesis (Table 9.9), inhibitors of CW assembly (Table 9.10), and the many advanced methods of studying CWs (Table 9.11). For scientists worldwide, this is an area of research that deserves much attention and enthusiasm, particularly by

Table 9.9 Genes involved in synthesis of cell wall chemicals.*

Chemicals	Some genes involved	Enzymes involved	Sites of synthesis and transport	References
Cellulose	Ces A		Plasma membrane "FRA1" kinesin-like protein is involved in the microtubule control of cellulose microfibril order	Zhong et al. (2002)
	Ces A-like (Csl)	Cellulose synthases		Doblin et al. (2002)
	Csl E1	Sucrose synthases		Suzuki et al. (2006)
	Csl H1	Korrigan synthases		Saxena and Brown (2007)
				Carpita et al. (2001)
Xyloglucan	Mur 2	α-Fucosyltransferase		Vanzin et al. (2002)
	AtXT1	β-Galactosyltransferase		Cavalier and Keegstra (2006)
	At5g			Caffall and Hohnen (2009)
Pectin	25 members of related gene family	Galacturonosyltransferase		Yin et al. (2010)
Proteins				
Expansins				
α	PttEXP2			Gray-Mitsume et al. (2004)
	PttEXP8			Asha et al. (2007)
β	MaEPA2,3,4,5			Gray-Mitsume et al. (2004)
	PttEXP			
Glycine-rich	86 identified			Fusaro et al. (2001)
Leucine-rich	LRX1			Baumberger et al. (2001)
Proline-rich	AtPRP			Fowler et al. (1999)

* In *Arabidopsis*, about 2500 genes are involved in plant cell wall biosynthesis (Carpita, 2011).

Table 9.10 Some inhibitors of plant cell wall components.

Inhibitor	Effect	Reference
Cytochalasin B, N-ethylmaleimide vinblastine, cycloheximide	Affects birefringency of a green alga cell wall	Noguchi and Ueda (1981)
Morlin	Inhibitor of cortical microtubule dynamics and cellulose synthase movement	DeBolt et al. (2007)
Thaxtomin A and synthetic herbicides	Affects the assembly of deposition of cellulose	Acebes et al. (2010)
Ferulate 5-hydrolyslase	Hydroxylates ferulate to 5-hydroxylferulate	
4-Coumarate: coenzyme A ligase	Catalyzes the formation of CoA thioesters	
Cinnamoly-coenzyme A reductase	Reduces hydroxycinnamoly-CoA thioesters to the appropriate aldehydes	
Cinnamyl alcohol dehydrogenase	Reduces hydroxycinnamaldehydes, hydroxycinnamal alcohols	
2-Galacturonsyltransferase (Gaut)		Caffall (2008) and Yin et al. (2010)

Table 9.11 Methods for analysis of plant cell walls.

Method	Reference
Acidic fingerprinting	Seveno et al. (2009)
Affinity labeling	Gibeaut and Capita (1994)
Atomic force microscopy	Marga et al. (2005)
Field emission scanning electron microscopy	Carpita et al. (2001)
Fourier-transform infrared spectroscopy	Wilson et al. (2000)
Gel permeation chromatography	Timpa (1991)
Gene expression profiling	Sanhu et al. (2009)
Genome and transcript analysis	Burton et al. (2006)
Inframicrospectroscopy	Carpita et al. (2001) Chen et al. (1997)
Mass spectrometry	Zablackis et al. (1996)
Nuclear magnetic resonance spectrometry	Zablackis et al. (1996)
Oligosaccharide mass profiling	Gunl et al. (2010)
HPLC fingerprinting plus mass spectrometry	https://www.ccrc.uga.edu/
Novel fingerprinting strategy—screens RG-II mutants	Seveno et al. (2009)
Proteomics	Jamet et al. (2006) Albenne et al. (2013)
Scanning probe microscopy	Radotic et al. (2012)
Resonance Raman spectroscopy	Barsberg et al. (2005)
Targeted genetic and enzymatic treatments	Burgert (2006)
Transcriptome sequencing	Wong et al. (2011)

Source: Adapted from https://www.ccrc.uga.edu/ as well as Sdeller and Ukviskou (2010).

plant biologists. After all, those who study the CW are investigating one of the most abundant and perhaps one of the most important biological complexities on earth!

References

Acebes, J.L., Encina, A., García-Angulo, P., Alonso-Simon, A., Melida, H. and Alvarez, J.M. (2010) Cellulose biosynthesis inhibitors: their uses as potential herbicides and as tools in cellulose and cell wall structural plasticity research. In: *Cellulose: Structure and Properties, Derivatives and Industrial Uses* (Lejeune, A. and Deprez, T., eds), Nova Publishers: New York, NY. pp. 39–73.

Albenne, C., Canut, H. and Jamet, E. (2013) Plant cell wall proteomics: the leadership of *Arabidopsis thaliana. Frontiers in Plant Science*, **4**, 111, http://doi.org/10.3389/fpls.2013.00111

Albersheim, P. (1975) The walls of growing plants. *Scientific American*, **232**, 80–95.

Albersheim, P., Darvill, A., Roberts, K., Sederoff, R. and Staehelin, A. (2010) *Plant Cell Walls*, Garland Publishing, New York, NY.

Asha, V.A., Sane, A.P. and Nath, P. (2007) Multiple forms of α-expansin genes are expressed during banana fruit ripening and development. *Postharvest Biology and Technology*, **45**, 184–192.

Atsushi, K., Azuma, J. and Koshijima, T. (1984) Lignin-carbohydrate complexes and phenolic acids in bagasse. *Holzforschung*, **38**, 141–149.

Bacic, A. (2006) Breaking an impasse in pectin biosynthesis. *Proceedings of the National Academy of Sciences of the United States of America*, **103**, 5639–5640.

Baldwin, T.C., Handford, M.G., Yuseff, M.I., Orellana, A. and Dupree, P. (2001) Identification and characterization of GONST1, a golgi-localized GDP-mannose transporter in Arabidopsis. *The Plant Cell*, **13**, 2283–2295.

Bar-Peled, M. and O'Neill, M. (2011) Plant nucleotide sugar formation, interconversion and salvage by sugar recycling. *Annual Review Plant Biology*, **62**, 127–155.

Barsberg, S., Matousek, P. and Towrie, M. (2005) Structural analysis of lignin by resonance Raman spectroscopy. *Macromolecular Bioscience*, **5**, 743–752.

Bauer, W.D., Talmadge, K.W., Keegstra, K. and Albersheim, P. (1973) The structure of plant cell walls. II. The hemicellulose of the walls of suspension-cultured sycamore cells. *Plant Physiology*, **51**, 174–187.

Baumberger, N., Ringli, C. and Keller, B. (2001) The chimeric leucine-rich repeat/extensin cell wall protein LRX1 is required for root hair morphogenesis in *Arabidopsis thaliana. Genes and Development*, **15**, 1128–1139.

Bidlack, J.E. and Buxton, D.R. (1992) Content and deposition rates of cellulose, hemicellulose and lignin during regrowth of forages grasses and legumes. *Canadian Journal of Plant Science* **72**, 809–818.

Bidlack, J.E. and Jansky, S.H. (2014) *Stern's Introductory Plant Biology*, 13th edn. McGraw-Hill Companies, New York, NY.

Bidlack, J., Malone, M. and Benson, R. (1992) Molecular structure and component integration of secondary cell walls in plants. *Proceedings of the Oklahoma Academy of Sciences of the United States of America*, **72**, 51–56.

Bolwell, G.P. (1988) Synthesis of cell wall components: aspects of control. *Phytochemistry*, **27**, 1235–1253.

Burgert, I. (2006) Exploring the micromechanical design of plant cell walls. *American Journal of Botany*, **93**, 1391–1401.

Burton, R.A., Wilson, S.M., Hrmova, M. *et al.* (2006) An enzyme identified in rice generates a complex sugar found in the cell walls of many grains that are as important as human and animal food. *Science*, **311**, 1940–1942.

Bouton S., Leboeuf, E., Mouille, G. *et al.* (2002) QUASIMODO1 encodes a putative membrane-bound glycosyltransferase required for normal pectin synthesis and cell adhesion in Arabidopsis. *The Plant Cell*, **14**, 2577–2590.

Brett, C. and Waldron, K. (1990) *Physiology and Biochemistry of Plant Cell Walls*, Unwin Hyman, London. 194 pp.

Brown, R.M. and Saxena, I.M. (2007) *Cellulose Molecular and Structural Biology*, Springer, New York, NY.

Bunzel, M., Ralph, J., Marita, J.M., Hatfield, R.D. and Steinhart, H. (2001) Diferulates as structural components in soluble and insoluble cereal dietary fibre. *Journal of the Science of Food and Agriculture*, **81**, 653–660.

Caffall K.H. and Mohnen, D. (2009) The structure, function, and biosynthesis of plant cell wall pectic polysaccharides. *Carbohydrate Research*, **344**, 1879–1900.

Caffall, K.H. (2008) *Expression and Characterization of galacturonosyltransferase-6 (gaut6) of the galac-turonosyltranserferase-1 (gaut1)-related Gene Family of Arabidopsis thaliana*, Ph.D. Dissertation, University of Georgia, Athens, GA, USA.

Carpita, N. (1983) Hemicellulosic polymers of cell walls of *Zea* coleoptiles. *Plant Physiology*, **72**, 515–521.

Carpita, N. (2011) Update on mechanisms of plant cell wall biosynthesis: how plants make cellulose and other $(1\rightarrow4)$-β-d-glycans. *Plant Physiology*, **155**, 171–184.

Carpita, N.C., Campbell, M. and Tierney, M. (2001) *Plant Cell Walls*, Springer, Berlin, Germany.

Carpita, N.C., Defernez, M., Findlay, K. *et al.* (2001) Cell wall architecture of the elongating maize coleoptile. *Plant Physiology*, **127**, 551–565.

Cassab, G.I. (1998) Plant cell wall proteins. *Annual Review Plant Physiology Molecular Biology*, **49**, 281–309.

Cavalier, D.M. and Keegstra, K. (2006) Two xyloglucan xylosyltransferases catalyze the addition of multiple xylosyl residues to cellohexaose. *The Journal of Biological Chemistry*, **281**, 34197–34207.

Chen, L., Wilson, R.H. and McCann, M.C. (1997) Infra-red microspectroscopy of hydrated biological systems: design and construction of a new cell with atmospheric control for the study of plant cell walls. *Journal of Microscopy*, **188**, 62–71.

Chen, Y.C. and Chiang, Y.M. (1995) Ultrastructure of cell wall regeneration from isolated protoplasts of *Grateloupia sparsa* (Halymeniaceae, Rhodophyta). *Botanica Marina*, **38**, 393–400.

Chesson, A., Gordon, A.H. and Lomax, J.A. (1983) Substituent groups linked by alkali-labile bonds to arabinose and xylose residues of legume, grass, and cereal straw cell walls and their fate during digestion by rumen microorganisms. *Journal of the Science of Food and Agriculture*, **34**, 1330–1340.

Cocking, E.C. (1972) Plant cell protoplasts – isolation and development. *Annual Review of Plant Physiology*, **23**, 29–50.

Cosgrove, D.J. (2000) Loosening of plant cell walls of expansins. *Nature*, **407**, 321–326.

Cosgrove, D.J. (2001) Wall structure and wall loosening. A look backwards and forwards. *Plant Physiology*, **125**, 131–134.

Dashek, W.V., and McMillin, D. (2009) *Biological Environmental Science*, Science Publishers. Enfield, NH, USA.

DeBolt, S., Gutierrez, R., Ehrhardt, D.W. *et al.* (2007) Morlin, an inhibitor of cortical microtubule dynamics and cellulose synthase movement. *Proceedings of the National Academy of Sciences of the United States of America*, **104**, 5854–5859.

Delmer, D.P. (1987) Cellulose biosynthesis. *Annual Review of Plant Physiology*, **38**, 259–290.

Dhugga, K.S. (2005) Plant Golgi cell wall synthesis: from genes to enzyme activities. *Proceedings of the National Academy of Sciences of the United States of America*, **102**, 1815–1816.

Diotallevi, F. and Mulder, B. (2007) The cellulose synthase complex: a polymerisation driven supramolecular motor. *Biophysical Journal*, **92**, 2666–2673.

Doblin, M.S., Kurek, I., Jacob-Wilk, D. and Delmer, D.P. (2002) Cellulose biosynthesis in plants: from genes to rosettes. *Plant Cell Physiology*, **43**, 1407–1420.

Doi, R.H. and Kosugi, A. (2004) Cellulosomes: plant-cell-wall-degrading enzyme complexes. *Nature Reviews Microbiology*, **2**, 541–551.

Ellis, M., Egelund, J., Schultz, C.J. and Bacic, A. (2010) Arabinogalactan-proteins: key regulators at the cell surface? *Plant Physiology*, **153**: 403–419.

Elstner, E.F. and Heupel, A. (1976) Formation of hydrogen peroxide by isolated cell walls from horseradish (Armoracia lapathifolia Gilib.). *Planta*, **130**, 175–180.

Esau, K. (1977) Cell Wall. In *Plant Anatomy*, John Wiley & Sons, Inc., New York, NY. pp. 43–60.

Evert, R.F. (2006). *Esau's Plant Anatomy*, John Wiley & Sons, Inc., Hoboken, NJ.

Fazary, A.E. and Ju, Y.-H. (2007) Feruloyl esterases as biotechnological tools: current and future perspectives. *Acta Biochimica et Biophysica Sinica*, **39**, 811–828.

Flowler, T.J., Bernhardt, C. and Tierney, M.L. (1999) Characterization and expression of four proline-rich cell wall protein genes in arabidopsis encoding two distinct subsets of multiple domain proteins. *Plant Physiology*, **121**, 1081–1091.

Freudenberg, K. and Neish, A.C. (1968) *Constitution and Biosynthesis of Lignin*, Springer-Verlag, Berlin/Heidelberg. 129 pp.

Fry, R.C., Smith, R.C., Renwich, K.F., Martin, D.J., Hodge, S.K. and Matthews, K.J. (1992) Xyloglucan endotransglycosylase, a new wall-loosening enzyme activity from plants. *Biochemical Journal*, **282**, 821–828.

Fry, S.C. (1995) Polysaccharide-modifying enzymes in the plant cell wall. *Annual Review of Plant Physiology and Plant Molecular Biology*, **46**, 497–520.

Fry, S.C. (1986) Cross-linking of matrix polymers in the growing cell walls of angiosperms. *Annual Review of Plant Physiology*, **37**, 165–186.

Fry, S.C., Addington, S., Hetherrington, P.R. and Aitken, J. (1993) Oligosaccharides as signals and substrates in the plant cell wall. *Plant Physiology*, **103**, 1–5.

Fusaro, A., Mangeon, A., Junqueira, R.M. *et al.* (2001) Classification, expression pattern and comparative analysis of sugarcane expressed sequences tags (ESTs) encoding glycine-rich proteins (GRPs). *Genetics and Molecular Biology*, **24**, 263–273.

Gibeaut, P.M. and Carpita, N.C. (1994) Biosynthesis of plant cell wall polysaccharides. *The FASEB Journal*, **8**, 904–915.

Giddings, T.H., Jr., Brower, D.L. and Staehelin, L.A. (1980) Visualization of particle complexes in the plasma membrane of *Micrasterias denticulata* associated with the formation of cellulose fibrils in primary and secondary cell walls. *Journal of Cell Biology*, **84**, 327–339.

Goodwin, T.W. and Mercer, E.I. (1983) The plant cell wall. In: *Introduction to Plant Biochemistry*, Pergamon Press, New York, NY. pp. 55–91.

Gray-Mitsumme, M., Mellerowicz, E.J., Hisashi, A. *et al.* (2004) Expansins abundant in secondary xylem belong to subgroup A of the α-expansin gene family. *Plant Physiology*, **135**, 1552–1564.

Grover, A. (2012) Plant chitinases: genetic diversity and physiological roles. *Critical Reviews in Plant Sciences*, **31**, 51–73.

Gruppen H., Hamer R.J. and Voragen A.G.J. (1992) Water-unextractable cell wall material from wheat flour. 1. Extraction of polymers with alkali. *Journal of Cereal Science*, **16**, 41–51.

Gu, X. and Bar-Peled, M. (2004) The biosynthesis of UDP-galacturonic acid in plants, functional cloning and characterization of arabidopsis UDP-d-glucuronic acid 4-epimerase. *Plant Physiology*, **136**, 4256–4264.

Gunl, M., Kraemer, F. and Pauly, M. (2010) Oligosaccharide mass profiling (OLIMP) of cell wall polysaccharides by MALDI-TOF/MS. *Methods in Molecular Biology*, **715**, 43–54.

Hahlbrock, K. and Grisebach, H. (1979) Enzymatic controls in the biosynthesis of lignin and flavonoids. *Annual Review of Plant Physiology*, **30**, 105–130.

Harholt, J., Suttangkakul, A. and Scheller, H.V. (2010) Biosynthesis of pectin. *Plant Physiology*, **153**, 384–395.

Harper, A.D. and Bar-Peled, M. (2002) Biosynthesis of UDP-xylose. Cloning and characterization of a novel *Arabidopsis* gene family, UXS, encoding soluble and putative membrane-bound UDP-glucuronic acid decarboxylase isoforms. *Plant Physiology*, **130**, 2188–2198.

Hartley, R.D. (1972) *p*-Coumaric and ferulic acid components of cell walls of ryegrass and their relationships with lignin and digestibility. *Journal of the Science of Food and Agriculture*, **23**, 1347–1354.

Hartley, R.D. and Jones, E.C. (1976) Diferulic acid as a component of cell walls of *Lolium multiflorum*. *Phytochemistry*, **15**, 1157–1160.

Hatfield, R.D. (1989) Structural polysaccharides in forages and their digestibility. *Agronomy Journal*, **18**, 39–46.

Hayashi, T. (2006) *The Science and Lore of the Plant Cell Wall: Biosynthesis Structure and Function*, Brown Walker Press, Boca Raton, FL.

Herth, W. (1985) Plasma-membrane rosettes involved in localized wall thickening during xylem vessel formation of *Lepidium sativum* (L.). *Planta*, **164**, 21–21.

Hirschberg, C.B., Robbins, P.W. and Abeijon, C. (1998). Transporters of nucleotide sugars, ATP and nucleotide sulfate in the endoplasmic reticulum and Golgi apparatus. *Annual Review of Biochemistry*, **67**, 49–69.

Huber, D.J. and Nevins, D.J. (1979) Partial purification of endo- and exo-β-D-glucanase-enzymes from *Zea mays* L. Seedlings and their involvement in cell wall autohydrolysis. *Planta*, **151**, 206–214.

Humphrey, T.V., Bonetta, D.T. and Goring, D.R. (2007) Sentinels at the wall: cell wall receptors and sensors. *New Phytologist*, **176**, 7–21.

Iiyama, B.-T., Lan, T. and Stone, B.A. (1994) Covalent cross-links in the cell wall. *Plant Physiology*, **104**, 315–320.

Jamet, E., Canut, H., Boudart, G. and Pont-Lezica, R.F. (2006) Cell wall proteins: a new insight through proteomes. *Trends in Plant Science*, **11**, 33–39.

Joshi, C.P., Bhandari, S., Ranjan, P. *et al.* (2004) Genomics of cellulose biosynthesis in poplars. *New Phytologist*, **164**, 53–61.

Jung, H.G. (1989) Forage lignins and their effects on fiber digestibility. *Agronomy Journal*, **81**, 33–38.

Keegstra, K. (2010) Plant cell wall. *Plant Physiology*, **154**, 483–486.

Keegstra, K., Talmadge, K.W., Bauer, W.D. and Albersheim, P. (1973) The structure of plant cell walls. III. A model of the walls of suspension-cultured sycamore cells based on the interconnections of the macromolecular components. *Plant Physiology*, **51**, 188–196.

Keller, C.P. and Van Volkenburgh, E. (1998) Evidence that auxin-induced growth of tobacco leaf tissues does not involve cell wall acidification. *Plant Physiology*, **118**, 557–564.

Kieliszewski, M.J. and Lamport, D.T.A. (1994) Extensin: repetitive motifs, functional sites, post-translational codes, and phylogeny. *The Plant Journal*, **5**, 157–172.

Labavitch, J.M. (1981) Cell wall turnover in plant development. *Annual Review of Plant Physiology*, **32**, 385–406.

Lai-Kee-Him. J., Chanzy, H., Muller, M., Putaux, J.L. Imai, T. and Bulone V. (2002) *In vitro* versus *in vivo* cellulose microfibrils from plant primary wall synthases: structural differences. *The Journal of Biological Chemistry*, **277**, 36931–36939.

Lamport D.T.A. and Catt J.W. (1981) Glycoproteins and enzymes of the cell wall. In: *Plant Carbohydrates II: Extracellular Carbohydrates* (The Encyclopedia of Plant Physiology), vol. **13 B**, Tanner W. and Loewus F.A. (eds), pp. 133–165, Springer, New York.

Lerouxel, O., Cavalier, D.M., Liepman, A.H. and Keegstra, K. (2006) Biosynthesis of plant cell wall polysaccharides—a complex process. *Current Opinion Plant Cell Biology*, **9**, 621–630.

Lewis, N.G. and Yamamoto, E. (1990) Lignin: occurrence, biogenesis and biodegradation. *Annual Review of Plant Physiology*, **41**, 455–496.

Li, Y., Jones, L. and McQueen-Mason, S. (2003) Expansins and cell growth. *Current Opinion in Plant Biology*, **6**, 603–610.

Lim, E.-K., Li, Y., Papp, A., Jackson, R., Ashford, D.A. and Bowles, D.J. (2001) Identification of glucosyltransferase genes involved in sinapate metabolism and lignin synthesis in Arabidopsis. *The Journal of Biological Chemistry*, **276**, 4344–4349.

Liszkay, S., Kenk, B. and Schopfer, P. (2003) Evidence for the involvement of cell wall peroxidase in the generation of hydroxyl radicals mediating extension growth. *Planta*, **217**, 658–667.

Liu Q.Y., Chen, L.C.M. and Taylor, A.R.A. (1992) Ultrastructure of cell wall regeneration by isolated protoplasts of *Palmaria palmate* (Rhodophyta). *Botanica Marina*, **35**, 21–33.

Liu, Y., Liu, D., Zhang, H. *et al.* (2006) Isolation and characterisation of six putative wheat cell wall-associated kinases. *Functional Plant Biology*, **33**, 811–821.

Mader, M. and Fussl, R. (1982) Role of peroxidase in lignification of tobacco cells. *Plant Physiology*, **70**, 1132–1134.

Malmström, S., Askerlund, P. and Palmgren M.G. (1997) A calmodulin-stimulated Ca^{2+}-ATPase from plant vacuolar membranes with a putative regulatory domain at its amino terminus. *FEBS Letters*, **400**, 324–328.

Marga, F., Grandbois, M., Cosgrove, D.J. and Baskin, T.I. (2005) Cell wall extension results in the coordinate separation of parallel microfibrils: evidence from scanning electron microscopy and atomic force microscopy. *The Plant Journal*, **43**, 181–190

Meychik, N. and Yermakov, I. (2001) Ion exchange properties of plant root cell walls. *Plant and Soil*, **234**, 181–193.

Mauseth, J.D. (1988) The cell. In: *Plant Anatomy*, Benjamin Cummings Publishing Company, Inc., Menlo Park, CA. pp. 13–40.

Minorsky, P.V. (2002) The wall becomes surmountable. *Plant Physiology*, **128**, 345–353.

Mohnen, D. (2008) Pectin structure and biosynthesis. *Current Opinion in Plant Biology*, **11**, 266–277.

Morrison, I.M. (1974) Structural Investigations on the lignin-carbohydrate complexes of *Lolium perenne*. *Biochemistry Journal*, **139**, 197–204.

Mousavi, A. and Hotta, Y. (2005) Glycine-rich proteins: A class of novel proteins. *Applied Biochemistry and Biotechnology*, **120**, 169–174.

Mueller-Harvey, I. and Hartley, R.D. (1986) Linkage of *p*-coumaroyl and feruloyl groups to cell-wall polysaccharides of barley straw. *Carbohydrate Research*, **148**, 71–85.

Narvaez-Vasquez J., Pearce, G. and Ryan, C.A. (2005) The plant cell wall matrix harbors a precursor of defense signaling peptides. *Proceedings of the National Academy of Sciences of the United States of America*, **102**, 12974–12977.

Nebenfuhr, A., Gallagher, L.A., Dunahay, T.G. *et al.* (1999) Stop-and-go movements of plant Golgi stacks are mediated by the acto-myosin system. *Plant Physiology*, **121**, 1127–1141.

Newby, V.K., Sablon, R.M., Synge, R.L.M., Casteel, K.V. and Van Sumere, C.F. (1980) Free and bound phenolic acids of lucerne (*Medicago sativa* cv Europe). *Phytochemistry*, **19**, 651–657.

Nishiyama, Y., Aoki, Y. and Yamaguchi, H. (1995) Morphological aspects of cell wall formation during protoplast regeneration in *Candida albicans*. *Journal of Electron Microscopy*, **44**, 72–78.

Noguchi, T. and Ueda, K. (1981) Effect of metabolic inhibitors on the formation of cell walls in a green alga, *Micrasterias crux-melitensis*. *Plant and Cell Physiology*, **22**, 1437–1445.

O'Neill, M. and York, W.S. (2003) The composition and structure of primary plant cell walls. In: *The Plant Cell Wall* (Rose, J.K.C., ed.), CRC Press, Boca Raton, FL. pp. 1–54.

Park, Y.B. and Cosgrove, D.J. (2015) Xyloglucan and its interactions with other components of the growing cell wall. *Plant and Cell Physiology*, **56**, 180–194.

Paradez A., Wright, A. and Ehrhardt, D.W. (2006) Microtubule cortical array organization and plant cell morphogenesis. *Current Opinion in Plant Biology*, **9**, 571–578.

Paredez, A.R., Somerville, C.R. and Ehrhardt, D.W. (2006) Visualization of cellulose synthase demonstrates functional association with microtubules. *Science*, **312**, 1491–1495.

Pauly, M. and Keegstra, K. (2010) Plant cell wall polymers as precursors for biofuels. *Current Opinion in Plant Biology*, **13**, 305–312.

Pedreno, M., RosBarcelo, A., Sabater, F. and Munoz, R. (1989) Control by pH of cell wall peroxidase activity involved in lignification. *Plant Cell Physiology*, **30**, 237–241.

Preston, R.D. (1979) Polysaccharide conformation and cell wall function. *Annual Review of Plant Physiology*, **30**, 55–78.

Perrin, R.M., DeRocher, A.E., Bar-Peled, M. *et al.* (1999) Xyloglucan fucosyltransferase, an enzyme involved in plant cell wall biosynthesis. *Science*, **284**, 1976–1979.

Rao, K.S. and Prakash, A.H. (1995) A simple method for the isolation of plant protoplasts. *Journal of Biosciences*, **20**, 645–655.

Radotic, K., Roduit, C., Simonovic, J. *et al.* (2012) Atomic force microscopy stiffness tomography on living *Arabidopsis thaliana* cells reveals the mechanical properties of surface and deep cell-wall layers during growth. *Biophysical Journal*, **103**, 386–394.

Rayle, D.L. and Cleland, R.E. (1992) The acid growth theory of auxin-induced cell elongation is alive and well. *Plant Physiology*, **99**, 1271–1274.

Rose, J. (2003) *The Plant Cell Wall*, Blackwell, Oxford, UK.

Salisbury, F.B. and Ross, C.W. (1992) Plant physiology and plant cells. In: *Plant Physiology*, Wadsworth, Inc., Belmont, CA. pp. 3–26.

Sardar, H.S. and Showalter, A.M. (2007) A cellular networking model involving interactions among glycosyl-phosphatidylinositol (GPI)-anchored plasma membrane arabinogalactan proteins (AGPs), microtubules and F-actin in tobacco BY-2 cells. *Plant Signaling and Behavior*, **2**, 8–9.

Saxena, I.M. and Brown, R.M. (2007) A perspective on the assembly of cellulose-synthesizing complexes: possible role of Korrigan and microtubules in cellulose synthesis in plants. In: *Cellulose: Molecular and Structural Biology* (Brown, R.M. and Saxena, I.M., eds), Springer, Netherlands, pp. 169–181.

Scalbert, A., Monties, B., Lallemand, J.Y., Guittet, E. and Rolando, C. (1985) Ether linkage between phenolic acids and lignin fractions from wheat straw. *Phytochemistry*, **24**, 1359–1362.

Sanhu, A.P.S., Ravhawa, G.S. and Dhugga, K.S. (2009) Plant cell wall matrix polysaccharide biosynthesis. *Molecular Plant*, **2**, 840–850.

Schachta, T., Ungerb, C., Pichc, A. and Wydraa, K. (2011) Endo- and exopolygalacturonases of Ralstonia solanacearum are inhibited by polygalacturonase-inhibiting protein (PGIP) activity in tomato stem extracts. *Plant Physiology and Biochemistry*, **49**, 377–387.

Scheller, H.V. and Ulviskov, P. (2010) Hemicelluloses. *Annual Review of Plant Biology*, **61**, 263–289.

Scheller, H.V., Jensen, J.K., Sørensen, S.O., Harholt, J. and Geshi, N. (2007) Biosynthesis of pectin. *Physiologia Plantarum*, **129**, 283–295.

Schopfer, P. (2001) Hydroxyl radical-induced cell-wall loosening *in vitro* and *in vivo*: implications for the control of elongation growth. *The Plant Journal*, **28**, 679–688.

Seifert, G.J. and Roberts, K. (2007) The biology of arabinogalactan proteins. *Annual Review of Plant Biology*, **58**, 137–161.

Selvendran, R.R. (1983) The chemistry of plant cell walls. In: *Dietary Fibre* (Birch, G.G., and Parker, K.J., eds), Applied Science Publ., London. pp. 95–147.

Seveno, M., Voxeur, A., Rihouey, C. *et al.* (2009) Structural characterisation of the pectic polysaccharide rhamnogalacturonan II using an acidic fingerprinting methodology. *Planta* **230**, 947–957.

Showalter, A.M. (1993) Structure and function of plant cell wall proteins. *The Plant Cell*, **5**, 9–23.

Sterling J.D., Atmodjo, M.A., Inwood, S.E. *et al.* (2006) Functional identification of an Arabidopsis pectin biosynthetic homogalacturonan galacturonosyltransferase. *Proceedings of the National Academy of Sciences of the United States of America*, **103**, 5236–5241.

Suslov, D., Verbelen, J.-P. and Vissenberg, K. (2010) Is acid-induced extension in seed plants only protein-mediated? *Plant Signaling and Behavior*, **5**, 757–759.

Suzuki, S., Li, L., Sun, Y.H. and Chiang, V.L. (2006) The cellulose synthase gene superfamily and biochemical functions of xylem-specific cellulose synthase-like genes in *Populus trichocarpa*. *Plant Physiology*, **142**, 1233–1245.

Talmadge, K.W., Keegstra, K., Bauer, W.D. and Albersheim, P. (1973) The structure of plant cell walls. I. The macromolecular components of the walls of suspension-cultured sycamore cells with a detailed analysis of the pectic polysaccharides. *Plant Physiology*, **51**, 158–173.

Takahisa, H., Wong, Y.-S. and Maclachlan, G. (1984) Pea xyloglucan and cellulose: II. Hydrolysis by pea endo-1,4-β-glucanases. *Plant Physiology*, **75**, 605–610.

Theander, O. and Aman, P. (1984) Anatomical and Chemical Characteristics. In: *Straw and Other Fibrous By-Products as Feed* (Sundstol, F., and Owen, E., eds), Elsevier, Amsterdam/Holland. pp. 45–78.

Timpa, J.D. (1991) Application of universal calibration in gel permeation chromatography for molecular weight determinations of plant cell wall polymers: cotton fiber. *Journal of Agricultural and Food Chemistry*, **39**, 270–275.

Vanzin, G.F., Madson, M., Carpita, N.C., Raikhel, N.V., Keegstra, K. and Reier, W.-D. (2002) The *mur2* mutant of *Arabidopsis thaliana* lacks fucosylated xyloglucan because of a lesion in fucosyltransferase *AtFUT1*. *Proceedings of the National Academy of Sciences of the United States of America*, **99**, 3340–3345.

Verma, D.P.S. (2001) Cytokinesis and building of the cell plate in plants. *Annual Review of Plant Physiology and Plant Molecular Biology*, **52**, 751–784.

Vincken, J.P., Beldman, G. and Voragen, A.G.J. 1994. The effect of xyloglucans on the degradation of cell-wall-embedded cellulose by the combined action of cellobiohydrolase and endoglucanases from *Trichoderma viride*. *Plant Physiology*, **104**, 99–107.

Vogel, J. (2008) Unique aspects of the grass cell wall. *Current Opinion in Plant Biology*, **11**, 301–307.

Wagner T.A. and Kohorn, B.D. (2001) Wall-associated kinases are expressed throughout plant development and are required for cell expansion. *The Plant Cell*, **13**, 303–318.

Whetten, R.W. and Sederoff, R.R. (1992) Phenylalanine ammonia-lyase from loblolly pine. Purification of the enzyme and isolation of complementary DNA clones. *Plant Physiology*, **98**, 380–386.

Wightman R. and Turner S. R. (2010) Trafficking of the plant cellulose synthase complex. *Plant Physiology*, **153**, 427–432.

Willats, W.G.T., Knox, J.P. and Mikkelsen, J.D. (2006) Pectin: new insights into an old polymer are starting to gel. *Trends in Food Science and Technology*, **17**, 97–104.

Willats, W.G.T., Orfila, C., Limberg, G. *et al.* (2001b) Modulation of the degree and pattern of methyl-esterification of pectic homogalacturonan in plant cell walls. Implications for pectin methyl esterase action, matrix properties, and cell adhesion. *The Journal of Biological Chemistry*, **276**, 19404–19413.

Wilson, R.H., Smith, A.C., Kacurakova, M., Saunders, P.K., Wellner, N. and Waldron, K.W. (2000) The mechanical properties and molecular dynamics of plant cell wall polysaccharides studied by Fourier-transform infrared spectroscopy. *Plant Physiology*, **124**, 397–406,

Wong, M.M., Cannon, C.H. and Wickneswari, R. (2011) Identification of lignin genes and regulatory sequences involved in secondary cell wall formation in *Acacia auriculiformis* and *Acacia mangium* via *de novo* transcriptome sequencing. *BMC Genomics*, **12**, 342, http://doi.org/10.1186/1471-2164-12-342.

Yin, Y., Chen, H., Hahn, M.G., Mohnen, D. and Xu, Y. (2010) Evolution and function of the plant cell wall synthesis-related glycosyltransferase family 8. *Plant Physiology*, **153**, 1729–1746.

Zablackis, E., York, W.S., Pauly, M. *et al.* (1996). Substitution of L-fucose by L-galactose in cell walls of *Aarabidopsis mur1*. *Science*, **272**, 1808–1810.

Zhang, G.F. and Staehelin, L.A. (1992) Functional compartmentation of the Golgi apparatus of plant cells: immunocytochemical analysis of high-pressure and freeze-substituted sycamore maple suspension culture cells. *Plant Physiology*, **99**, 1070–1083.

Zhong, R. and Ye, Z.-H. (2009) Secondary cell walls. In: *Encyclopedia of Life Source*, John Wiley & Sons, Ltd, Chichester, UK.

Zhong, R., Burke, D.H., Morrison, W.H. and Ye, Z.-H. (2002) A kinesin-like protein is essential for oriented deposition of cellulose microfibrils and cell wall strength. *The Plant Cell*, **14**, 3101–3117.

Further reading

Grison, M.S., Brocard, L., Fouillen, L. *et al.* (2015) Specific membrane lipid composition is important for plasmodesmata function in Arabidopsis. *The Plant Cell*, doi: http://dx.doi.org/10.1105/tpc.114.135731.

Jackson, D. (2015) Plasmodesmata spread their influence. *F1000Prime Reports*, **7**, 25. doi: 10.12703/P7-25.

Jarvis, M.C. (2011) Plant cell walls: supramolecular assemblies. *Food Hydrocolloids*, **25**, 2, 257–262.

Kitagawa, M., Paultre, D. and Rademaker, H. (2015) Intercellular communication via plasmodesmata, EMBO Workshop on Intercellular Communication in Plant Development and Disease (2014) Bischoffsheim, France, *New Phytologist*, **205**, 970–972.

Tavares, E.Q.P. and Buckeridge, M.S. (2015) Do plant cell walls have a code? *Plant Science*, **241**, 286–294.

Vogler, H., Felekis, D., Nelson, B.J. and Grossniklaus, U. (2015) Measuring the mechanical properties of plant cell walls. *Plants*, **4**, 2, 167–182.

Xu, M., Cho, E., Burch-Smith, T.M. and Zambryski, P.C. (2012) Plasmodesmata formation and cell-to-cell transport are reduced in decreased size exclusion *limit* 1 during embryogenesis in Arabidopsis. *Proceedings of the National Academy of Sciences of USA*, **109**, 13.

CHAPTER 10

Plastid structure and genomics

Gurbachan S. Miglani

School of Agricultural Biotechnology, Punjab Agricultural University, Ludhiana, India

The first idea for the existence of extranuclear genetic factors came from non-Mendelian inheritance described by E. Baur in 1908 and C. Correns in 1909. They provided the first genetic evidence for the existence of cytoplasmic genes regulating chloroplast development. All genetic factors that lie outside the nucleus are called "extranuclear genetic factors" or "extranuclear genomes." Extranuclear genomes are present in mitochondria, chloroplasts, kinetoplasts, and centrioles. Some authors have erroneously been using the term "non-Mendelian genes" to describe cytoplasmic or organelle genes. These genes were better described by R. Sager in 1977 after the name of the organelle—for example, chloroplast genes and mitochondrial genes. The best insights into the behavior of chloroplast genes have come from an integrated approach combining genetic and molecular studies of chloroplast DNAs. Understanding of the genomic function is dependent upon the identification of individual chloroplast genes.

Plastid structure

Plastid is a general name for a cell organelle of plants that is enclosed by a double membrane and contains a series of internal membranes and vesicles. Plastids are best known as sites of photosynthesis, the process discussed in Chapter 11. In addition to this role, plastids manufacture and store a myriad of biomolecules, including carbohydrates, amino acids, fatty acids, plant hormones, and nucleotides (Buchanan *et al.*, 2000). Sulfur and nitrogen are assimilated into biologically available forms in plastids, and antioxidants such as ascorbic acid (vitamin C) and other secondary metabolites are also produced there. Plastids are believed to have originated from the engulfment of a photosynthetic bacterium by a eukaryotic cell that already contained mitochondria, which are the by-products of a separate endosymbiotic event. Plastids are thought to have originated from endosymbiotic cyanobacteria (Chan and Bhattachatrya, 2010). They developed around 1.5×10^9 years ago and allowed eukaryotes to carry out oxygenic photosynthesis. Due to a split into three evolutionary lineages, the plastids are named differently: "chloroplasts" in green algae and plants, "rhodoplasts" in red algae, and "cyanelles" in the glaucophytes. The plastids differ not only by their pigmentation, but also in ultrastructure. Many authors use terms "plastid" and "chloroplast" synonymously.

Different forms of plastids

Proplastids

In plants, plastids may differentiate into several forms, depending upon which function they need to play in the cell (Figure 10.1). All plastids are derived from proplastids, the precursors of high differentiated plastids (Lopez-Jurez and Pylse, 2005; Evert, 2006), which are present in the meristematic regions of the plant. Proplastids and young chloroplasts commonly divide by binary fission, but more mature chloroplasts also have this capacity.

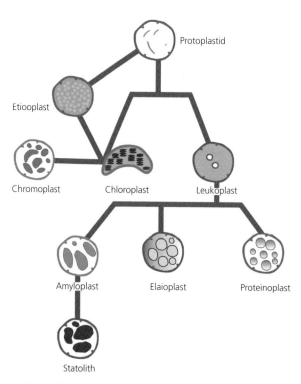

Figure 10.1 Several forms of plastids and their origin. Source: Redrawn from http://www.eplantscience.com/index/plastid.php.

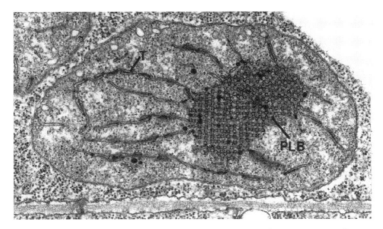

Figure 10.2 Thin section electron micrograph of a pea etioplast during a very early stage of light-induced conversion into a chloroplast. Thylakoid (T), prolamellar body (PLB). Source: Staehelin (2006).

Etioplasts

Etioplasts are chloroplasts that have not been exposed to light. They are usually found in flowering plants (angiosperms) grown in the dark. If a plant is kept out of light for several days, its normal chloroplasts will actually convert into etioplasts. Electron micrograph of a pea etioplast is presented in Figure 10.2. Mackender (1978) examined etioplasts in dark-grown leaves of *Zea mays* L. These etioplasts contained prolamellar bodies (PLBs) and lacked lamellae. PLBs are semicrystalline tubular structures. These structures have also been observed in Arabidopsis etioplasts (Sperling *et al.*, 1998). Upon exposure to light, etioplasts can be converted to chloroplasts. Etioplasts have been shown to contain monogalactosyl and digalactosyldiglycerides, phosphatidyl choline, phosphatidylglycerol, and phosphatidylinositol (Mackender, 1978). Proteome analysis of rice etioplasts yielded 240 proteins (von Zychlinski *et al.*, 2005). Certain of these protein appear to be involved in the "plastid gene expression machinery."

Undifferentiated plastids (etioplasts, proplastids) may develop into any of the plastids, namely, chromoplasts, chloroplasts, and leucoplasts (Wise and Hoober, 2006).

Chromoplasts

Chromoplasts are plastids, other than chloroplasts, that produce and store pigments. Found in flowers, leaves, roots, and ripe fruits, they contain carotenoids (lipid-soluble pigments ranging from yellow to red in color), which lend color to the plant tissues containing them. Chromoplasts are responsible for pigment synthesis and storage. A photomicrograph of chromoplast and fine structure of a chromoplast from the yellow part of fruit immediately after flowering are presented in Figure 10.3a and b, respectively.

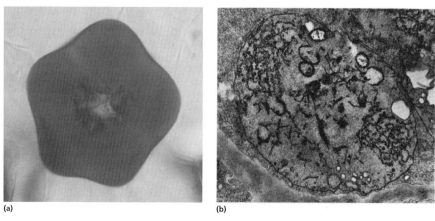

(a)   (b)

Figure 10.3 (a) Photomicrograph of chromoplasts. Source: http://www.gettyimages.in/detail/ photo/light-micrograph-of-a-section-of-begonia-high-res-stock-photography/128611402. (b) Chromoplast from the yellow part of the fruit immediately after flowering. Besides two tubular complexes (tc) localized peripherically, there are only few single tubules and hylakoids. Source: Ljubešic (1970).

In *Crocus sativus* L. Iridaceae chromoplasts a reticulo-tubular structure is apparent (Caiola and Canini, 2004). The reticular contains plastoglobules. In addition to the reticular type, there are glandular, membranous, and crystalline chromoplasts (Evert, 2006). The globular type chromoplast may contain carotenoid-containing plastoglobules. In contrast, the membranous type chromoplast houses double-carotenoid-containing membranes. The crystalline chromoplast possesses crystalline carotene (Evert, 2006). Chromoplasts accumulate elevated amounts of carotenoids (Howitt and Pogson, 2006) within plastoglobules. These pigments are derivatives of isoprene (Liu *et al.*, 2007; Breitmaier, 2006). Mettal *et al.* (1988) reported the enzymology of monoterpene synthesis. Tandem mass spectrometry coupled with database searches revealed 151 proteins in *Capsicum annum* L. chromoplasts. Among the most abundant proteins were capsanthin/capsorubin synthase and fibrillin (Siddique *et al.*, 2006). These proteins may be involved in carotenoid synthesis. Zeng *et al.* (2011) identified 493 proteins from purified *Citrus sinensis* L. asbeck chromoplasts. Certain of these proteins function in carbohydrate metabolism, amino acid and protein synthesis and secondary metabolism.

Chloroplasts
Chloroplast structure
Chloroplasts are organelles found in plant cells and eukaryotic algae. Photomicrograph of cells of an elodea leaf showing chloroplasts is presented in Figure 10.4a. Chloroplasts are similar to mitochondria but are found only in plants. In green plants, chloroplasts are surrounded by two lipid bilayer membranes, with an intermembrane space, now thought to correspond to the outer

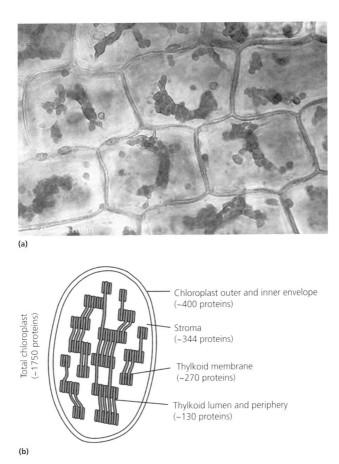

(a)

(b)

Chloroplast outer and inner envelope
(~400 proteins)

Stroma
(~344 proteins)

Thylkoid membrane
(~270 proteins)

Thylkoid lumen and periphery
(~130 proteins)

Total chloroplast
(~1750 proteins)

Figure 10.4 (a) Photomicrograph of cells of an elodea (*Elodea sp.*) leaf, showing chloroplasts. Courtesy: J. C. Revy. Source: http://dkphoto.photoshelter.com/gallery-image/Microscopic-Plant-Cells/ G00005l.MbuYRzEQ/I0000kf6l9zujQ68/C0000oyPxKwu0APU. (b) Compartments of chloroplasts and numbers of experimentally identified chloroplast proteins. Redrawn from Armbruster *et al.* (2011).

and inner membranes of the ancestral cyanobacterium. The genome is considerably reduced compared to that of free-living cyanobacteria, but the parts that are still present show clear similarities. The fluid within the chloroplast is called the "stroma," corresponding to the cytoplasm of the bacterium, and contains tiny circular DNA and ribosomes, though most of their proteins are synthesized by the cell nucleus. Within the stroma are stacks of thylakoids, the suborganelle where photosynthesis actually takes place. This stack of thylakoids is called a "granum." A thylakoid looks like a flattened disk (stacks of coins), and inside is an empty area called the "thylakoid space" or "lumen." The photosynthesis reaction takes place on the surface of the thylakoid. Chloroplasts contain chlorophyll and are responsible for the green coloration in most plants. Figure 10.4b shows compartments of chloroplasts and numbers of experimentally identified chloroplast proteins.

Function of chloroplasts

Chloroplasts convert light energy from the Sun into adenosine triphosphate (ATP) through a process called "photosynthesis" (see Chapter 11). The energy produced by chloroplasts can be used by cells. Thus chloroplasts are sites of photosynthesis. The photosynthetic proteins in the membrane bind chlorophyll, which is present with various accessory pigments, and which give chloroplasts their green color. Algal chloroplasts may be golden, brown, or red and show variation in the number of membranes and the presence of thylakoids. Efforts to identify the entire complement of chloroplast proteins and their interactions are progressing rapidly, making the organelle a prime target for systems biology research in plants (Armbruster et al., 2011).

Kleffmann et al. (2004) identified 690 different proteins from purified Arabidopsis chloroplasts. In the same year, Friso et al. (2004) reported the proteome analysis of "Arabidopsis" thylakoid membranes. They identified 154 proteins of which 76 were α-helical integral membrane proteins. Some of these were rubredoxins, a metallochaperone and a Dnal-like protein suggesting that these proteins function in thylakoid biogenesis. The chloroplast cytochrome system consists of the cytochrome b6/f complex (Hoober, 2006). This complex contains four different integral proteins.

Biogenesis of chloroplasts

Chloroplasts can originate by fission, that is, division into equal halves. In meristematic cells, division of proplastids occurs; in mature cells, division of mature plastids takes place (Evert, 2006). Plastid division is initiated by a constriction in the middle of the plastid. The constriction narrows between daughter plastids by an isthmus. Dynamins and dynamin-like proteins are mechanochemical proteins (large GTPases) which play role in functioning of transport vesicles in chloroplast and mitochondrial divisions (McFadden and Ralph, 2003). The flattened vesicles originate from the inner envelope membrane and develop into grana and stroma thylakoids. To identify the nuclear genes required for early chloroplast development, de la Luz Gutiérrez-Nava et al. (2004) generated a collection of Arabidopsis photosynthetic mutants.

Plastids are produced by division of existing plastids. In chloroplast division, the plastid-dividing (PD) ring is a main structure of the PD machinery and is a universal structure in the plant kingdom. By proteomic analysis of PD machineries isolated from Cyanidioschyzon merolae, Yoshida et al. (2010) identified the glycosyltransferase protein PD ring 1 (PDR1) which constructs the PD ring and is widely conserved from red alga to land plants. Electron microscopy showed that the PDR1 protein forms a ring with carbohydrates at the chloroplast division site. Fluorometric saccharide ingredient analysis of purified PD ring filaments showed that only glucose was included, and downregulation of PDR1-impaired chloroplast division. Thus, chloroplasts are divided by the PD ring, which is a bundle of PDR1-mediated polyglucan filaments.

Gerontoplasts
These plastids are derived from chloroplasts during leaf senescence (Krupinska, 2006; Biswal and Raval, 2012). The transition is characterized by extensive alteration of thylakoids.

Phycobilisomes
These are present in red algae and contain the phycobilin pigments: phyco-erythrin and phycocyanin. The phycobilisomes are arranged on thylakoids (Rosinski *et al.*, 1981), and they contain a light-harvesting antenna complex (Liu *et al.*, 2013).

Plastoglobules
Plastoglobules are lipoprotein particles inside chloroplasts. Their numbers have been shown to increase during the upregulation of plastid lipid metabolism in response to oxidative stress and during senescence. Austin *et al.* (2006) charac-terized the structure and spatial relationship of plastoglobules to thylakoid membranes in developing, mature, and senescing chloroplasts and demon-strated that plastoglobules are attached to thylakoids through a half-lipid bilayer that surrounds the globule contents and is continuous with the stroma-side leaflet of the thylakoid membrane. Plastoglobules function as both lipid biosynthesis and storage subcompartments of thylakoid membranes. The per-manent structural coupling between plastoglobules and thylakoid membranes, shown in Figure 10.5, suggests that the lipid molecules contained in the

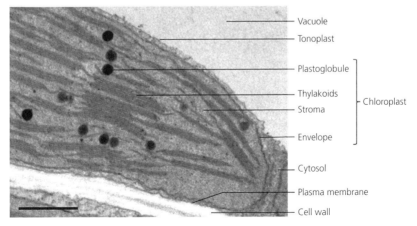

Figure 10.5 Electron micrograph of Arabidopsis chloroplast showing plastoglobules physically linked to thylakoids. Source: Nacir and Bréhélin (2013).

plastoglobule cores (carotenoids, plastoquinone, and tocopherol (vitamin E]) are in a dynamic equilibrium with those located in the thylakoid membranes (Bréhélin and Kessler, 2008).

Leucoplasts

Leucoplasts are the sites for monoterpene synthesis. Leucoplasts are food-storage bodies, which store large amounts of starch. Leucoplasts sometimes differentiate into more specialized plastids, namely, amyloplasts, elaioplasts, and proteinoplasts.

Amyloplasts

Amyloplasts are specialized leucocytes found in the roots of many plants and are used for starch storage and detecting gravity. Figure 10.6a shows a potato cell containing (amyloplasts). Ultrastructure of *Psilotum nudum* amyloplast containing starch grains (S) is given in Figure 10.6b.

These are types of leucoplasts that synthesize and store starch which is synthesized from glucose-6-phosphate (Wischmann *et al.*, 1999). Electron micrographs revealed that these plastids are encased in a membranous envelope and contain multiple starch grains. In some organisms, the amyloplasts contain a single starch grain. Starch is present as grains possessing a series of concentric rings. Amyloplasts occur mainly in tubers and seed endosperm. Amyloplasts function in gravitropism (a response to gravity). The amyloplasts contain either a single grain or multiple starch grains. The envelope houses a triose phosphate transporter which imports triose phosphate during starch synthesis. The amyloplasts also contain ADPG/C pyrophosphatase (AGPase) which utilizes glucose-1-phosphate and ATP during starch synthesis (Wischmann *et al.*, 1999). The potato tuber possesses starch synthases I and II, alpha glucan dikase, and enzymes for the assimilation of nitrogen and sulfur (Stensballe *et al.*, 2008).

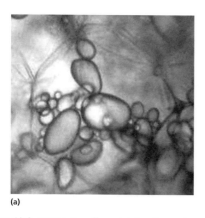 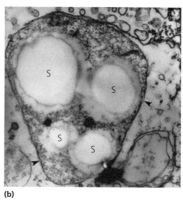

(a)   (b)

Figure 10.6 (a) Potato cell containing leucoplasts (amyloplasts). Courtsey: E.M. McCarthy. HYPERLINKhttp://www. macroevolution. net/leucoplasts.html. (b) Electron micrograph of *Psilotum nudum* leucoplast (amyloplast) showing plastid envelope (arrows) and starch (S). Source: http://users.humboldt.edu/ dkwalker/images/P05-Amyloplast(labels)t.jpg.

Some amyloplasts differentiate into **statoliths**. Statoliths are enmeshed in a web of actin, and it is thought that their sedimentation transmits the gravitropic signal by activating mechanosensing channels. The gravitropic signal then leads to reorientation of auxin efflux carriers and subsequent redistribution of auxin streams in root cap and root as a whole. The changed relations in concentration of auxin lead to differential growth of the root tissues. Taken together, the root then turns following the gravity stimuli.

Elaioplasts

Elaioplasts are another type of leucocytes used for storing fat and are found in seeds. They are nonpigmented and fall into the much broader organelle category of plant plastids. For ultrastructure of an elaioplast from *Brassica napus* tapetal tissue, see Figure 10.7 in which liquid droplets are shown. These are lipid-storing plastids containing steryl esters as droplets (Hsieh and Huang, 2005). Elaioplasts have been most studied in developing pollen. Helianthus elaioplasts are composed of numerous lipid bodies surrounded by endoplasmic reticulum, microtubules, a few mitochondria, Golgi bodies, and microbodies (Kwiatkowska *et al.*, 2010). *Arabidopsis thaliana* elaioplasts contain steryl esters and plastid-associated proteins (Hsieh and Huang, 2007). These investigators also reported the presence of Rubisco subunits. In *Brassica napus*, elaioplasts were composed of sterol esters and triacylglycerols as well as 34 and 36 kD proteins (Wu *et al.*, 1999).

Proteinoplasts

Proteinoplasts (sometimes called "proteoplasts," "aleuroplasts," or "aleuronaplasts") are also a specialized type of leucocytes which store and modify proteins within the plant cell. Figure 10.8 presents a photomicrograph of proteinoplast from *helleborus*

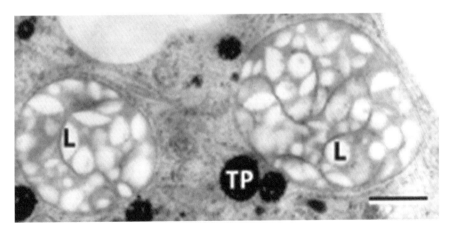

Figure 10.7 Elaioplasts from *Brassica napus* tapetal tissue. Numerous lipid droplets (L) appear clear. Several tapetosomes (Tp) are seen in the cytoplasm. Micrograph courtesy of Dr. Denis Murphy, University of Glamorgan, UK.

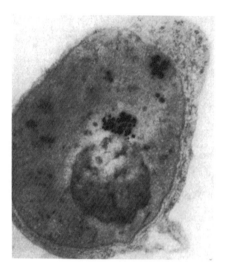

Figure 10.8 Ultrastructure of proteinoplast of *Helleborus corsicus*. Source: Salema and Badenhuizen (1969).

corsicus. These plastids consist of a three-dimensional network of fenestrated tubules encased in an envelope (Vigil and Ruddat, 1985). Proteinoplasts contain protein storage structures such as crystals. Earlier, Hurkman and Kennedy (1976) showed that proteinoplasts of developing primary leaves of mung bean contained large protein inclusions exhibiting granular matrix.

Plastid stromules

Stromules are thin projections from plastids that are generally longer and more abundant on nongreen plastids than on chloroplasts. Gunning (2005) studied through video microscopy outgrowth, retraction, tensioning, anchoring, branching, bridging, and tip-shedding aspects of plastid stromules. Figure 10.9 shows structure of a plastid stromule joining two plastids. Occasionally, stromules can be observed to connect two plastid bodies with one another. Stromules have role in trafficking of material from one plastid to another (Hanson and Sattarzadch, 2013).

Chlorophyll biosynthesis

In plants, chlorophyll may be synthesized from succinyl-CoA and glycine, although the immediate precursor to chlorophyll a and b is protochlorophyllide (for details, see Chapter 11). In Angiosperm plants, the last step, conversion of protochlorophyllide to chlorophyll, is light-dependent and such plants are pale (etiolated) if grown in the darkness. Nonvascular plants and green

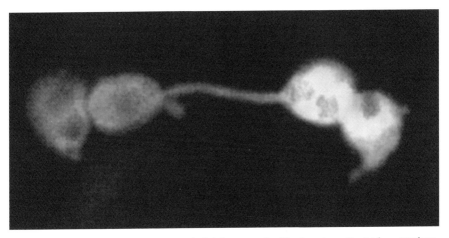

Figure 10.9 Micrograph showing stromule joining two plastids. Courtsey: Dr. Jaideep Mathur, University of Guelph.

algae have an additional light-independent enzyme and grow green in the darkness instead.

In the first phase of chlorophyll biosynthesis, the amino acid glutamic acid is converted to 5-aminolevulinic acid (ALA). This reaction is unusual in that it involves a covalent intermediate in which the glutamic acid is attached to a transfer RNA molecule. This is one of a very small number of examples in biochemistry in which a tRNA is utilized in a process other than protein synthesis. Two molecules of ALA are then condensed to form porphobilinogen (PBG), which ultimately form the pyrrole rings in chlorophyll. The next phase is the assembly of a porphyrin structure from four molecules of PBG. This phase consists of six distinct enzymatic steps, ending with the product protoporphyrin IX. All the biosynthesis steps up to this point are the same for the synthesis of both chlorophyll and heme. But here, the pathway branches, and the fate of the molecule depends on which metal is inserted into the center of the porphyrin. If magnesium is inserted by an enzyme called "magnesium chelatase," then the additional steps needed to convert the molecule into chlorophyll take place; if iron is inserted, the species ultimately becomes heme.

The next phase of the chlorophyll biosynthetic pathway is the formation of the fifth ring (ring E) by cyclization of one of the propionic acid side chains to form protochlorophyllide. The pathway involves the reduction of one of the double bonds in ring D, using nicotinamide adenine dinucleotide phosphate. This process is driven by light in angiosperms and is carried out by an enzyme called "protochlorophyllide oxidoreductase".

Despite versatile functions, plastid genomes encode fewer than 100 proteins (Sato *et al.*, 1999). Rather, the vast majority of plastid proteins are encoded by nuclear genes. It has been estimated that approximately 3000 proteins are transcribed in the nucleus, translated in the cytoplasm, and then imported into plastids. After endosymbiosis, most genes were either lost or transferred to the nucleus, where they acquired the regulatory sequences for expression, as well as sequences that encode a transit peptide for targeting proteins to the plastid where they are imported and cleaved (Kleine *et al.*, 2009). Once these proteins reach their proper compartment within the plastid, many are incorporated into multisubunit complexes that include components encoded in plastid genomes. Thus, signaling between plastids and the nucleus is required to maintain plastid biological functions. Signaling between chloroplasts and the nucleus is bidirectional. In a process called "anterograde regulation," the nucleus encodes regulators that convey information about cell type and expresses proteins that are appropriate for plastid functions within that particular cell type (Kleine *et al.*, 2009).

Plastid genomics

In 1963, R. Sager and M.R. Ishida described a procedure for isolating relatively intact chloroplasts from the alga *Chlamydomonas*. These chloroplasts contain about 3% of the cellular DNA, of which 25–40% consists of a satellite band with a buoyant density of 1.702 g/cm^2 and guanine–cytosine (GC) content of 39.3%. The enrichment of satellite DNA in the chloroplast extract sevenfold over its concentration in total cell extracts provides strong evidence for its association with the chloroplast *in vivo*.

Plastid DNA exists as large protein–DNA complexes associated with the inner envelope membrane and is called "plastid nucleoids." Each plastid creates 10–1000 copies of the circular 75–250 kilobase (kb) plastome. The proplastid contains a single nucleoid located in the center of the plastid. The number of genome copies per plastid is flexible, ranging from more than 1000 in rapidly dividing cells, which generally contain few plastids, to 100 or fewer in mature cells, where plastid divisions give rise to a large number of plastids. The plastome contains about 100 genes encoding ribosomal ribonucleic acids (rRNAs) and transfer ribonucleic acids (tRNAs) as well as proteins involved in photosynthesis and plastid gene transcription and translation. However, these proteins only represent a small fraction of the total protein necessary to build and maintain the structure and function of a particular type of plastid. Nuclear genes encode the vast majority of plastid proteins, and the expression of plastid genes and nuclear genes is tightly co-regulated to allow proper development of plastids in relation to cell differentiation. During the development of proplastids to chloroplasts, and when plastids convert from one type to another, nucleoids change in morphology, size, and location within the organelle. The remodeling of nucleoids is believed to occur by modifications of the composition and abundance of nucleoid proteins.

The chloroplast DNA differs from nuclear DNA of same species in nucleotide composition. Plastid DNA is shown to replicate through a semiconservative mode in *Chlamydomonas*. It higher plant species, chloroplast DNA replicates by the rolling circle method. There are no histones associated with the chloroplast DNA. The organization of the chloroplast genome of the *Marchantia polymorpha* is shown in Figure 10.10. The known genes and open reading frames (ORFs) are shown as boxes except for the small tRNA genes that are indicated by single lines. The tRNA genes that contain inrons are thus larger and are denoted here by triangles extending from the circular map. The existence of extranuclear genetic systems raises a question as to their need for such separate systems along with the one present in the nucleus. P. Borst and L.A. Grivell suggested in 1981 that it was simply an evolutionary dead end—a relic of the putative symbiotic bacterium.

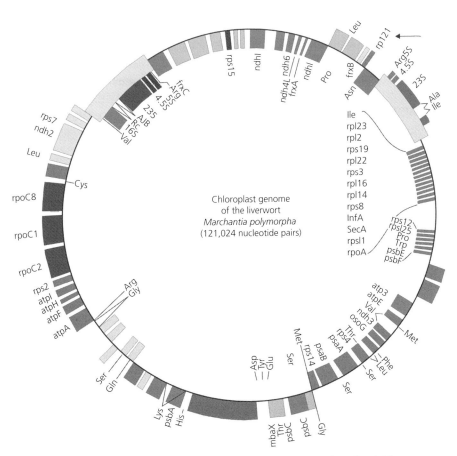

Figure 10.10 Organization of the genome of *Marchantia polymorpha* chloroplast DNA.
Source: Adapted with permission from Gardner *et al.* (2005).

Evidence that chloroplast DNA is maternally inherited

To demonstrate that a trait is maternally inherited, specific crosses need to be made to generate the required offspring. A number of crosses were made between cultivated tomato (*Lycopersicon esculentum*) and a number of wild species. Chloroplast DNA is 150 kb in size and a circular molecule. Digestion of this molecule produces relatively few fragments. Within the plant cell, chloroplast DNA represents a significant portion (15%) of the DNA. This is a result of the large number of chloroplasts per cell (50) and the large number of chloroplast DNA molecules per chloroplast (150). Chloroplast DNA was obtained from F_1 plants of a number of crosses in which *L. esculentum* was the female in the cross. In each case, the F_1 restriction fragment pattern was found to be identical to the female parent *L. esculentum* (Figure 10.11). This experiment by Palmer and Zamir (1982) provided conclusive evidence that chloroplast DNA is inherited in a maternal manner.

Paternal inheritance

In the case of *Plumbago zeylanica*, cytoplasmic genes are transferred to the egg through the male parent (Russel, 1980). The fertilized egg has organelles from the male as well as female parent. This type of inheritance is termed "biparental cytoplasmic inheritance." Male cytoplasmic organelles were identified in both the egg and the central cell immediately after entry of sperm nuclei. Sperm plastids are distinguished by their smaller size, elongated shape, inflated internal lamellae, and relatively electron-dense stromata. Egg plastids are larger, circular to elliptical in section, less electron-dense, and occasionally contain starch grains. Mitochondria of male origin can be distinguished from those of female gametophyte by their size and cellular location. Sperm mitochondria when compared to

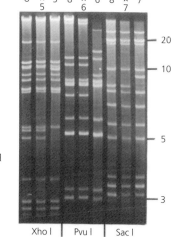

Figure 10.11 Evidence that chloroplast DNA is inherited in a maternal manner. Lanes 8 × 5, 8 × 6, and 8 × 7 indicate chloroplast DNA from F_1 hybrids for which sample 8 was maternal parent and samples 5, 6, and 7, respectively, were paternal parents. Source: Adapted with permission from Palmer and Zamir (1982).

sperm nuclei are circular, and female gametophyte mitochondria are elliptical. The presence of extranuclear sperm organelles in the cytoplasm of the egg and central cell supports the view that syngamy in angiosperms is initiated by cellular fusion. Both male organelles and nuclei may contribute to the development of the embryo and endosperm. This is referred to as "paternal inheritance." The genetic impact of male cytoplasmic inheritance during development on the mature plant depends on whether the male organelles remain viable, reproduce, and transmit genetic information within the embryo.

Sequenced plastomes

A plastome is the genome of a plastid, a type of organelle found in plants and in a variety of protoctists. The flowering plants are especially well represented in complete chloroplast genomes with more than 100 plastomes sequenced so far.

General features of plastomes

Chloroplasts contain unique DNA genomes that are transcribed within the organelle. Chloroplast genomes function in the biogenesis of chloroplasts. The chloroplast genes are mappable by recombination analysis. Almost all chloroplast DNAs carefully examined so far exist as a single, more or less homogeneous, size class of circular molecules (Figure 10.12). The existence and properties of the chloroplast genome were established by a combination of genetic methods, which identified chloroplast mutations and placed them into a linear sequence or map, and by chemical methods—cesium chloride (CsCl) density gradient ultracentrifugation and base analysis, which identified nonnuclear DNA extracted from isolated chloroplasts.

The studies carried out in the 1950s and 1960s primarily with *Chlamydomonas* laid the framework for distinguishing plastid and nuclear genomes. Based on this, the coding and regulatory functions of three genomes—nuclear, chloroplast, and mitochondrial—are being addressed in modern plant molecular biology. Chloroplast DNA also varies in size and molecular weight. The chloroplast DNA unlike mitochondrial DNA is always circular. Their size varies from 11 to 63 μm with a molecular weight from 22 to 1500 Md and number of base pairs ranges from 1.3×10^5 to 1.5×10^6. There are some striking features of the chloroplast DNA. All identified proteins encoded by chloroplast DNA are parts of protein complexes that have at least one component encoded by nuclear DNA. This may be a regulatory feature. There is clustering and co-transcription of functionally related genes, as in *Escherichia coli*. There are several examples of overlapping genes, a prokaryotic feature. Gene encoding for ribosomal protein S12 has exons located in different strands of DNA. This suggests the existence of trans-splicing mechanism in chloroplasts. Pribnow box—a relatively variable six-nucleotide sequence, TATAAT, located in prokaryotic promoters upstream (at position −10)

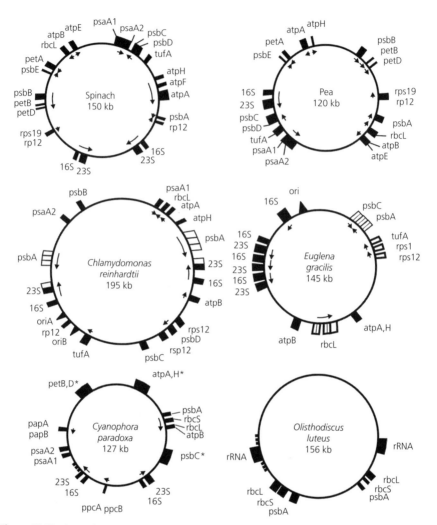

Figure 10.12 Physical and genetic maps of chloroplast DNAs representing six major lines of chloroplast evolution. Heavy lines centered on the circle indicate the extent of major repeat elements in the genomes. Filled boxes indicate location of exon sequences, while open boxes indicate the location of introns for all mapped genes encoding rRNAs and proteins. Transfer RNA genes are not shown. The arrows indicate the direction of transcription.
Source: Reproduced with permission of Palmer (1985).

from the start codon—is present in chloroplast genes. Upstream transcription termination sequences are also seen in chloroplast genes. Some of the species studied in detail for their chloroplast genomes are described later.

Comparative studies on wheat, rice, and maize plastomes
Structural features of the wheat plastome were clarified by comparison of the complete sequence of wheat chloroplast DNA with those of rice and maize

chloroplast genomes. The wheat plastome consists of a 134 545-bp circular molecule with 20 703-bp inverted repeats (IRs) and the same gene content as the rice and maize plastomes. Rice (*Oryza sativa* indica) plastome has 134 496 bp and maize (*Zea mays*) plastome has 140 384 bp. However, some structural divergence was found even in the coding regions of genes. These alterations are due to illegitimate recombination between two short direct repeats and/or replication slippage. The overall comparison of chloroplast DNAs among the three cereals indicated the presence of some hot-spot regions for length mutations. Whereas the region with clustered tRNA genes and that downstream of rbcL showed divergence in a species-specific manner, the deletion patterns of ORFs in the inverted repeat regions and the borders between the IRs and the small single-copy (SSC) region support the notion that wheat and rice are related more closely to each other when compared to maize.

Comparative genomics of liverworts and tobacco chloroplasts

In 1986, two Japanese laboratories have completely sequenced the chloroplast DNA of liverworts (*M. polymorpha*) (K. Ohyama and coworkers) and a higher plant *Nicotiana tobaccum* (K. Shinozaki and coworkers). The chloroplast DNAs of *M. polymorpha* and tobacco are 121 024 and 155 844 nucleotide pairs in length, respectively. The size, gene content, and organization of both these genomes are remarkably similar. Their genomes differ in "inverted repeat" regions only. There is no evidence of any abnormal use of genetic code, as found in bacteria. Therefore, chloroplast DNA seems to obey the "universal rules." Plastid DNA, like bacterial and mitochondrial DNA, is organized into protein–DNA complexes (called *nucleoids*). Jeong *et al.* (2003) have characterized the coiled-coil DNA-binding protein MFP1 (MAR binding filament-like protein 1, where MAR stands for matrix attachment regions [MARs]) as a protein associated with nucleoids and with the thylakoid membranes in mature chloroplasts. MFP1 is located in plastids in both suspension culture cells and leaves and is attached to the thylakoid membrane with its C-terminal DNA-binding domain oriented toward the stroma. It has a major DNA-binding activity in mature *Arabidopsis* chloroplasts and binds to all tested chloroplast DNA fragments without detectable sequence specificity. Its expression is tightly correlated with the accumulation of thylakoid membranes. Importantly, it is associated *in vivo* with nucleoids, suggesting a function of MPF1 at the interface between chloroplast nucleoids and the developing thylakoid membrane system.

Chloroplast genome of *Welwitschia mirabilis*

The chloroplast genome of *Welwitschia mirabilis* comprises 119 726 bp and exhibits large single-copy (LSC) and SSC regions and two copies of the large IR (McCoy *et al.*, 2008). Only 101 unique gene species are encoded. The *Welwitschia* plastome is the most compact photosynthetic land plant plastome sequenced to date: 66% of the sequence codes for product. The genome also exhibits a slightly expanded IR, a minimum of nine inversions that modify gene order, and 19 genes that are lost or present as pseudogenes.

Chloroplast genome of *Lolium arundinaceum*

Cahoon *et al.* (2004) described the complete chloroplast genome of tall fescue, *Lolium arundinaceum*. The fescue plastome is 136 048 bp with a typical quadripartite structure and a gene order similar to other grasses; 56% of the plastome is coding region comprising 75 protein-coding genes, 29 tRNAs, 4 rRNAs, and 1 hypothetical coding region (*ycf*).

Chloroplast genome of *Coffea arabica*

The chloroplast genome sequence of coffee, *Coffea arabica* L., the first sequenced member of the fourth largest family of angiosperms, *Rubiaceae*, was reported by Samson *et al.* (2007). The genome is 155 189 bp in length, including a pair of IRs of 25 943 bp. Of the 130 genes present, 112 are distinct and 18 are duplicated in the IR. The coding region comprises 79 protein genes, 29 transfer RNA genes, 4 ribosomal RNA genes, and 18 genes containing introns (3 with three exons). Repeat analysis revealed five direct and three IRs of 30 bp or longer with a sequence identity of 90% or more.

Chloroplast genome of *Phoenix dactylifera*

Yang *et al.* (2010) reported a complete sequence of the date palm (*Phoenix dactylifera* L.) chloroplast genome based on pyrosequencing. The date palm chloroplast genome is 158 462 bp in length and has a typical quadripartite structure of the LSC (86 198 bp) and SSC (17 712 bp) regions separated by a pair of IRs (27 276 bp). Similar to what has been found among most angiosperms, the date palm chloroplast genome harbors 112 unique genes and 19 duplicated fragments in the IR regions. The junctions between LSC/IRs and SSC/IRs show different features of sequence expansion in evolution. They also identified 78 single-nucleotide polymorphisms (SNPs) as major intravarietal polymorphisms within the population of a specific chloroplast genome, most of which were located in genes with vital functions. Based on RNA-sequencing data, they also found 18 polycistronic transcription units and 3 highly expression-biased genes—*atpF*, *trnA-UGC*, and *rrn23*. Unlike most monocots, date palm has a typical chloroplast genome similar to that of tobacco—with little rearrangement and gene loss or gain.

Chloroplast genome of *Heterosigma akashiwo*

Heterosigma akashiwo strain CCMP452 (West Atlantic) chloroplast DNA is 160 149 bp in size with a 21 822-bp IR, whereas NIES293 (West Pacific) chloroplast DNA is 159 370 bp in size and has an IR of 21 665 bp (Cattolico *et al.*, 2008). Both strains contain multiple small inverted and tandem repeats, nonrandomly distributed within the genomes. Although both CCMP452 and NIES293 chloroplast DNAs contain 197 genes, multiple nucleotide polymorphisms are present in both coding and intergenic regions. Several protein-coding genes contain large, in-frame inserts relative to orthologous genes in other plastids. These inserts are maintained in mRNA products. Two genes of interest in *H. akashiwo*, not previously reported in any chloroplast genome, include *tyr*C, a tyrosine recombinase, which they hypothesized might be a result of a lateral gene transfer event, and an unidentified 456 amino acid protein, which they hypothesized serves as a G-protein-coupled receptor. The *H. akashiwo* chloroplast genomes share little synteny—co-localization of genes on chromosomes of different species—with other algal chloroplast genomes sequenced to date.

Chloroplast genome of *Glycine max*

Saski *et al.* (2005) sequenced the soybean chloroplast genome and compared it with the other completely sequenced legumes: *Lotus* and *Medicago*. The chloroplast genome of *Glycine* is 152 218 bp in length, including a pair of IRs of 25 574 bp of identical sequence separated by an SSC region of 17 895 bp and an LSC region of 83 175 bp. The genome contains 111 unique genes, and 19 of these are duplicated in the IR. Comparisons of *Glycine*, *Lotus,* and *Medicago* confirm the organization of legume chloroplast genomes. Gene content of the three legumes is nearly identical. The *rpl22* gene is missing from all three legumes, and Medicago is missing *rps16* and one copy of the IR. Gene order in *Glycine*, *Lotus*, and *Medicago* differs from the usual gene order for angiosperm chloroplast genomes by the presence of a single, large inversion of 51 kb. Detailed analyses of repeated sequences indicate that many of the *Glycine* repeats that are located in the intergenic spacer regions and introns occur in the same location in the other legumes and in *Arabidopsis*, suggesting that they may play some functional role. The presence of small repeats of psbA and rbcL in legumes that have lost one copy of the IR indicates that this loss has only occurred once during the evolutionary history of legumes.

Chloroplast genome of *Chara vulgaris*

Turmel *et al.* (2009) have determined the complete chloroplast genome sequence (184 933 bp) of a representative of the Charales, *Chara vulgaris*, and compared this genome to those of *Mesostigma* (Mesostigmatales), *Chlorokybus* (Chlorokybales), *Staurastrum,* and *Zygnema* (Zygnematales), *Chaetosphaeridium* (Coleochaetales), and selected land plants. The phylogenies they inferred from 76 chloroplast DNA-encoded proteins and genes using various methods favor the hypothesis that the Charales diverged before the Coleochaetales and Zygnematales.

Chloroplast genome of *Lolium perenne* L.

Diekmann *et al.* (2009, 2010) sequenced, assembled, and annotated the entire chloroplast genome of *Lolium perenne* "Cashel," and searched it for SNPs and chloroplastic simple sequence repeats (cpSSRs). The chloroplast genome sequence of *L. perenne* is 135 282 bp long with a typical quadripartite structure and encodes 128 genes. It contains genes for 76 unique proteins, 30 tRNAs, and 4 rRNAs. As in other grasses, the genes *accD*, *ycf1*, and *ycf2* are absent. Forty SNPs were found within the sequenced sample of "Cashel." Thirty mononucleotide cpSSRs of more than 10-bp length were detected. Genome size differences are mainly due to length variations in noncoding regions. However, considerable length differences of 1–27 codons in comparison of *L. perenne* to other Poaceae and 1–68 codons among all Poaceae were also detected. Within the chloroplast genome of this outcrossing cultivar, 10 insertion/deletion polymorphisms and 40 SNPs were detected. Two of the polymorphisms involve tiny inversions within putative hairpin structures. By comparing the genome sequence with real-time polymerase chain reaction (RT–PCR) products of transcripts for 33 genes, 31 mRNA editing sites were identified, five of them unique to *Lolium*.

Chloroplast genome of *Alsophila spinulosa*

Gao *et al.* (2009) have sequenced the complete chloroplast genome of a scaly tree fern *Alsophila spinulosa* (Cyatheaceae). The *Alsophila* chloroplast genome is 156 661 bp in size, and has a typical quadripartite structure with LSC (86 308 bp) and SSC (21 623 bp) regions separated by two copies of IRs (24 365 bp each). This genome contains 117 different genes encoding 85 proteins, 4 rRNAs, and 28 tRNAs. Pseudogenes of ycf66 and trnT-UGU are also detected in this genome. A unique trnR-UCG gene (derived from trnR-CCG) is found between rbcL and accD.

Chloroplast genome of *Boea hygrometrica*

Zhang *et al.* (2012) determined the complete nucleotide sequences of the chloroplast genome of resurrection plant *Boea hygrometrica* (*Bh*) with the length of 153 493 bp. The chloroplast genome contains 147 genes with a 72% coding sequence. Similar to other seed plants, the *Bh* chloroplast genome has a typical quadripartite organization with a conserved gene in each region. Compared to other angiosperms, one remarkable feature observed of the *Bh* mitochondrial genome is the frequent transfer of genetic material from the chloroplast genome during recent *Bh* evolution. The chloroplast-derived sequences including tRNAs found in angiosperm mitochondrial genomes support the conclusion that frequent gene transfer events may have begun early in the land plant lineage.

Promiscuous DNA

Promiscuous DNA was defined by J. Ellis in 1982–1983 as a nucleotide sequence that occurs in more than one of the three membrane-bound organellar genetic systems of eukaryotic cells and should be distinguished from a wide range of prokaryotic cells. The phenomenon of this DNA transfer is known as intracellular promiscuity, and it diminishes the separate identity of each genome.

Various examples clearly demonstrate that the movement of DNA between the various genomes has occurred. In 1983, J.N. Timmis and N.S. Scott found that the nuclear DNA of spinach contains integrated sequences homologous with most of the chloroplast genome. Several lines of evidences have indicated that relocation of some organelle genes could have been more recent. In 1990, S.L. Baldauf and J.D. Palmer presented gene sequence and molecular phylogenetic evidence for the transfer of chloroplast *tufA* gene to the nucleus in the green algal ancestor of land plants. The *tufA* gene encodes chloroplast protein synthesis elongation factor Tu (EF-Tu). A 12-kilobase DNA sequence has been identified by D.B. Stern and D.M. Lonsdale in 1982 in the maize mitochondrial genome, which is homologous to part of the IR of the maize chloroplast genome. In chloroplasts, the sequence contains a 16S rRNA gene, and also the coding sequences for tRNA^{Ile} and tRNA^{Val}.

The information that has been obtained from complete sequences of the chloroplast genomes is helping us understand not only how chloroplasts function within present-day plants, but may also yield insights into the evolutionary relationships of photosynthetic organisms and into the gene movement that has occurred among the various genetic compartments of eukaryotic cells. Chloroplast and mitochondrial sequences have been found in the nucleus of plant cells. Further, chloroplast sequences have been observed in the mitochondria of plant cells. This is clear evidence that genetic control of certain biochemical functions was relinquished (or taken from) the progenitor cells. Figure 10.13 shows the flow of genetic information from organelles in a plant cell. Chloroplast and mitochondrial DNA before invading a eukaryotic cell were free-living primitive prokaryotes. In due course of time, these entrapped organisms lost their independence by losing genetic information to nuclear genome. This promiscuity of DNA gave further support to the endosymbiont hypothesis given by L. Margulis in 1970. There are two possible methods of transfer of promiscuous DNA. First, some kind of vector or transposon or transducing phage may be involved. Second, two organelles might have undergone fusion at some stage. The second mechanism is preferred.

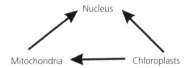

Figure 10.13 The flow of genetic information from organelles in a plant cell. Source: Adapted with permission from www.ndsu.edu/.../plsc431/maternal/maternal3.htm. © 1997 Phillip McClean.

Plastid genome organization

Most plastid genomes are circular and range in size from 120 to 217 kbp. Variation in genome size is related primarily to the size of an IR in the plastid genome and limited differences in coding capacity. Plastids contain from 22 to over 300 genome copies, depending on the developmental stage. The genome is membrane associated, complexed with protein, and resembles bacterial nucleoids. Plastid DNA encodes approximately 135 genes that fall into three major functional categories: (i) genes encoding proteins and RNAs involved in transcription and translation of the plastid genome (RNA polymerase subunits, tRNAs, rRNAs, ribosomal proteins, initiation factor 1); (ii) genes encoding proteins of the photosynthetic apparatus (ribulose-1,5-bisphosphate carboxylase/oxygenase or RuBisCo, PSI, PSII, ATP synthase, Cyt complex); and (iii) genes encoding proteins of the NADH oxido-reductase complex. Whereas plastids function in nonphotosynthetic tissues, the plastid genome is clearly specialized for the production of the photosynthetic apparatus. Most plastid genes are organized in complex operons that are conserved among plastid genomes.

Chloroplast genome differences with respect to introns

The most common structural differences found among individual chloroplast genes involve introns. *Euglena gracilis* contains at least 50 introns (accounting for over 20% of the genome). There are perhaps as few as six introns in broad bean. Most *Euglena* protein-encoding genes contain multiple introns, while with only three known exceptions, protein-encoding genes in angiosperms lack introns. Conversely, tRNA genes lack introns in *Euglena* but contain large introns (451–949 bp) in angiosperms (Figure 10.14). *Chlamydomonas reinhardtii* is somewhat intermediate in intron content, containing the only chloroplast tRNA intron yet identified and four large introns in a single one (*psbA*) of several protein-encoding genes examined. Intron differences (optional introns) have also been found among more closely related taxa. For example, spinach *rpl2* lacks a large intron found in the tobacco gene, while the 1.1 kb intron 3 of *psbA* from *C. reinhardtii* is completely absent from *psbA* of the interfertile species *Caryophyllia smithii*.

Extreme reconfiguration of plastid genomes

Geraniaceae plastid genomes (plastomes) have experienced a remarkable number of genomic changes. The plastomes of *Erodium texanum*, *Geranium palmatum*, and *Monsonia speciosa* were sequenced and compared with other rosids and the previously published *Pelargonium hortorum* plastome (Guisinger *et al.*, 2011). Geraniaceae plastomes were found to be highly variable in size, gene content and order, repetitive DNA, and codon usage. Several unique plastome rearrangements include the disruption of two highly conserved operons (*S10* and *rps2-atpA*), and

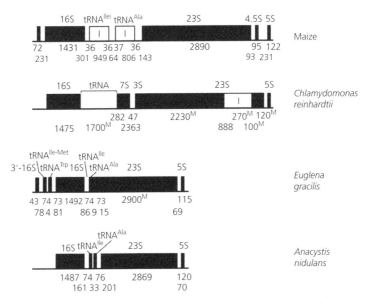

Figure 10.14 Ribosomal RNA transcription units of chloroplasts and cyanobacteria. Filled boxes indicate exons, while open boxes indicate introns. Numbers immediately below the maps indicate sizes of the coding regions, while numbers next below indicate sizes of spacers and introns. Source: Reproduced with permission of Palmer (1985).

the IR region in *M. speciosa* does not contain all genes in the ribosomal RNA operon. The sequence of *M. speciosa* is unusually small (128 787 bp); among angiosperm plastomes sequenced to date, only those of nonphotosynthetic species and those that have lost one IR copy are smaller. In contrast, the plastome of *P. hortorum* is the largest, at 217 942 bp. These genomes have experienced numerous gene and intron losses and partial and complete gene duplications. Some of the losses are shared throughout the family (e.g., *trnT-GGU* and the introns of *rps16* and *rpl16*); however, other losses are homoplasious (similar due to convergent evolution) (e.g., *trnG-UCC* intron in *G. palmatum* and *M. speciosa*). IR length is also highly variable. The IR in *P. hortorum* was previously shown to be greatly expanded to 76 kb, and the IR is lost in *E. texanum* and reduced in *G. palmatum* (11 kb) and *M. speciosa* (7 kb). Geraniaceae plastomes contain a high frequency of large repeats (>100 bp) relative to other rosids. Within each plastome, repeats are often located at rearrangement end points and many repeats shared among the four Geraniaceae flank rearrangement end points. GC content is elevated in the genomes and also in coding regions relative to other rosids. Codon usage per amino acid and GC content at third position sites are significantly different for Geraniaceae protein-coding sequences relative to other rosids. It has been suggested that relaxed selection and/or mutational biases lead to increased GC content, and this in turn alters codon usage. Increases in genomic rearrangements, repetitive DNA, nucleotide substitutions, and GC content may be caused by relaxed selection resulting from improper DNA repair.

Repeated sequences in chloroplast genomes
Size variation in IRs

All angiosperm and land plant chloroplast DNAs are between 120 and 160 kbp in size. However, there are three known exceptions found among angiosperms— *N. accuminata* (genome size 171 kbp), duckweed (*Spirodela oligorhiza*) (genome size (180 kbp), and geranium (*Pelargonium horiorum*) (genome size 217 kbp). Almost two-thirds of 97-kbp size variation found among angiosperm chloroplast DNAs can be accounted for by changes in the size of the large, rRNA-encoding IR. A comparison of gene order and IR size in geranium and spinach cpDNAs has been made in Figure 10.15. At one extreme, geranium chloroplast DNA genome posses a greatly enlarged IR of 76 kbp almost three times the size of any other angiosperm IR. This increase is the result of spreading of the IR into both the SSC and LSC regions, producing duplicate genes for *rbc*L, *pet*A, *psb*B, *pet*B, and *pet*D, which are single copy in all other angiosperms. Simple spreading for inversion repeat cannot explain all the differences in gene order and orientation present in geranium; several inversions have to be postulated. In addition, some other rearrangements (deletions, duplications, inversions, etc.) and mutations are also responsible for these differences.

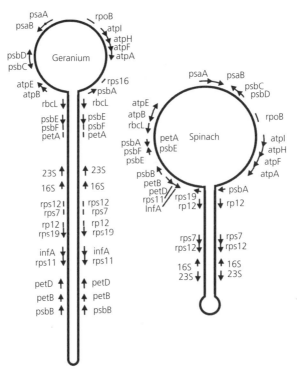

Figure 10.15 Comparison of gene order and inverted repeat size in geranium and spinach cpDNAs. Source: Reproduced with permission of Palmer et al. (1987).

Nature of repeated sequences

Chloroplast genomes generally have few repeated sequences that often dominate the landscape of the genome and are often associated with recombinational and evolutionary properties. Aside from short repeated sequences of less than 100 bp, which have been described in a number of sequencing studies, only five repeat families have been found in chloroplast genomes. One of these is organized as a two-copy IR, two as tandem repeats, and two as dispersed repeats. The most widespread of these is a large (10–76 kb) inverted duplication found in chloroplast DNAs from almost all land plants and from several major lineages of algae. Large IRs in the chloroplast DNAs of all land plants and algae contain a complete set of rRNA genes. Figure 10.16 shows through physical maps the arrangement of homologous sequences in the 22-kb inversion region of the lettuce and either *Barnadesia* or *Vernonia* chloroplast genomes. Aside from the presence of rRNA genes, there are no features common to the IRs. Large differences are apparent in positioning of the repeat segments in *Chlamydomonas, Cyanophora,* and *Olisthodiscus* and in the gene content of the repeat. There are overall differences in the organization of these genomes, and the IRs have the ability to spread and shrink in size. Maybe these species are of single common origin but are highly altered in present structure owing to subsequent rearrangements. Separate endosymbiotic events are an alternate possibility.

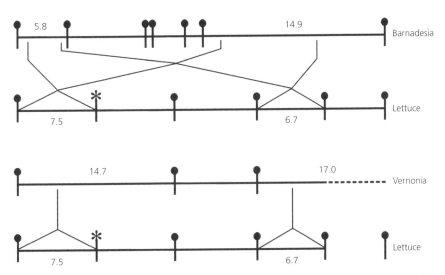

Figure 10.16 Physical maps showing the arrangement of homologous sequences in the 22-kb inversion region of the lettuce and either *Barnadesia* or *Vernonia* chloroplast genomes. The *Barnadesia* and *Vernonia Sac* I fragments in which the 7.5-kb *Sac* I-*Sal* I and 6.7-kb *Sac* I probes hybridize are indicated by lines leading from lettuce fragments: the *Barnadesia* and *Vernonia* fragments. Numbers indicate fragment sizes in kilobases. Restriction sites are shown: ●, *Sac* I; *, *Sal* I. Source: Reproduced with permission of Jansen and Palmer (1987).

Small single-copy regions

In 1988, T. Kohchi and coworkers characterized the genes in the regions of large IRs (IR$_A$ and IR$_B$, 10058-bp each) and an SSC (19813 bp) of chloroplast DNA from *M. polymorpha*. The IR regions contain genes for four ribosomal RNAs (16S, 23S, 4.5S, and 5S rRNAs) and five transfer RNAs (valine tRNA^{GAC}, isoleucine tRNA^{GAU}, alanine tRNA^{UGC}, arginine tRNA^{ACG}, and asparagine tRNA^{GUU}). The gene organization of the IR regions in the liverwort chloroplast genome is conserved, although the IR regions are smaller (10058 bp) than any reported in higher plant chloroplasts. The SSC (19813 bp) region encoded genes for 17 ORFs, a leucine tRNA^{UAG}, and a proline tRNA^{GGG}-like sequence. They identified 12 ORFs by homology of their coding sequences to a 4Fe-4S-type ferredoxin protein, a bacterial nitrogenase reductase component (Fe-protein), five human mitochondrial components of NADH dehydrogenase (ND1, ND4, ND4L, ND5, and ND6), two *E. coli* ribosomal proteins (S15 and L21), two putative proteins encoded in the kinetoplast maxicircle DNA of *Leishmania tarentolae* (LtORF 3 and LtORF 4), and a bacterial permease inner membrane component (encoded by *malF* in *E. coli* or *hisQ* in *Salmonella typhimurium*).

Tandem repeats and large insertion/deletions

Plants in the family Solanaceae are used as model systems in comparative and evolutionary genomics. The complete chloroplast genomes of seven solanaceous species have been sequenced. Jo *et al.* (2010) analyzed the complete chloroplast genome sequence of the hot pepper, *Capsicum annuum*. The pepper chloroplast genome was 156 781 bp in length, including a pair of IRs of 25 783 bp. The content and the order of 133 genes in the pepper chloroplast genome were identical to those of other solanaceous plastomes. To characterize pepper plastome sequence, they performed comparative analysis using complete plastome sequences of pepper and seven solanaceous plastomes. Frequency and contents of large indels and tandem repeat sequences and distribution pattern of genome-wide sequence variations were investigated. In addition, a phylogenetic analysis using concatenated alignments of coding sequences was performed to determine evolutionary position of pepper in Solanaceae. These results revealed two distinct features of pepper plastome compared to other solanaceous plastomes. First, large indels, including insertions on *accD* and *rpl20* gene sequences, were predominantly detected in the pepper plastome compared to other solanaceous plastomes. Second, tandem repeat sequences were particularly frequent in the pepper plastome.

Partially overlapping genes in chloroplasts

The chloroplast NAD(P)H dehydrogenase (NDH) C (*ndhC*) and *ndhK* genes partially overlap and are co-transcribed in many plants. The tobacco *ndhC/K* genes are translationally coupled but produce NdhC and NdhK, subunits of the NDH complex, in similar amounts. Generally, translation of the downstream cistron in overlapping mRNAs is very low. It has been suggested that the *ndhK* cistron is translated not only from the *ndhC* 5′-untranslated region (UTR) but also by an additional pathway. Using an *in vitro* translation system from tobacco chloroplasts, Yukawa and Sugiura (2013) observe that free ribosomes enter, with formylmethionyl-tRNA$_f^{Met}$, at an internal AUG start codon that is located in frame in the middle of the upstream *ndhC* cistron; translate the 3′ half of the *ndhC* cistron; reach the *ndhK* start codon; and that, at that point, some ribosomes resume *ndhK* translation. They detected a peptide corresponding to a 57-amino-acid product encoded by the sequence from the internal AUG to the *ndhC* stop codon. It seems that the internal initiation site AUG is not designed for synthesizing a functional isoform but for delivering additional ribosomes to the *ndhK* cistron to produce NdhK in the amount required for the assembly of the NDH complex. This pathway is a unique type of translation to produce protein in the needed amount with the cost of peptide synthesis.

Organization and molecular evolution of organelle genomes from *Marchantia polymorpha*

The complete nucleotide sequence of chloroplast DNA from liverwort *M. polymorpha* has made clear the entire gene organization of the chloroplast genome. Quite a few genes encoding components of photosynthesis and protein synthesis machinery have been identified by comparative computer analysis. The liverwort chloroplast DNA contains 121 024 bp, consisting of a set of large IRs (IR$_A$ and IR$_B$, each of 10 058 bp) separated by an SSC (19 813 bp) region and an LSC (81 095 bp) region. They detected 128 possible genes throughout the liverwort chloroplast genome, including coding sequences for four kinds of ribosomal RNAs, 32 species of transfer RNAs, and 55 identified ORFs for proteins, which are separated by short A + T-rich spacers. A total of 20 genes (8 encoding tRNAs and 12 encoding proteins) contain introns in their coding sequences. These introns can be classified as belonging to either group I or group II. Interestingly, seven of the identified ORFs show high homology to unidentified reading frames (URFs) found in human mitochondria. The complete nucleotide sequence from chloroplast DNA comprises 20 introns (19 group II and 1 group I) in 18 different genes. One of the chloroplast group II introns separated the ribosomal protein gene in the trans-position. The mitochondrial genome also has 32 introns (25 group II and 7 group I) in the coding regions of 17 genes.

Plastid gene organization, expression, and regulation

The chloroplast DNA contains essential genes for its maintenance and operation. Several components of the photosystems and proteins involved in biosynthetic pathways are also encoded by the chloroplast genome. Exploring the genetic repository of this organelle is vital due to its conserved nature, small size, persistent gene organization, and promising ability for transgenic expression. Therefore, chloroplast DNA sequence information has been instrumental in phylogenetic studies and molecular taxonomy of plants. Chloroplast genome sequencing efforts have been initiated with conventional cloning and chain-termination sequencing technologies.

The gene order among all higher plant chloroplast genomes is essentially conserved. Many of chloroplast genes encode proteins that are involved in photosynthesis. In total, the genome appears to encode for a complete set of rRNA genes (16S-spacer-23S-spacer-5S), tRNA genes (25–45), and 45 protein products, including larger subunit of ribulose-1,5-bisphosphate (RuBP) carboxylase; thylakoid membrane protein; adenosine triphosphate (ATP) synthase; cytochrome b oxidase; cytochrome C oxidase I, II, and III; ATPase-6; NADH dehydrogenase; RNA polymerase; ribosomal protein genes; ferredoxin; etc.

The organization of the single-copy genes in the chloroplast genomes of liverworts and tobacco is remarkably similar considering that they are evolutionarily very distant from each other. The major difference between these two chloroplast DNAs is that the inverse repeat region containing the rRNA genes are considerably larger in tobacco. The best estimates of chloroplast DNA gene number are 136 in *Marchantia* and 150 in tobacco.

Co-transcription linkage of chloroplast genes

The plastid operons show that genes with common functions are often co-localized in operons. For example, the *rpoB–rpoC1–rpoC2* operon encodes subunits of the plastid-encoded RNA polymerase, whereas the *psbI–psbK–psbD–psbC* operon encodes proteins of PSII. Some operons contain genes involved only in transcription (rpoB operon), translation (rrn operon), or transcription and translation ($23 operon), but not photosynthesis. The localization of genes encoding proteins of a single protein complex in operons facilitates coordinated and stoichiometric accumulation of subunits. This organization also offers the opportunity to differentially regulate the transcription of some genes involved in transcription/translation independently of genes for photosynthesis. Other operons contain mixtures of genes encoding different functions. For example, the gene *rpsl4* coding for a ribosomal protein is co-transcribed with *psaA–psaB* (PSI subunits), *rpoA* gene coding for RNA polymerase subunit with genes for ribosomal proteins, and *psbB/psbH* (PSII subunits) with *petB/petD* (Cyt complex), and tRNAs are included in several operons. Expression of specific genes within mixed function operons involves multiple levels of regulation including complex promoter/terminator combinations, selective RNA processing, differential RNA stability, regulated translation, and protein turnover. Not surprisingly, complex RNA populations are associated with most plastid operons. In some cases, the RNA complexity is due to the action of multiple promoters that allow differential transcription of selected genes within the same operon. For example, at least 12 different RNAs are produced from the barley *psbI–psbK–psbD–psbC–orf6–tmG* operon through the action of four different promoters and several RNA cleavage events.

Several sets of two or three genes are closely linked in two or more chloroplast genomes. Often, these sets of genes are known to be co-transcribed in at least one of the chloroplast genomes, as well as in the putative cyanobacterial ancestors of chloroplasts. An example of this type of co-transcriptional linkage is the ribosomal RNA operon, which has the same basic structure and transcriptional order (*16S–tRNA^{Ile}–tRNA^{Ala}–23S*) in all examined chloroplast genomes and in the cyanobacterium *Anacystis nidulans*). A point to be noted here is that this transcriptional linkage remains unaffected by the variable presence of large introns in the spacer tRNA genes and 23S gene, by splitting of small RNA species from the 5′-end (7S and 3S rRNA in *Chlamydomonas*) and 3′-end (4.5S rRNA in angiosperms) of the 23S rRNA gene, and by the duplicational insertion of part of the middle of the operon into the 16S leader region in *Euglena*. There are many other cases of co-transcriptional linkage in chloroplasts and cyanobacteria.

Changing perspectives of plastid transcription

Early studies of chloroplast genes revealed −10 (TATAA) and −35 (GTGACA) transcription promoter elements and putative ribosome binding sites (GGAGG) that resembled prokaryotic transcription and translation elements. These sequence elements were consistent with the plastid prokaryotic-like RNA polymerase and 70S ribosomes. Sequence homology between plastid and bacterial genes provides additional evidence of common origin. Even more striking and important relative to gene regulation is the presence of conserved plastid operons, often having the same gene order as their counterparts in *E. coli*. Although some features of chloroplast gene expression, such as RNA splicing, are less commonly associated with bacterial systems, it was generally expected that regulation of chloroplast gene expression would follow bacterial paradigms. Thus, when differential expression of *rbcL* was observed in mesophyll versus bundle sheath cells of maize and selective accumulation of *psbA* mRNA was documented in illuminated maize leaves, differential transcription was suggested as the likely mechanism involved. Determination of relative mRNA levels and transcription rates during light-induced chloroplast development in barley showed that changes in *psaA–psaB* and *afpB* mRNA levels were paralleled by changes in transcription rate. However, for other genes such as *psbA*, changes in transcription and mRNA levels were not coupled, suggesting that mRNA stability contributed significantly to the determination of *psbA* mRNA levels.

Regulation of chloroplast transcription

Plastids are ubiquitous DNA-containing organelles present in plant cells. These organelles carry out numerous metabolic functions some of which are highly expressed in differentiated plastids. Although basal levels of plastid gene expression are observed in all plastid types, a large increase in plastid transcription is observed during chloroplast biogenesis. This suggests that one central regulatory point of chloroplast differentiation is the activation of chloroplast transcription. However, chloroplast development requires coordinated expression of plastid and nuclear genes; therefore, activation of plastid gene transcription must be coordinated with the activation of nuclear gene expression by signals that can be transmitted between plastids and the nucleus. The study of chloroplast gene expression also provides an opportunity to understand how plants sense and alter gene expression in response to light. Light is a substrate for photosynthesis, a trigger of plant developmental processes, and a potential source of injury that modulates chloroplast transcription.

Posttranscriptional control of chloroplast gene expression

Chloroplasts contain their own genome, organized as operons, which are generally transcribed as polycistronic transcriptional units. These primary transcripts are processed into smaller RNAs, which are further modified to produce functional RNAs. The RNA processing mechanisms remain largely unknown and represent an important step in the control of chloroplast gene expression. Such mechanisms include RNA cleavage of preexisting RNAs, RNA stabilization, intron splicing, and RNA editing. Recently, several nuclear-encoded proteins that participate in diverse plastid RNA processing events have been characterized. Many of them seem to belong to the pentatricopeptide repeat (PPR) protein family that is implicated in many crucial functions including organelle biogenesis and plant development.

RNA self-splicing mechanism in *Chlamydomonas* chloroplasts

The 23S rRNA gene of the *C. reinhardtii* chloroplast contains an 88-base pair intron with structural features characteristic of group I introns. The nuclear, chloroplast ribosome-deficient mutant of *C. reinhardtii, ac20*, overaccumulates approximately 3.6-kb unspliced 23S preRNA compared to wild-type cells. In 1990, D.L. Herr and coworkers used [α-^{32}P]GTP labeling of total RNA preparations from ac20 to rapidly determine that 23S preRNA is capable of self-splicing. The ability of the 23S intron (with flanking exon sequences) to correctly catalyze its own splicing was confirmed using RNA produced by *in vitro* transcription of cloned DNA. These results identify the first example of a self-splicing RNA of chloroplast origin.

Translational regulation of chloroplast genes

In 1996, C.R. Hauser and coworkers examined the effects of illumination, carbon source, and levels of chloroplast protein synthesis on trans-acting proteins that bind to the leaders of five representative chloroplast mRNAs. The accumulation of these five chloroplast mRNAs and the proteins they encode were measured in cells grown under identical conditions. Extracts from all cell types examined contain a minimum set of six chloroplast 5'-UTR-binding proteins (81, 62, 56, 47, 38, and 15 kDa). Fractionation results suggest that multiple forms of the 81-, 62-, and 47-kDa proteins may exist. A 36-kDa protein was found in all cells except those deficient in chloroplast protein synthesis. Binding of the 81-, 47-, and 38-kDa proteins to the rps12 leader is effectively competed by the atpB or rbcL 5'-UTRs, indicating that the same proteins bind to all three leaders. In contrast, these three proteins do not bind to the nuclear-encoded α-1 tubulin leader, which bound novel proteins of 110, 70, and 43 kDa. Cis-acting sequences within the 5'-UTRs of two chloroplast mRNAs (rps7 and atpB) have been identified, which are protected from digestion by RNase T1 by extracts enriched for the 81-, 47-, and 38-kDa proteins.

Protein stoichiometry, mRNA abundance, and transcription rates

Segregation of genes encoding proteins of different functions into separate operons could be related, in part, to a requirement for different levels of protein production. The abundance of plastid-encoded proteins varies over 1000-fold, with the large subunit of RuBisCo being the most abundant protein, followed by proteins involved in electron transport, ribosomes, and subunits of the plastid-encoded RNA polymerase. Among plastid genes, mRNA abundance also varies approximately 1000-fold and, in general, mRNA abundance parallels protein abundance. A 300-fold range of transcription rate is observed among these same genes, and transcription activity varies in parallel with mRNA level and protein abundance. Thus, transcription plays a central role in establishing the levels of many plastid mRNAs and proteins. Variation in plastid mRNA stability also significantly influences plastid gene expression.

Plastid mRNA stability

Recent assays of plastid mRNA stability revealed half-lives ranging from 6 h to over 40 h. In contrast, the stability of most mRNAs in *E. coli* ranges from 20 s to several minutes. Thus, the stability of many plastid mRNAs is similar to the stability of mRNAs found in eukaryotic organisms. Long plastid mRNA half-lives mean that RNA stability is an important determinant of plastid mRNA levels. In addition, high mRNA stability limits the rate with which plastids can alter mRNA levels through a change in transcription rate.

Systems biology approach in understanding chloroplast development

Chloroplasts not only receive signals from the nucleus, but chloroplasts also send signals to the nucleus via chloroplast-to-nucleus retrograde signaling.

Anterograde regulation

Anterograde regulation is the regulation that occurs or moves in the normal or forward direction of conduction or flow. Anterograde regulation coordinates expression levels of nuclear genes encoding photosynthetic proteins and plastid gene expression machineries. Gene expression profiling, RNA target identification, and plastid proteomics provide a systematic view of anterograde regulation.

The plastid proteome varies with its differentiated form, which is associated with its various functions within plant cells. Some of these plastid types are not very prevalent or are present only in cell types that are difficult to isolate. Thus, the question of how the plastid proteome varies during development or under a specific environmental condition is still a challenge. Chloroplasts are the only plastid types for which there is useful proteomic information. Because chloroplasts are green and abundant and mesophyll protoplasts are easy to isolate, the chloroplast proteome has been heavily interrogated. Considering the fact that most chloroplast proteins are encoded in the nucleus, major questions are as follows: What are the regulators of nuclear genes for plastid-destined proteins? How do nuclear genes regulate plastid transcription and translation? How are the expression levels of these nuclear genes regulated by the environment?

When a seedling first encounters light, there is a massive reprogramming of the nuclear genome that must be coordinated with plastid function. After exposure to light, plastids are differentiated from proplastids or etioplasts to chloroplasts. During a short time, photosynthetic complexes are synthesized and assembled into thylakoid membranes, and proteins required for chloroplast DNA replication, transcription, and translation are synthesized and imported into the developing chloroplast. The protein complexes, especially photosynthetic protein complexes, consist of components encoded in the nuclear genome with a small but critical contribution of the chloroplast genome. Anterograde regulation coordinates levels of proteins encoded in the chloroplast and nuclear genomes. To better understand the effects of the environment and genetic background on plastid composition, a set of 3000 nuclear genes predicted to encode proteins that function in plastids was interrogated on microarrays. Expression levels of these genes were determined under 101 different conditions, which included changing environment or genetic backgrounds, and groups of co-regulated genes were identified. This study concluded that, in the nucleus, photosynthesis-related gene expression is co-regulated with chloroplast gene transcription and translation factors.

Expression profiling of nuclear and chloroplast genes encoding chloroplast ribosomal proteins showed that expression levels of these genes in two compartments simultaneously increased during tobacco (*N. tabacum*) seedling development. During the etioplast-to-chloroplast transition, proteins related to plastid protein translation, as well as photosynthesis-related proteins, were up-regulated, which is consistent with the gene expression profiling. Isolation of etioplast protein complexes followed by component identification determined the presence of such complexes as the ATPase, cytochrome b6f, and partially assembled RuBisCo. These results indicated that the accumulation of chloroplast- or nuclear-encoded subunits of multisubunit complexes was coordinated from the earliest stages of plant development.

Chloroplast localized RNA-binding proteins are known to have a prominent role in chloroplast gene expression. They are required for plastid transcript maturation through splicing of introns, formation of correct 3' ends, and RNA editing. RNA binding also regulates RNA stability and translation initiation of chloroplast mRNA. RNA-mediated control of plant development and stress responses is widespread. It seems likely that cytoplasmic and/or nuclear proteins, which are involved in RNA metabolism, may also contribute to signaling between chloroplasts and the nucleus.

Retrograde regulation

Signaling pathways leading from the organelles (chloroplasts, mitochondria) to the nucleus are referred to as "retrograde signaling" and were first discovered in the yeast *Saccharomyces cerevisiae*. Retrograde signaling is broadly defined as cellular responses to alterations in the functional state of organelles. Retrograde signaling regulates nuclear gene expression depending on the developmental and functional state of plastids. Gene expression profiling, protein-interacting networks, and changes in the metabolome of chloroplasts have shed light on systems biological analysis of responses of retrograde signaling.

Effect of light quality on gene expression is shown in Figure 10.17. The three cell compartments are depicted: chloroplast, cytosol, and nucleus. The integrated cellular network consists of two subnetworks. One network is the signaling transduction pathways in the cytosol and the other network is the gene regulatory network for transcriptional regulation in nucleus. An important mediator for plastoquinone (PQ) pool redox states is the thylakoid kinase STN7. STN7 kinase transmits the decisive signal to the nucleus, resulting in the ensuing regulation of the relative amount of each of the photosystems. However, the mechanism by which the redox signal is transmitted from the chloroplast double membrane into the cytosol is poorly understood. In this study, the gene regulatory network (the protein–DNA interaction) in the nucleus is constructed under different retrograde signals originating from different redox states in the PQ pool. The light environment is perceived by cytosolic photoreceptors. Although both PQ pool and photoreceptor systems report changes of ambient light environment by different signal transduction pathways to the nucleus, some common TFs in nucleus may be employed simultaneously in different light-related systems to respond to the prevailing environment.

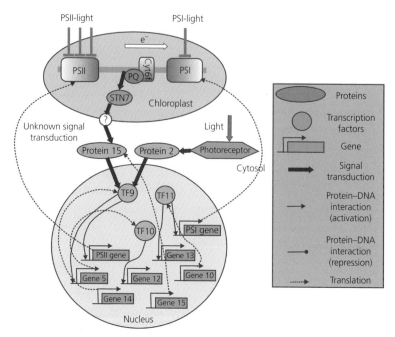

Figure 10.17 Effect of light quality on gene expression. Source: Reproduced with permission of Yao *et al.* (2011).

Plastid-to-nucleus retrograde coordination of nuclear gene expression with chloroplast function

Chloroplasts contain about 3000 proteins, of which more than 95% are encoded by nuclear genes. Plastid-to-nucleus retrograde coordinates nuclear gene expression with chloroplast function and is essential for photoautotrophic lifestyle of plants. Three retrograde signals have been described. One pathway involves accumulation of Mg-protoporphyrin IX (Mg-Proto IX), which is a biosynthetic intermediate. Genes *GUN2*, *GUN3*, *GUN4*, and *GUN5* are required to maintain its levels. Second pathway represses *Lheb* expression in response to inhibition of plastid gene expression and requires gene *GUN1*. The third signaling pathway mediates signals derived from the reduction/oxidation (redox) state of the photosynthetic electron transfer chain. Plant chloroplastids respond to environmental and developmental stress through a signaling pathway that controls gene expression in the nucleus (Zhang, 2007). These three chloroplast signals generate a common signal. GUN1 may either generate this common signal or perceive it. GUN1 then communicates with the nucleus (by an unknown pathway) to control gene expression by the transcription factor AB14.

Koussevitzsky *et al.* (2007) have proposed a model in which multiple indicators of aberrant plastid function in *Arabidopsis* are integrated upstream of *GUN1* within plastids, which lead to AB14-mediated repression of nuclear-encoded genes. Plastid-to-nucleus retrograde signaling coordinates nuclear gene expression with chloroplast function and is essential for the photoautotrophic lifestyle of plants. They have shown that GUN1, a chloroplast-localized pentatricopeptide-repeat protein, and ABI4, an Apetala 2 (AP2)-type transcription factor, are common to all three pathways describe before. ABI4 binds the promoter of a retrograde-regulated gene through a conserved motif found in close proximity to a light-regulatory element.

Plant cells coordinately regulate the expression of nuclear and plastid genes that encode components of the photosynthetic apparatus. Nuclear genes that regulate chloroplast development and chloroplast gene expression provide part of this coordinate control. There is evidence that information also flows in the opposite direction, from chloroplasts to the nucleus. They have shown that the tetrapyrrole intermediate Mg-Proto IX acts as a signaling molecule in one of the signaling pathways between the chloroplast and nucleus. Accumulation of Mg-Proto IX is both necessary and sufficient to regulate the expression of many nuclear genes encoding chloroplastic proteins associated with photosynthesis.

Tetrapyrrole biosynthetic pathway

The tetrapyrrole biosynthetic pathway leads to the synthesis of a number of important products including the chlorophylls and hemes. The enzymatic steps of the pathway are well characterized, and genes for virtually all of the enzymes have been identified in higher plants. The tetrapyrrole pathway in plants is shown in Figure 10.18. Genes coding for different enzymes and proteins involved in tetrapyrrole synthesis in higher plants are *GTS*, glutamate-tRNA synthetase; *HEMA1*, glutamyl-tRNA reductase 1; *HEMA2*, glutamyl-tRNA reductase 2;

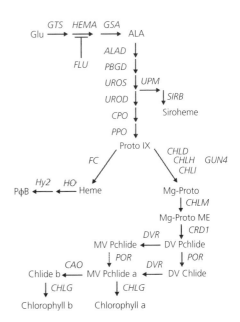

Figure 10.18 The tetrapyrrole pathway in plants showing intermediates and genes analyzed in this study. Intermediates: Glu, glutamate; ALA, 5-aminolevulinic acid; Proto IX, protoporphyrin IX; Mg-proto, Mg-protoporphyrin; Mg-proto ME, Mg-protoporphyrin monomethyl ester; DV Pchlide, divinyl protochlorophyllide; MV Chlide, monovinyl chlorophyllide. Source: Reproduced with permission of Moulin *et al.* (2008).

HEMA3, glutamyl-tRNA reductase 3; *GSA1*, glutamate-1-semialdehyde-2, 1-aminotransferase 1; *GSA2*, glutamate-1-semialdehyde-2,1-aminotransferase 2; *FLU*, regulator of ALA synthesis; *ALAD1*, ALA dehydratase 1; *ALAD2*, ALA dehydratase 2; *PBGD*, porphobilinogen deaminase; *UROS*, uroporphyrinogen-III synthase; *UROD1*, uroporphyrinogen-III decarboxylase; *UROD2*, uroporphyrinogen-III decarboxylase; *CPO1*, coproporphyrinogen oxidase 1; *PPO1*, protoporphyrinogen oxidase 1 (Cp-localized); *PPO2*, protoporphyrinogen oxidase 2 (Mt-localized, possibly also Cp); *CHLI1*, Mg-chelatase subunit I1; *CHLI2*, Mg-chelatase subunit I2; *CHLD*, Mg-chelatase subunit D; *CHLH*, Mg-chelatase subunit H; *CHLM*, Mg-protoporphyrin IX methyltransferase; *CRD1*, Mg-protoporphyrin IX monomethylester cyclase; *PORA*, NADPH: protochlorophyllide oxidoreductase A; *PORB*, NADPH: protochlorophyllide oxidoreductase B; *PORC*, NADPH: protochlorophyllide oxidoreductase C; *DVR*, divinyl chlorophyllide reductase; *CHLG*, chlorophyll synthase; *CHLP*, geranylgeranyl reductase; *CAO*, chlorophyllide *a* oxygenase; *GUN4*, regulator of Mg-porphyrin synthesis; *FCI*, ferrochelatase 1 (Cp-, Mt-localized); *FCII*, ferrochelatase 2 (Cp-localized); *HO1*, heme oxygenase 1; *HO2*, heme oxygenase 2; *HO3*, heme oxygenase 3; *HO4*, heme oxygenase 4; *HY2*, phytochromobilin synthase; *UPM1*, uroporphyrinogen III methylase; and *SIRB*, sirohydrochlorin ferrochelatase.

Chloroplast development-dependent nuclear gene expression profiles

In the absence of chloroplast development, expression of nuclear genes encoding chloroplast proteins, such as the *Lhcb* gene for a light harvesting complex protein, is repressed. To date, molecular genetic studies in *C. reinhardtii* and *Arabidopsis* have suggested that perturbations in the tetrapyrrole pathway around the chlorophyll/heme branch point may be involved in generating a distress signal from plastids. The signal may be a tetrapyrrole intermediate, or a secondary signal may be generated that exits the plastid. To identify a comprehensive set of nuclear genes under the control of retrograde pathways, Koussevitzky *et al.* (2007) analyzed the global gene expression response of the wild-type and two genomes uncoupled (*gun*) mutants defective in retrograde signaling, *gun1* (wild-type of which encodes a chloroplast PPR protein) and *gun5* (wild-type of which encodes a Mg-chelatase subunit). About half of these genes were repressed by norflurazon (an inhibitor of phytoene desaturase that causes photooxidative damage in plastids) treatment in the wild-type, with 330 of these having expression in both gun1 and gun5. This statistical analysis indicates that retrograde signaling regulates a large number of nuclear genes with diverse function. When the promoters of these 330 genes were systematically interrogated, an element, ACGT, was found to be significantly overrepresented (Priest *et al.*, 2009). ACGT is the core sequence of ABA response elements, implicating components of the abscisic acid (ABA) response pathway in retrograde signaling. Indeed, ABI4, an AP2-type transcription factor, was then shown to be a component of the plastid retrograde signaling pathway. ABI4 has been shown to be regulator of mitochondrial retrograde signaling as well, providing a point of convergence in the plastid and mitochondrial retrograde signaling pathways (Giraud *et al.*, 2009).

Cross talk between chloroplasts and mitochondria

Several pieces of evidence demonstrate the existence of cross talk between chloroplasts and mitochondria. Mutants either affected in photorespiration activity or having lesions in a gene encoding a component of mitochondrial protein complexes, such as NADH dehydrogenase, showed altered chloroplast functional states, including photosynthetic activity, while, in chloroplast development mutants (i.e., albostrians) of *Arabidopsis*, mitochondria development was also affected. Furthermore, chloroplast–mitochondria cross talk seems to contribute to chloroplast-to-nucleus retrograde signaling. The *immutans* mutant (has green and white sectoring due to the action of a nuclear recessive gene, *immutans*) of *Arabidopsis* has a variegated phenotype, having green and albino leaf sectors. In contrast to the complete arrest of chloroplast development seen in norflurazon-treated plants, white and green sectors of *immutans* provide genetically identical sectors of developmentally arrested and wild-type chloroplasts, respectively. Gene expression profiles were compared (Aluru *et al.*, 2009). These experiments demonstrated significant nuclear gene expression changes in the white versus green sectors.

Chloroplast functional state-dependent nuclear gene expression profiles

In addition to chloroplast developmental state, nuclear gene expression is regulated by the functional state of chloroplasts via retrograde signaling. In contrast to chloroplast development-dependent signaling studies, intact chloroplasts from mature leaves are the source of tissue in functional state-dependent signaling studies. To alter the chloroplast functional state, environmental conditions, such as very strong light, photosystem-specific light, or electron transport inhibitors, have been used. Very strong light results in a reduced state of all photosynthetic components in chloroplasts, and the reduced state changes nuclear gene expression. In general, expression levels of genes encoding proteins involved in photosynthesis are decreased, whereas those of stress response genes are increased. The up-regulated stress-response genes encode antioxidant proteins, chaperones, and enzymes involved in biosynthesis of antioxidants. It has been proposed that the redox state of the PQ pool may regulate the very strong light-driven gene expression, but only 8% of the induced genes are under the control of PQ redox state.

Decoding the signaling pathway(s) between chloroplasts and the nucleus

Protein–protein interactions and subcellular localization play a critical role in signal transduction and are sure to play a role in retrograde signaling. A few proteins are good candidates to begin biochemical analyses, including the chloroplast localized PPR protein encoded by *GUN1*, the nuclear localized transcription factor, ABI4, and cryptochrome 1 identified in multiple genetic screens. Computational approaches have been employed to identify genome-wide protein–protein interactions (Morsy *et al.*, 2008). A group of chloroplast localized proteins were predicted and used to specifically predict protein–protein interactions.

Dramatic increase in plastid DNA copy number during chloroplast development

Plastid DNA copy number and ribosome abundance increased dramatically during chloroplast development; in 1987, A.J. Bendich hypothesized that rRNA synthesis was template limited and that the increase in plastid DNA copy number during chloroplast development was needed to activate rRNA synthesis. The hypothesis suggested that expression of other plastid genes (not encoding rRNA) was not limited by transcription and that regulation of expression occurred post-transcriptionally at the levels of mRNA stability, translation, or protein turnover. It is now known that transcription rates vary among plastid genes over 300-fold and that transcription rates are predictive of mRNA levels and protein abundance for many plastid genes.

Overall dynamics of chloroplast transcription

Overall transcription rates change dramatically during chloroplast development, and in sorghum, RNA polymerase levels increase in parallel with transcription activity. However, despite the global changes in transcription rates, the relative ratio of transcription of many plastid genes is constant in different tissues and developmental stages. This observation is not surprising for three reasons. First, the coding capacity of the plastid genome is limited, thus precluding the need for regulation in situations where plastid genes are not relevant. Second, with a few exceptions (i.e., *ndh* genes), plastid genes can be classified into two major groups: genes encoding proteins and RNAs directly involved in gene expression, that is, transcription and translation) and genes encoding the photosynthetic apparatus. Expression of these two groups of genes is sequentially linked during chloroplast development in that high levels of the plastid DNA transcription and translation machinery are required only during synthesis and assembly of the photosynthetic apparatus. Third, large numbers of plastid genes encode proteins that accumulate in fixed stoichiometries in a few large protein complexes (RNA polymerase, ribosomes, and photosynthetic electron transport complexes). Therefore, constant ratios of transcription of genes encoding a basic set of plastid proteins needed in fixed stoichiometry is not surprising.

Differential transcription during chloroplast development

The events described in the following text are based on the study of cells and plastids as they move from the meristematic zone in the monocot leaf base, through a zone of cell enlargement, followed by final maturation of cells and chloroplasts in older apical regions of the leaf. The first step in chloroplast development in monocot leaves involves the activation of plastid DNA synthesis and plastid replication. This occurs while cells are in the leaf basal meristem and when it is required to maintain plastid number and DNA content in the dividing cells. At present, no plastid genes involved in plastid replication or DNA synthesis have been identified, indicating that this early phase of chloroplast development is controlled through nuclear gene expression. Plastid transcription activity and RNA levels remain low in cells of the leaf basal meristem, but they dramatically increase when cells enter the zone of cell enlargement. In particular, transcription and RNA levels for *rpoB–rpoC1–rpoC2*, *rps6* rRNA, and some tRNAs are differentially elevated at this stage relative to genes encoding proteins of the photosynthetic apparatus. This step in chloroplast development is followed shortly by the activation of plastid and nuclear genes encoding proteins of the photosynthetic apparatus. Coordinated expression of nuclear and plastid genes at this stage of development involves a "plastid signal," which is required for high expression of nuclear genes encoding proteins of the photosynthetic apparatus. If chloroplast development is blocked either by inhibitors of translation or through inhibition of transcription, then production of the plastid signal is limited. During normal development, activation of genes encoding the photosynthetic apparatus leads to the synthesis and assembly of the photosynthetic apparatus. Finally, as chloroplasts mature, overall transcription rates decline. However, the need for continued synthesis of the PSII reaction center proteins (Dl/D2) remains high due to light-mediated damage of these proteins. Differential light-induced transcription of *psbA* and *psbD* genes helps sustain synthesis of these proteins in mature chloroplasts.

Chloroplast biogenesis and regulation of photosynthesis

Photosynthetic complexes each consist of nucleus- and chloroplast-encoded subunits. The former are synthesized as precursors on cytosolic 80S ribosomes and targeted to the chloroplast (Figure 10.19a). The latter are synthesized on chloroplast 70S ribosomes. Several posttranscriptional steps in the chloroplast, such as RNA stability, processing, splicing, editing and translation, and the assembly of the protein complexes, require the action of numerous nucleus-encoded factors. In a reciprocal manner, the state of the chloroplast is perceived by the nucleus through retrograde signaling chains.

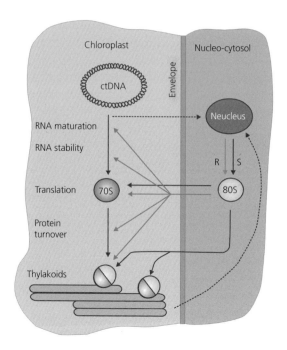

(a)

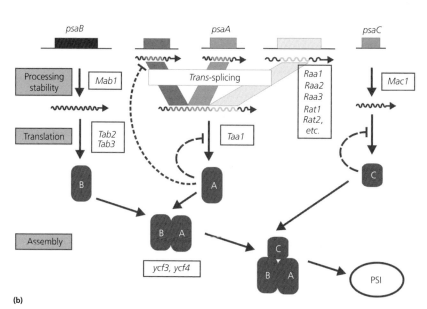

(b)

Figure 10.19 (a) Chloroplast biogenesis and regulation of photosynthesis. Source: Reproduced with permission of Dr. J.-D. Rochaix and N. Roggli, University of Geneva. (b) Regulation of chloroplast gene expression. Source: Reproduced with permission of Dr. M. Goldschmidt-Clermont, University of Geneva.

Genetic studies in the green unicellular alga, *C. reinhardtii*, have revealed many nucleus-encoded factors that are imported in the chloroplast where they are required for plastid gene expression. They are involved in posttranscriptional steps such as RNA splicing, RNA processing, or translation (Figure 10.19b). Each of these factors is surprisingly specific for a single gene in the chloroplast or a small set of genes.

The mRNA of *psaA* is assembled from three separate transcripts encoded at different loci in the chloroplast genome. The maturation of *psaA* mRNA requires two steps of splicing in trans, which depend on at least fourteen nuclear genes. Studies on some of these genes (*Raa1*, *Raa2*, and *Raa3*) have shown that the corresponding splicing factors belong to large ribonucleoprotein complexes. In mutants deficient for *psaA* trans-splicing, the precursor of *psaA*-exon1 is overexpressed. This overaccumulation is also apparent with chimeric reporter genes under the control of the promoter and 5'-UTR of *psaA*-exon1. The studies on other proteins that are imported into the chloroplast and are necessary for the maturation of *psaB* mRNA (*Mab1*) or for its translation (*Tab2*, *Tab3*) and for the translation of *psaA* (*Taa1*) are being taken up.

A nuclear-encoded plastid-localized RNA polymerase

Differential transcription of *rpoB–rpoC1–rpoC2* and other genes early in chloroplast development has been hypothesized to be caused by increased expression of a nuclear-encoded plastid-localized RNA polymerase. It has been suggested that expression of the nuclear-encoded plastid-localized RNA polymerase is activated early in chloroplast development. This results in the differential transcription of the *rpoB* operon and other loci involved in plastid gene expression early in chloroplast development. Transcription of the *rpoB* operon leads to increased levels of the plastid-encoded RNA polymerase, which, in turn, further activates transcription of genes needed for synthesis and assembly of the photosynthetic apparatus. This sequential cascade model for the activation of plastid gene transcription during chloroplast development can be directly tested once the nuclear-encoded plastid-localized RNA polymerase is isolated and characterized.

Chaperones import proteins into chloroplasts for their development

Molecular chaperones are required for the translocation of many proteins across organellar membranes, presumably by providing energy in the form of ATP hydrolysis for protein movement. In the chloroplast protein import system, a heat shock protein 100 (Hsp100), known as Hsp93, is hypothesized to be the chaperone providing energy for precursor translocation. There is little direct evidence for this hypothesis. To know the possible function of Hsp93 during protein import into chloroplasts, Constan *et al.* (2004) isolated knockout mutant lines that contain T-DNA disruptions in either *atHSP93-V* or *atHSP93-III*, which encode the two *Arabidopsis thaliana* homologs of Hsp93. The *atHsp93-V* mutant plants are much smaller and paler than wild-type plants. In addition, mutant chloroplasts contain less thylakoid membrane when compared to the wild-type. Plastid protein composition, however, seems to be largely unaffected in *atHsp93-V* knockout plants. Chloroplasts isolated from the *atHsp93-V* knockout mutant line are still able to import a variety of precursor proteins, but the rate of import of some of these precursors is significantly reduced. These results indicate that atHsp93-V has an important, but not essential, role in the biogenesis of *Arabidopsis* chloroplasts. In contrast, knockout mutant plants for atHsp93-III, the second *Arabidopsis* Hsp93 homolog, had a visible phenotype identical to the wild-type, suggesting that atHsp93-III may not play as important a role as atHsp93-V in chloroplast development and/or function.

Chloroplast development regulated by plant hormones

The development and function of the chloroplast are regulated by plant hormones, particularly cytokinin as a positive regulator and brassinosteroid as a negative regulator. The molecular mechanism of chloroplast development regulated by these two phytohormones is discussed here.

Cytokinin

Cytokinin is an adenine-derived plant growth regulator with important biological activities, including the promotion of cell division and chloroplast development. When dicot plants were grown in a medium containing cytokinin, the plants had short hypocotyls and highly greened leaves that correspond to the "accelerated photomorphogenesis" phenotype. The fundamental nature of these cytokinin effects suggests that changes in gene expression may be required in mediating the response of chloroplast development to cytokinin. Although previous studies have shown that most photosynthetic genes were induced by cytokinin, key genes on the upstream of the cytokinin signal transduction that can affect chloroplast development and functions have never been identified.

Brassinosteroid

Brassinosteroid is a sterol-derived plant growth regulator with important biological activities, including the promotion of cell elongation and negative regulation of chloroplast development. In 1999, J. Li and J. Chory studied *Arabidopsis de-etiolated 2* (*DET2*) and some recently identified mutants concerned with brassinosteroid biosynthesis, and the possible *Arabidopsis* brassinosteroid receptor *brassinosteroid-insensitive dwarf 1* (*BRI1*) suggested that deficiency in brassinosteroid causes dwarf with highly greened leaves. In 1998, T. Asami and S. Yoshida synthesized brassinosteroid biosynthesis inhibitor, brz. The brz-treated plants exhibited the same dwarf phenotype similar to *DET2* and *BRI1* mutants, which corresponded to the "accelerated photomorphogenesis" phenotype. The brz-treated plants also had enhanced chloroplast development compared to the non-treated plants that could be result of the chloroplastic gene expression that encoded on nuclear and chloroplast genomes.

Nakano *et al.* (2001) determined that cultured green tobacco cells, *N. tabacum* cv. Samsun NN, are suitable for studying cytokinin signal transduction because cell growth and chloroplast development were up-regulated by a medium supplemented with cytokinin. Based on the developed fluorescent differential display using cultured green tobacco cells, the expression of three cytokinin-inducible genes (*cig*) during the earliest stages of chloroplast development were identified.

Repression of chloroplast development

The phytochrome-interacting factor PIF3 has been proposed to act as a positive regulator of chloroplast development. Stephenson *et al.* (2009) showed that the *pif3* mutant has a phenotype that is similar to the *pif1* mutant, lacking the repressor of chloroplast development PIF1, and that a *pif1pif3* double mutant has an additive phenotype in all respects. The *pif* mutants showed elevated protochlorophyllide levels in the dark, and etioplasts of *pif* mutants contained smaller prolamellar bodies and more prothylakoid membranes than corresponding wild-type seedlings, similar to previous reports of constitutive photomorphogenic mutants. Consistent with this observation, *pif1*, *pif3*, and *pif1pif3* showed reduced hypocotyl elongation and increased cotyledon opening in the dark. Transfer of 4-day-old dark-grown seedlings to white light resulted in more chlorophyll synthesis in *pif* mutants over the first 2 h, and analysis of gene expression in dark-grown *pif* mutants indicated that key tetrapyrrole regulatory genes such as *HEMA1* encoding the rate-limiting step in tetrapyrrole synthesis were already elevated 2 days after germination. Circadian regulation of *HEMA1* in the dark also showed reduced amplitude and a shorter, variable period in the *pif* mutants, whereas expression of the core clock components *TOC1*, *CCA1*, and *LHY* was largely unaffected. Expression of both *pif1* and *pif3* was circadian regulated in dark-grown seedlings. PIF1 and PIF3 are proposed to be negative regulators that function to integrate light and circadian control in the regulation of chloroplast development.

Development of the photosynthetic apparatus

In 1972, N.I. Bishop and H. Senger reported that mutant C-2A' of *Scenedesmus obliquus* formed only traces of chlorophyll and showed no detectable photosynthesis when grown heterotrophically. When transferred to light, this mutant developed chlorophyll and its photosynthetic capacity was established. Following a short initial lag phase, both photosynthetic capacity and total light absorption of the cells reached saturation more rapidly than did the rate of chlorophyll synthesis. Consequently, the quantum requirement of photosynthesis showed a rapid decline during the initial 6 h of greening, Subsequently, a slow increase occurred as additional chlorophyll was synthesized. Behavior parallel to that of the quantum requirement was also noted for the relative fluorescence yield and for the onset of the 520 nm light-induced absorbance change. Of the two photosystems, PS-I seemed to develop more rapidly than did PS-II. The appearance of PS-II activity appeared to accompany linkage of the two photosystems, as revealed by analysis of the variable-yield fluorescence and the kinetics of the 520 nm light-induced absorbance change. During the phase of greening at which photosynthetic capacity developed its maximum quantum efficiency, no significant changes in type or content of the various chloroplast cytochromes were detected. Analysis of the ratio of chlorophyll/PQ, however, showed that changes in this value followed more closely the observed increase in the quantum efficiency of photosynthesis and the other parameters of photosynthesis examined.

Chloroplast RNA-binding and PPR proteins

Chloroplast gene expression is mainly regulated at the post-transcriptional level by numerous nuclear-encoded RNA-binding protein factors. Two RNA-binding proteins, chloroplast ribonucleoprotein (cpRNP) and PPR, are suggested to be major contributors to chloroplast RNA metabolism. Tobacco cpRNPs are composed of five different proteins containing two RNA-recognition motifs and an acidic N-terminal domain. The cpRNPs are abundant proteins and form heterogeneous complexes with most ribosome-free mRNAs and the precursors of tRNAs in the stroma. The complexes could function as platforms for various RNA-processing events in chloroplasts. It has been demonstrated that cpRNPs contribute to RNA stabilization, 3'-end formation, and editing. The PPR proteins occur as a superfamily only in the higher plant species. They are predicted to be involved in RNA/DNA metabolism in chloroplasts or mitochondria. Nuclear-encoded HCF152 is a chloroplast-localized protein that usually has 12 PPR motifs. The null mutant of *Arabidopsis*, *hcf152*, is impaired in the 5'-end processing and splicing of *petB* transcripts. HCF152 binds the *petB* exon–intron junctions with high affinity. The number of PPR motifs controls its affinity and specificity for RNA. It has been suggested that each of the highly variable PPR proteins is a gene-specific regulator of plant organellar RNA metabolism.

Nuclear genes required for early chloroplast development

Cell autonomous trait is genetic in multicellular organisms in which only genotypically mutant cells exhibit the mutant phenotype. Conversely, a non-cell-autonomous trait is one in which genotypically mutant cells cause other cells (regardless of their genotype) to exhibit a mutant phenotype. Cell autonomous and non-cell-autonomous actions of certain mutants have shown that nuclear genes are required for early chloroplast development.

In order to identify nuclear genes required for early chloroplast development, a collection of photosynthetic pigment mutants of *Arabidopsis* was assembled and screened for lines with extremely low levels of chlorophyll. Nine chloroplast biogenesis (*clb*) mutants that affect proplastid growth and thylakoid membrane formation and result in an albino seedling phenotype were identified. These mutations identify six new genes as well as a novel allele of *cla1*. The *clb* mutants have less than 2% of wild-type chlorophyll levels, and little or no expression of nuclear and plastid-encoded genes required for chloroplast development and function. In all but one mutant, proplastids do not differentiate enough to form elongated stroma thylakoid membranes. Analysis of mutants during embryogenesis allows differentiation between *clb* genes that act non-cell-autonomously, where partial maternal complementation of chloroplast development is observed in embryos, and those that act cell autonomously, where complementation during embryogenesis is not observed. Molecular characterization of the non-cell-autonomous *clb4* mutant established that the *clb4* gene encodes for hydroxy-2-methyl-2-(*E*)-butenyl 4-diphosphate synthase (HDS), the next to the last enzyme of the methylerythritol 4-phosphate (MEP) pathway for the synthesis of plastidic isoprenoids. The non-cell-autonomous nature of the *clb4* mutant suggests that products of the MEP pathway can travel between tissues, and provides *in vivo* evidence that some movement of MEP intermediates exists from the cytoplasm to the plastid.

Chloroplast differentiation and dedifferentiation

Arabidopsis seed formation is coupled with two plastid differentiation processes. Chloroplast formation starts during embryogenesis and ends with the maturation phase. It is followed by chloroplast dedifferentiation/degeneration that starts at the end of the maturation phase and leads to the presence of small non-photosynthetic plastids in dry seeds. Allorent *et al.* (2013) have analyzed mRNA and protein levels of nucleus- and plastid-encoded (NEP and PEP) components of the plastid transcriptional machinery; mRNA and protein levels of some plastid RNA polymerase target genes; changes in plastid transcriptome profiles; and mRNA and protein levels of some selected nucleus-encoded plastid-related genes in developing seeds during embryogenesis, maturation, and desiccation. As expected, most of the mRNAs and proteins increase in abundance during maturation and decrease during desiccation, when plastids dedifferentiate/degenerate. In contrast, mRNAs and proteins of components of the plastid transcriptional apparatus do not decrease or even still increase during the period of plastid dedifferentiation. The proteins of the plastid transcriptional machinery are specifically protected from degradation during the desiccation period and conserved in dry seeds to allow immediate regain of plastid transcriptional activity during stratification/germination. Furthermore, accumulation and storage of mRNAs coding for RNA polymerase components and sigma factors in dry seeds has been observed. These mRNAs seem to provide immediate-to-use templates for translation on cytoplasmic ribosomes in order to enhance RNA polymerase protein levels and to provide regulatory proteins for stored PEP to guaranty efficient plastid genome transcription during germination.

Chloroplast genetic engineering

Procedure of chloroplast genetic engineering

A novel system that appears to circumvent the concerns about nuclear modification is genetic engineering of chloroplast DNA. In chloroplast genetic engineering, the recombinant DNA plasmid is bound to small gold nanoparticles that are then injected into the chloroplasts of a leaf using a "gene gun". This device uses high pressure to insert the plasmid-coated particles into the cell. These plasmids contain multiple important genes: the therapeutic gene, a gene for antibiotic resistance, a gene that increases expression of the therapeutic gene, and two flanking sequences that ensure that the plasmid is not randomly integrated into the chloroplast genome. The flanking sequences guide the human recombinant DNA into a specific place on the chloroplast genome by binding to corresponding parts on the genome. The leaf is then grown on a plate containing an antibiotic, which ensures that the only surviving plant cells will be those that contain the gene for antibiotic resistance and, therefore, contain the therapeutic gene as well. These cells are then exposed to regenerative factors that induce them to start sprouting shoots and grow into full plants that express the desired protein. Figure 10.20 provides details of the procedure of chloroplast genetic engineering.

Elimination of marker genes from the plastid genome

Incorporation of a selectable marker gene during transformation is essential to obtain transformed plastids. However, once transformation is accomplished, having the marker gene becomes undesirable. Corneille *et al.* (2001) reported on adapting the P1 bacteriophage Cre-*lox* site-specific recombination system for the elimination of marker genes from the plastid genome. The system was tested by the elimination of a negative selectable marker, *codA*, which is flanked by two directly oriented *lox* sites (*codA*). Highly efficient elimination of *codA* was triggered by introduction of a nuclear-encoded plastid-targeted Cre by *Agrobacterium* transformation or via pollen. Excision of *codA* in tissue culture cells was frequently accompanied by a large deletion of a plastid genome segment that includes the tRNA-Val^{UAC} gene. However, the large deletions were absent when *cre* was introduced by pollination. Thus, pollination is our preferred protocol for the introduction of *cre*. Removal of the *codA* coding region occurred at a dramatic speed, in striking contrast to the slow and gradual buildup of transgenic copies during plastid transformation. The nuclear *cre* gene could subsequently be removed by segregation in the seed progeny. The Cre/*lox* system requires additional retransformation and sexual crossing for introduction and subsequent removal of the CRE recombinase. The modified Cre/*lox* system will be a highly efficient tool to obtain marker-free transplastomic plants.

Another method used to remove the selection gene from higher plant plastids is based on loop-out-type recombination, a process difficult to control because

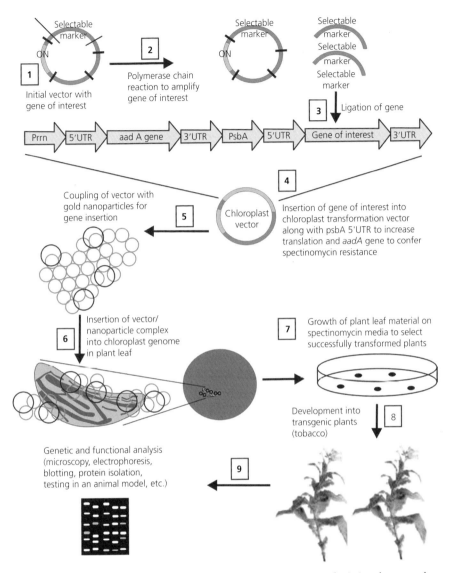

Figure 10.20 Procedural schematic of chloroplast genetic engineering, depicting the general procedure used to insert transgenes into chloroplasts of tobacco plants. Source: Gaglani (2006). Reproduced with permission of Dr. S. Gaglani.

selection of homoplastomic transformants is unpredictable. In 2002, S.M. Klaus and coworkers described the generation of marker-free chloroplast transformants in tobacco using the reconstitution of wild-type pigmentation in combination with plastid transformation vectors, which prevent stable integration of the kanamycin selection marker. One benefit of a procedure using mutants is that marker-free plastid transformants can be produced directly in the first generation (T_0) without retransformation or crossing.

Marker-free chloroplast genetic engineering

Chen and Melis (2013) applied a transgene expression method based on the replacement of an inactive *rbcL* gene as the selection marker in *C. reinhardtii* chloroplasts. The native *rbcL* gene in strain CC2653 has a point mutation that causes early translation termination, thus resulting in a photosynthesis mutant. Recovery of rbcL function for selection is offered along with the heterologous expression of the alcohol dehydrogenase *ADH1* gene from *S. cerevisiae* in the *Chlamydomonas* chloroplast. The *CrCpADH1* gene was inserted via double homologous recombination in the *psaB–rbcL* chloroplast intergenic region of recipient strain, using the *psaB* and *rbcL* gene sequences for the double homologous recombination. This transformation conferred a functional *rbcL* gene and expression of the *CrCpADH1* transgene in the recipient strain. This method alleviated the need to use antibiotics for selection, resulting in a negligible number of false positives during screening and attaining a high transformation efficiency. The approach also ensured segregation of chloroplast DNA copies, so as to achieve homoplasmy of the transformant chloroplast DNA, with a concomitant elimination of recipient strain Cp DNA. High levels of steady-state *CrCpADH1* transcripts were detected in the homoplasmic transformants. However, CrCpADH1 protein levels were attenuated under continuous illumination growth conditions due to oxygen accumulation in the cells. Under conditions of low oxygen partial pressure, or anoxia, accumulation of CrCpADH1 protein in the cells and ethanol in the growth medium was observed. A metabolic pathway for ethanol production has been proposed in *Chlamydomonas*, mediated by the chloroplast-localized CrCpADH1 transgenic enzyme.

Genetically engineered chloroplasts
Chloroplast transformation in oilseed rape

The chloroplast transformation vector pNRAB carries two expression cassettes for the spectinomycin resistance gene *aadA* and the insect resistance gene *cry1Aa10*. The two cassettes are situated between the *rps7* and *ndhB* targeting fragments. Biolistic delivery of the vector DNA, followed by spectinomycin selection, yielded chloroplast transformants at a frequency of four in 1000 bombarded cotyledon petioles. PCR analysis and Southern blot of PCR products confirmed the site-specific integration of *aadA* and *cry1Aa10* into the chloroplast genomes of transgenic oilseed rape. When transgenic oilseed rape leaves were fed to second instar *Plutella xylostera* larvae, 47% mortality was observed against this insect, and the surviving larvae had significantly lower weight than the control. This is the first report of chloroplast transformation in oilseed rape and the introduction of novel genes between the *rps7* and *ndhB* genes in the chloroplast genome.

Chloroplast transformation in tobacco

In 1999, Kavanagh and coworkers achieved efficient plastid transformation in *N. tabacum* using cloned plastid DNA of *Solanum nigrum* carrying mutations conferring spectinomycin and streptomycin resistance. The use of the incompletely homologous (homeologous) *Solanum* plastid DNA as donor resulted in a *Nicotiana* plastid transformation frequency comparable with that of other experiments where completely homologous plastid DNA was introduced. Physical mapping and nucleotide sequence analysis of the targeted plastid DNA region in the transformants demonstrated efficient site-specific integration of the 7.8-kb *Solanum* plastid DNA and the exclusion of the vector DNA. The integration of the cloned *Solanum* plastid DNA into the *Nicotiana* plastid genome involved multiple recombination events as revealed by the presence of discontinuous tracts of *Solanum*-specific sequences that were interspersed between *Nicotiana*-specific markers.

Huang *et al.* (2002) reported on the development of a new dominant selection marker for plastid transformation in higher plants using the aminoglycoside phosphotransferase gene *aphA-6* from *Acinetobacter baumannii*. Vectors containing chimeric *aphA-6* gene constructs were introduced into the tobacco chloroplast using particle bombardment of alginate-embedded protoplast-derived microcolonies or polyethylene glycol (PEG)-mediated DNA uptake. Targeted insertion into the plastome was achieved via homologous recombination, and plastid transformants were recovered on the basis of their resistance to kanamycin. Variations in kanamycin resistance in transplastomic lines were observed depending on the 5′ and 3′ regulatory elements associated with the *aphA-6* coding region. Transplastomic plants were fertile and showed maternal inheritance of the transplastome in the progeny.

Advantages of chloroplast genetic engineering
Environment-friendly technique
Unlike nuclear transformation, this method ensures that the recombinant transgenes are contained within the chloroplast and therefore will not spread to other plants. Chloroplasts (and the genes they contain) are not passed into the sperm (i.e., pollen) of a plant, so they cannot be spread by pollination. Researchers demonstrated that, even though chloroplasts in leaves were modified to express an insecticidal protein, called CRY, at very high levels (47% total soluble protein), the pollen did not contain any traces of the protein. This signifies that the recombinant genes and proteins are contained within the chloroplast, so this technique is environmental-friendly.

Large-scale protein production
Chloroplast engineering also allows for large-scale protein production. The levels of pharmaceutical proteins produced in nuclear-modified plants are less than 1% of the levels needed for the purified protein to be commercially feasible. Each plant cell contains approximately 100 chloroplasts and each chloroplast contains about 100 copies of its genome. So, chloroplast genetically engineered plants have high levels of integration of transgenes—up to 10 000 copies per cell—which elevate expression levels of recombinant proteins (up to 47% of the plant's total soluble protein).

Applications of chloroplast genetic engineering
Therapeutic proteins produced

A number of therapeutic proteins have been produced using the chloroplast genetic engineering system. These include human somatotropin (growth hormone), serum albumin (blood protein), insulin-like growth factor (hormone), antimicrobial peptides (proteins that kill pathogens), interferon alpha/gamma (cytokines in the immune system which are effective against hepatitis and leukemia), monoclonal antibodies (immune system molecules that fight off invading pathogens and toxins), and vaccines to cholera, plague, canine parvovirus (dog virus) and anthrax. Each of these proteins is clinically relevant and has not been produced efficiently in nuclear modified plants. Somatotropin, also known as human growth hormone, is used as therapy for stunted growth and even to help maintain muscle mass in patients. Serum albumin is the most widely used intravenous protein that is administered to replace blood volume since it accounts for 60% of blood protein composition. Insulin is a crucial hormone that regulates carbohydrate metabolism and therefore energy production. It was the first marketed therapeutic protein produced through genetic engineering (the pharmaceutical company Eli Lilly sold it as Humulin beginning in 1982). Most of the therapeutic proteins produced by chloroplast genetic engineering are still in the developmental stage and need to be tested in humans. Chlorogen Inc. is working to commercialize this technology and bring the plant-produced therapeutic proteins to the market. According to Chlorogen's site, their first product will be human serum albumin for the non-therapeutic market (Chlorogen Inc., URL: http://www.chlorogen.com). More work is required before chloroplast genetic engineering can be applied commercially. This work will probably include modifying more types of crops and plants as well as ensuring the functionality of the resultant therapeutic proteins in humans. But it may not be too far in the future when mothers may nag their children not only to eat their broccoli, but to eat their transgenes.

Vaccine production

Vaccines are, of course, needed to immunize people from harmful pathogens, such as the polio virus, but many times there is a shortage of the amount of vaccine available. Plague vaccine, which immunizes against *Yersinia pestis*, has been expressed in transgenic tobacco plants at commercially feasible levels. In addition, the canine parvorvirus vaccine (CPV), which protects dogs against CPV and stomach complications, has been expressed highly. Recently, a team of scientists working on chloroplast genetic engineering reported achieving such high levels of expression of the anthrax protective antigen that, according to their extrapolation, one acre of transgenic tobacco could produce about 400 million doses of the vaccine. This is a crucial property of vaccine production due to modern concerns over viral epidemics such as avian flu.

The possibility to directly manipulate chloroplast genome-encoded information has paved the way to detailed *in vivo* studies of virtually all aspects of plastid gene expression. Moreover, plastid transformation technologies have been intensely used in functional genomics by performing gene knockouts and site-directed mutagenesis of plastid genes. These studies have contributed greatly to our understanding of the physiology and biochemistry of bioenergetic processes inside the plastid compartment.

Plastid transformation is becoming more popular and an alternative to nuclear gene transformation because of various advantages like high protein levels, the feasibility of expressing multiple proteins from polycistronic mRNAs, and gene containment through the lack of pollen transmission. Recently, much progress in plastid engineering has been made. In addition to model plant tobacco, many transplastomic crop plants have been generated that possess higher resistance to biotic and abiotic stresses and molecular pharming. For some more examples of transplastomic plants developed so far through plastid engineering and the various applications and advantages of plastid transformation, refer to Wani *et al.* (2010).

Recent trends in chloroplast research

Genetic and genomic technologies have greatly boosted the rate of discovery and functional characterization of chloroplast proteins during the past decade. Indeed, data obtained using high-throughput methodologies, in particular proteomics and transcriptomics, are now routinely used to assign functions to chloroplast genes (see Chapter 13).

A generalized map of the chloroplast genome

Elegant molecular biology techniques including shotgun sequencing, rolling circle amplification (RCA), amplification, sequencing and annotation of plastome (ASAP), and next-generation sequencing are being used to accelerate data output. Owing to manifold increase in submission of chloroplast DNA sequences in nucleotide databases, challenges of in-depth data analysis stimulated the emergence of devoted annotation, assembling, and phylogenetic software. Recently, reported bioinformatics software for chloroplast genome studies comprise DOGMA for annotation, SCAN-SE, ARAGON, and PREP suit for RNA analyses and CG viewer for circular map construction/comparative analysis. Faster algorithms for gene-order-based phylogenetic reconstruction and bootstrap analysis have attracted the attention of research community. Current trends in sequencing strategies and bioinformatics with reference to chloroplast genomes hold great potential to illuminate more hidden corners of this ancient cell organelle. A generalized map of the chloroplast genome is given in Figure 10.21.

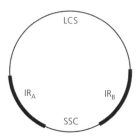

Figure 10.21 Generalized map of the chloroplast genome. IR_A, inverted repeat region A; IR_B, IR region B; LSC, large single-copy region; SSC, small single-copy region. Source: Khan et al. (2010). Reproduced with permission of *African Journal of Biotechnology*.

Establishment of redox markers and *in vivo* sensors

The area of redox regulation of chloroplast functions is emerging as a hot topic in chloroplast research, making it necessary to establish redox markers and *in vivo* sensors in chloroplast research. Redox-sensitive GFP (roGFP) has already been successfully targeted as sensor of the glutathione redox potential to chloroplasts of *Arabidopsis*. HyPer, a hydrogen peroxide sensor tested in cytosol and mitochondria of HeLa cells, and Redoxfluor, a redox sensor for cytosol and peroxisomes of yeast and Chinese hamster ovary (CHO) cells, might represent alternative genetically based redox sensors.

While reversible phosphorylation of thylakoid proteins is a well-characterized posttranslational modification in chloroplasts, protein *S*-nitrosylation has emerged as the most important mechanism for transduction of the bioactivity of nitric oxide, and also several chloroplast proteins have been described to become *S*-nitrosylated. Glutathionylation is a more recently described redox posttranslational modification and represents the major form of *S*-thiolation in cells by formation of a mixed disulfide between a free thiol on a protein and a molecule of glutathione. Glutathionylation is thought to occur under oxidative stress and can protect cysteine residues from irreversible oxidation, and alter positively or negatively the activity of diverse proteins. In *A. thaliana*, several chloroplast proteins have been described to be glutathionylated, including thioredoxin f during redox signaling. A systematic proteomic approach in *C. reinhardtii* identified 25 glutathionylation targets, mainly chloroplastic, involved in various metabolic processes.

Riboswitches

The further improvement of methods to detect *S*-nitrosylation and glutathionylation with high specificity and sensitivity will be crucial for the full dissection of the impact of these posttranslational modifications on the regulation of chloroplast processes. Application of deep sequencing approaches toward transcript quantification may provide a cheaper alternative to hybridization-based microarray platforms (Lister *et al.*, 2009). The analysis of inducible systems that allow one to transiently generate or complement lesions in organellar properties, thus enabling reconstruction of effects on nuclear and plastid gene expressions with high temporal resolution, promises to extend the power of transcriptomics. One emerging system is represented by riboswitches, which are natural RNA sensors that control gene expression via their capacity to bind small molecules (metabolites). They fold into RNA secondary structures whose conformation switches between an "on" state and an "off" state in response to ligand binding. Recently, a synthetic translational riboswitch controlled by the ligand theophylline was successfully employed in tobacco chloroplasts (Verhounig *et al.*, 2010).

Chloroplast interactome

Interactome is defined as the whole set of molecular interactions in cells. It is usually displayed as a directed graph. Molecular interactions can occur between molecules belonging to different biochemical families (proteins, nucleic acids, lipids, carbohydrates, etc.) and also within a given family. When spoken in terms of proteomics, interactome refers to protein–protein interaction network (PPIN), or protein interaction network (PIN). Another extensively studied type of interactome is the protein–DNA interactome (network formed by transcription factors and DNA or chromatin regulatory proteins) and their target genes. The word "interactome" was originally coined in 1999 by a group of French scientists headed by Bernard Jacq.

It has been suggested that the size of an organism's interactome correlates better than genome size with the biological complexity of the organism. Although PPI maps containing several thousands of binary interactions are now available for several organisms, none of them are presently complete and the size of interactomes is still a matter of debate. In 2010, the most complete gene interactome produced to date was compiled from 54 million two-gene comparisons to describe the interaction profiles for nearly 75% of all genes in the budding yeast, with 170 000 gene interactions (Costanzo *et al.*, 2010).

Identification of PPIs typically relies on purification of a bait protein and interacting prey proteins, using either a custom antibody specific for the bait or a commercially available antibody to a peptide or protein epitope tag fused to the N or C terminus of the bait. Interaction networks can be predicted on the premise that orthologous proteins that are known to interact in one organism can interact in the system under study. This approach was used recently to predict the interactome of *Arabidopsis*. Predictions are still at an early stage but are expected to improve rapidly as more interactome data become available in plants. The Search Tool for the Retrieval of Interacting Genes/Proteins (STRING) database combines physical and functional PPIs from 630 organisms (Jensen *et al.*, 2009).

Reconstruction of regulatory networks

The combinatorial analysis of multiple (transcriptomics and other "omics") datasets will be essential for the *in silico* reconstruction of regulatory networks. Eventually, a chloroplast development or functional state-driven nuclear gene expression network would be built based on genome-wide identification of transcription factors via yeast one-hybrid assays. Additionally, their direct and indirect target genes will be identified using chromatin immunoprecipitation with an antibody against such transcription factors followed by a deep sequencing approach and identification of genes whose expression levels in the loss-of-function mutant background are affected.

A novel rewired chloroplast-signaling pathway

Photosynthetic electron transport regulates chloroplast gene transcription through the action of a bacterial-type sensor kinase known as chloroplast sensor kinase (CSK). CSK represses photosystem I (PS-I) gene transcription in PS-I light and thus initiates photosystem stoichiometry adjustment. In cyanobacteria and in nongreen algae, CSK homologs coexist with their response regulator partners in canonical bacterial two-component systems. In green algae and plants, however, no response regulator partner of CSK is found. Yeast two-hybrid analysis has revealed interaction of CSK with sigma factor 1 (SIG1) of chloroplast RNA polymerase. Recently, Puthiyaveetil *et al.* (2013) have presented further evidence for the interaction between CSK and SIG1. CSK interacts with quinone. *Arabidopsis* SIG1 becomes phosphorylated in PS-I light, which then specifically represses transcription of PS-I genes. In view of the identical signaling properties of CSK and SIG1 and of their interactions, it has been suggested that CSK is a SIG1 kinase. The selective repression of PS-I genes arises from the operation of a gene-regulatory phosphoswitch in SIG1. The CSK–SIG1 system thus represents a novel, rewired chloroplast-signaling pathway created by evolutionary tinkering. This regulatory system supports a proposal for the selection pressure behind the evolutionary stasis of chloroplast genes.

Tailored Chloroplasts

The ability to transform chloroplasts by homologous transformation and drive high-level expression of transgenes in chloroplasts, coupled with their maternal mode of inheritance in most species of interest, makes chloroplasts a prime target for biotechnological improvement of crop plants. However, commercial varieties harboring transgenic chloroplasts have not been generated yet. One promising target for modifying chloroplast functions is photorespiration. Photorespiration results from the oxygenase reaction catalyzed by RuBisCo and serves as a carbon recovery system. It comprises enzymatic reactions distributed in chloroplasts, peroxisomes, and mitochondria (Maurino and Peterhansel, 2010; Armbruster *et al.*, 2011). The *E. coli* glycolate catabolic pathway has been introduced into *A. thaliana* chloroplasts to reduce the loss of fixed carbon and nitrogen that occurs in C3 plants when phosphoglycolate, an inevitable by-product of photosynthesis, is recycled by photorespiration. The resulting transgenic plant produced more biomass, giving rise to the hope that the manipulation of photorespiration can indeed be used to improve agronomic performance.

Summary

In plants, plastids may differentiate into several forms, depending upon which function they need to play in the cell. Plastids are thought to have originated from endosymbiotic cyanobacteria. Promiscuity of DNA has provided support to the endosymbiont hypothesis. Chloroplasts convert light energy from the sun into ATP through a process called "photosynthesis." Efforts to identify the entire complement of chloroplast proteins and their interactions are progressing rapidly, making the organelle a prime target for systems biology research in plants. Plastids contain DNA and are produced by division of existing plastids. In chloroplast division, the plastid-dividing (PD) ring is a main structure of the PD machinery. Cyanelle DNA resembles both chloroplast DNA and cyanobacteriun DNA. Cyanelles may thus represent a bridge between cynobacteria and chloroplasts. All cyanobacteria and most algae possess an inorganic carbon-concentrating mechanism (CCM) that involves a microcompartment—carboxysomes in prokaryotes and pyrenoids in eukaryotes. Plastid DNA exists as large protein–DNA complexes associated with the inner envelope membrane and is called "plastid nucleoids." Each plastid creates 10–1000 copies of the circular 75–250 kbp plastome. There is conclusive evidence at molecular level that chloroplast DNA is inherited in a maternal manner. In the case of *P. zeylanica*, cytoplasmic genes are transferred to egg through the male parent. More than 100 plastomes have been sequenced so far. DNA sequencing of plastids is helping a great deal in understanding interesting features of chloroplast genome and gene organization, which in turn is playing a role in understanding the systems biology in chloroplast-to-nucleus signals during chloroplast development. Chloroplasts not only receive signals from the nucleus but also send signals to the nucleus via chloroplast-to-nucleus retrograde signaling.

The plastid operons show that genes with common functions are often co-localized in operons. Chloroplast development requires coordinated expression of plastid and nuclear genes, and, therefore, activation of plastid gene transcription must be coordinated with the activation of nuclear gene expression by signals that can be transmitted between plastids and the nucleus. The study of chloroplast gene expression also provides an opportunity to understand how plants sense and alter gene expression in response to light. Light is important for photosynthesis, a trigger of plant developmental processes, and a potential source of injury that modulates chloroplast transcription. Several nuclear-encoded proteins that participate in diverse plastid RNA processing events have been characterized. Many of them seem to belong to the PPR protein family. The CSK–SIG1 represents a novel, rewired chloroplast-signaling pathway created by evolutionary tinkering. This regulatory system supports a proposal for the selection pressure behind the evolutionary stasis of chloroplast genes.

Posttranscriptional and translation regulatory mechanisms have been found to exist in chloroplasts. Transcription plays a central role in establishing the levels of many plastid mRNAs and proteins. Variation in plastid mRNA stability also significantly influences plastid gene expression. The stability of many plastid mRNAs is similar to the stability of mRNAs found in eukaryotic organisms. A nuclear-encoded plastid-localized RNA polymerase has been discovered recently. Chaperones import proteins into chloroplasts for their development. This development is regulated by plant hormones such as cytokinin and brassinosteroid. The phytochrome-interacting factor PIF3 has been proposed to act as a positive regulator of chloroplast development. Of the two photosystems, PS-I seemed to develop more rapidly than PS-II. Some nuclear genes have been identified that are required for early chloroplast development. Accumulation and storage of mRNAs coding for RNA polymerase components and sigma factors in dry seeds has been observed. These mRNAs seem to provide immediate-to-use templates for translation on cytoplasmic ribosomes in order to enhance RNA polymerase protein levels and to provide regulatory proteins for stored PEP to guaranty efficient plastid genome transcription during germination.

A novel system that appears to circumvent the concerns about nuclear modification is genetic engineering of chloroplast DNA. There are many advantages to chloroplast genetic engineering. The technique is environmental-friendly. It allows large-scale protein production. Plastid transformation is becoming more popular and an alternative to nuclear gene transformation. Several therapeutic proteins and vaccines have been produced. Recently, a generalized map of the chloroplast genome has been established. The area of redox regulation of chloroplast functions is emerging as a hot topic in chloroplast research. A synthetic translational riboswitch controlled by the ligand theophylline was successfully employed in chloroplasts. PPIs involved in chloroplast development and function are being identified. A chloroplast development or functional state-driven nuclear gene expression network is being constructed. Through chloroplast genetic engineering, manipulation of photorespiration can indeed be used to improve agronomic performance of plants.

References

Allorent, G., Courtois, F., Chevalier, F. and Lerbs-Mache, S. (2013) Plastid gene expression during chloroplast differentiation and dedifferentiation into non-photosynthetic plastids during seed formation. *Plant Molecular Biology*, **82**, 59–70.

Aluru, M.M.R., Zola, M.J., Foudree, M.A. and Rodermel, M.S.R. (2009) Chloroplast photooxidation-induced transcriptome reprogramming in *Arabidopsis* immutans white leaf sectors. *Plant Physiology*, **150**, 904–923.

Armbruster, U., Pesaresi, P., Pribil, M. *et al.* (2011) Update on chloroplast research: New tools, new topics, and new trends. *Molecular Plant*, **4**, 1–16.

Austin, J.R. II, Frost, E., Vidi, P.A. et al. (2006). Plastoglobules are lipoprotein subcompartments of the chloroplast that are permanently coupled to thylakoid membranes and contain biosynthetic enzymes. *The Plant Cell*, **18** (7), 1693–1703.

Biswal, B. and Raval, M.K. (2003). *Chloroplast Biogenesis from Proplastid to Gerontoplast*. Kluwer Academic Publishers, Dordrecht.

Bréhélin, C. and Kessler, F. (2008). The plastoglobule: a bag full of lipid biochemistry tricks. *Photochemistry and Photobiology*, **84** (6), 1388–1394.

Breitmaier, E. (2006). *Terpenes, Flavors, Fragrences, Pharmaca, Pheromones*. Wiley-WCH, Weinheim.

Buchanan, B.B., Gruissem, W., and Jones, R.L. (2000). *Biochemistry and Molecular Biology of Plants*, American Society of Plant Physiologists, Rockville, MD.

Cahoon, A.B., Sharpe, R.M., Mysayphonh, C. *et al.* (2004) The complete chloroplast genome of tall fescue (*Lolium arundinaceum*; Poaceae) and comparison of whole plastomes from the family Poaceae. *Molecular Biology and Evolution*, **21**, 1445–1454.

Caiola, M.G. and Canini, A. (2004). Ultrastructure of chromoplasts and other plastids in *Crocus stivus L.* (Iridaceae). *Plant Biosystems*, **138**, 43–52.

Chan, C.X. and Bhattacharya, D. (2010). The origins of plastids. *Nature Education*, **3** (9), 84.

Cattolico, R.A., Jacobs, M.A., Zhou, Y. *et al.* (2008) Chloropalst genome sequencing analysis of *Heterosigma akashiwo* CCMP 452 (West Atlantic) and NIES 293 (West Pacific) strains, *BMC Genomics*, **9**, 211.

Chen, H.-C. and Melis, A. (2013) Marker-free genetic engineering of the chloroplast in the green microalga *Chlamydomonas reinhardtii*. *Plant Biotechnology Journal*, **11**, 818–828.

Constan, D., Froehlich, J.E., Rangarajan, S. and Keegstra, K. (2004) A stromal Hsp100 protein is required for normal chloroplast development and function in *Arabidopsis*. *Plant Physiology*, **136**, 3605–3615.

Corneille, S., Lutz, K., Svab, Z. and Maliga, P. (2001) Efficient elimination of selectable markers from the plastid genome by VRE-lox site specific recombination system. *The Plant Journal*, **27**, 171–178.

Costanzo, M., Baryshnikova, A., Bellay, J. *et al.* 2010, The genetic landscape of a cell. *Science*, **327**, 425–431.

de la Luz Gutiérrez-Nava, M., Gillmor, C.S., Jimenez, L.F. et al. (2004). Chloroplast biogenesis genes act cell and non cell autonomously in early development. *Plant Physiology*, **135**, 471–482.

Diekmann, K., Hodkinson, T.R., Wolfe, K.H. *et al.* (2009) Complete chloroplast genome sequence of a major allogamous forage species, perennial ryegrass (*Lolium perenne* L.). *DNA Research*, **16** (3), 165–176.

Diekmann, K., Hodkinson, T.R., Wolfe, K.H. *et al.* (2010) The complete chloroplast genome sequence of perennial ryegrass (*Lolium perenne* L.) reveals useful polymorphisms among European ecotypes. In: Huyghe, C. (ed.), *Sustainable Use of Genetic Diversity in Forage and Turf Breeding*, Part 5. Springer, London, pp. 409–411.

Evert, R.F. (2006). *Esau's Plant Anatomy, Meristems, Cells, and Tissues of the Plant Body: Their Structure, Function, and Development*, 3rd edn. John Wiley & Sons, Inc., Hoboken, NJ.

Friso, G., Giacomelli, L., Ytterberg, A.J. et al. (2004). In-depth analysis of the thylakoid membrane proteome of *Arabidopsis thaliana* chloroplasts: new protein, new functions and a plastid proteome database. *The Plant Cell*, **16**, 478–499.

Gaglani, S. 2006, Chloroplast genetic engineering. *Harvard Science Review*, **37**(Fall), 36–39.

Gao, L., Yi, X., Yang, Y.-X. *et al.* (2009) Complete chloroplast genome sequence of a tree fern *Alsophila spinulosa*: insights into evolutionary changes in fern chloroplast genomes. *BMC Evolutionary Biology*, **9**, 130.

Gardner, E.J., Simmons, M.J., and Snustad, D.P. (2005) *Principles of Genetics*, 8th edn., John Wiley & Sons, Singapore.

Giraud, E., Van Aken, O., Ho, L.H.M. and Whelan, J. (2009) The transcription factor ABI4 is a regulator of mitochondrial retrograde expression of *ALTERNATIVE OXIDASE1a*. *Plant Physiology*, **150**, 1286–1296.

Guisinger, M.M., Kuehl, J.V., Boore, J.L. and Jansen, R.K. (2011) Extreme reconfiguration of plastid genomes in the angiosperm family geraniaceae: rearrangements, repeats, and codon usage. *Molecular Biology and Evolution*, **28**, 583–600.

Gunning, B.E.S. (2005). Plastid stromules: video microscopy of their outgrowth, retraction, tensioning, anchoring, branching, bridging and tip-shedding. http://citeseerx.ist.psu.edu/viewdoc/download?doi=10.1.1.132.3222&rep=rep1&type=pdf (accessed August 1, 2016).

Hanson, M.R. and Sattarzadeh, A. (2013). Trafficking of proteins through plastid stromules. *The Plant Cell*, doi: http://dx.doi.org/10.1105/tpc.113.112870.

Hoober, J.K. (2006). Photosynthesis. In: *Plant Cell Biology*, Dashek, W.V. and Harrison, M., (eds) Science Publishers, Enfield, NH.

Howitt, C.A. and Pogson, B.J. (2006). Carotenoid accumulation and function in seeds and non-green tissues. *Plant Cell and Environment*, **29**, 435–445.

Hsieh, K. and Huang, A.H.C. (2005). Lipid-rich tapetosomes in *Brassica tapetum* are composed of oleosin-coated oil droplets and vesicles, both assembled and then detached from the endoplasmic reticulum. *The Plant Journal*, **43**, 889–899.

Hsieh, K. and Huang, A.H.C. (2007). Tapetosomes in brassica tapetum accumulate endoplasmic reticulum–derived flavonoids and alkanes for delivery to the pollen surface. *The Plant Cell*, **19** (2), 582–596.

Huang, F.C., Klaus, S.M.J., Herz, S. *et al.* (2002) Efficient plastid transformation in tobacco using the *apha-6* gene and kanamycin selection. *Molecular Genetics and Genomics*, **268**, 19–27.

Hurkman, W.J. and Kennedy, G.S. (1976). Fine structure and development of proteoplasts in primary leaves of mung bean. *Protoplasma*, **89** (1), 171–184.

Jansen, R.K. and Palmer, J.D. (1987) A chloroplast DNA inversion marks an ancient evolutionary split in the sunflower family (Asteraceae). *Proceedings of the National Academy of Sciences of the United States of America*, **84**, 5818–5822.

Jensen, L.J., Kuhn, M., Stark, M. *et al.* (2009) STRING 8: a global view on proteins and their functional interactions in 630 organisms. *Nucleic Acids Research*, **37**, D412-D416.

Jeong, S.Y., Rose, A. and Meier, I. (2003) MFP1 is a thylakoid-associated, nucloid-binding protein with a coiled-coil structure. *Nucleic Acids Research*, **31**, 5175–5185.

Jo, Y.D., Park, J., Kim, J. *et al.* (2010) Complete sequencing and comparative analyses of the pepper (*Capsicum annuum* L.) plastome revealed high frequency of tandem repeats and large insertion/deletions in pepper plastome. *Plant Cell Reports*, **30**, 2, 217–229.

Khan, A., Khan, I.A., Asif, H. and Azim, M.K. (2010) Current trends in chloroplast genome research. *African Journal of Biotechnology*, **9**, 3494–3500.

Kleffmann, T., Russenberger, D., Von Zychlinski, A. et al. (2004). The chloroplast proteome reveals pathway abundance and novel protein functions. *Current Biology*, **14**, 354–362.

Kleine, T., Maier, U.G. and Leister, D. (2009). DNA transfer from organelles to the nucleus: the idiosyncratic genetics of endosymbiosis. *Annual Review of Plant Biology*, **60**, 115–138.

Koussevitzky, S., Nott, A., Mockler, T.C. *et al.* (2007) Multiple signals from damaged chloroplasts converge on a common pathway to regulate nuclear gene expression. *Science*, **316**, 715–719.

Krupinska, K. (2006). Fate and activity of plastids during leaf senescence. In: *The Structure and Function of Plastids*, Wise, J.K. and Hoober, K. (eds) Springer, Dordrecht.

Kwiatkowska, M., Stepinski, D., Poplonska, K. et al. (2010). Elaioplasts of *Haemanthus albiflos* are true leptubuloids: cytoplasmic domains rich in lipid bodies entwined by microtubules. *Acta Physiologial Plantaium*, **32**, 1189–1196.

Lister, R., Gregory, B.D. and Ecker, J.R. (2009) Next is now: new technologies for sequencing of genomes, transcriptomes, and beyond. *Current Opinions in Plant Biology*, **12**, 107–118.

Liu, R.H. (2007). Whole grain phytochemicals and health. *Journal of Cereal Science*, **46** (3), 207–219.

Liu, H., Zhang, H., Niedzwiedski, D.M. *et al.* (2013). Phycobilisomes supply excitations to both photosystems in a megacomplex in cyanobacteria. *Science*, **342**, 1104–1107.

Ljubešic, N. (1970) Fine structure of developing chromoplasts in outer yellow fruit parts of *Cucurbita pepo* cv. Pyriformis. *Acta Botanica Croatica*, **29**, 51–56.

Lopez-Jurez, E. and Pylse, K.A. (2005). Plastids unleashed: their development and their integration in plant development. *Developmental Biology*, **49**, 557–577.

Mackender, R.O. (1978). Etioplast development in dark-grown leaves of *Zea mays*. *Plant Physiology*, **62**, 499–505.

McFadden, G.I. and Ralph, S.A. (2003). Dynamin: the endosymbiosis ring of power. *Proceedings of National Academy of Sciences of the United States of America*, **100**, 3557–3559.

Maurino, V.G. and Peterhansel, C. (2010) Photorespiration: current status and approaches for metabolic engineering. *Current Opinions in Plant Biology*, **13**, 249–256.

McCoy, S.R., Kueh, J.V., Boore, J.L. and Raubeson, L.A. (2008) The complete plastid genome sequence of *Welwitschia mirabilis*: an unusually compact plastome with accelerated divergence rates. *BMC Evolutionary Biology*, **8**, 130.

Mettal, U., Boland, W., Beyer, P., and Kleinig, H. (1988). Biosynthesis of monoterpene hydrocarbons by isolated chromoplasts from daffodil flowers. *European Journal of Biochemistry*, **170**, 613–616.

Morsy, M., Gouthu, S., Orchard, S. *et al.* (2008) Charting plant interactomes: possibilities and challenges. *Trends in Plant Science*, **13**, 183–191.

Moulin, M., McCormac, A.C., Terry, M.J. and Smith, A.G. (2008) Tetrapyrrole profiling in Arabidopsis seedlings reveals that retrograde plastid nuclear signaling is not due to Mg-protoporphyrin IX accumulation. *Proceedings of the National Academy of Sciences of the United States of America*, **105**, 15178–15183.

Nacir, H. and Bréhélin, C. (2013) When proteomics reveals unsuspected roles: the plastoglobule example. *Frontiers in Plant Science*, **4**, 114, doi:10.3389/fpls.2013.00114.

Nakano, T., Kimura, T., Kaneko, I. *et al.* (2001) Molecular mechanism of chloroplast development regulated by plant hormones. *RIKEN Review*, **41**, 86–87.

Palmer, J.D. 1985, Comparative organization of chloroplast genomes. *Annual Review of Genetics*, **19**, 325–354.

Palmer, J.D. and Zamir, D. (1982) Chloroplast DNA evolution and phylogenetic relationships in *Lycopersicon*. *Proceedings of the National Academy of Sciences of the United States of America*, **79**, 5006–5010.

Palmer, J.D., Nugent, J.M., and Herbon, L.A. (1987) Unusual structure of geranium chloroplast DNA: a triple-sized inverted repeat, extensive gene duplications, multiple inversions, and two repeat families. *Proceedings of the National Academy of Sciences of the United States of America*, **84**, 769–773.

Priest, H.D., Filichkin, S.A. and Mockler, T.C. (2009) Cis-regulatory elements in plant cell signaling. *Current Opinions in Plant Biology*, **12**, 643–649.

Puthiyaveetil, S., Ibrahim, I.M. and Allen, J.F. (2013) Evolutionary rewiring: a modified prokaryotic gene-regulatory pathway in chloroplasts. *Philosophical Transactions of the Royal Society London Biological Sciences B*, **368**, 1622.

Rosinski, J., Hainfeld, F., Rigbi, M., and Siegelman, H.W. (1981). Phycobilisome ultrastructure and chromatic adaptation in *Fremyella diplosiphon*. *Annals of Botany*, **47**, 1–12.

Russel, S.D. (1980) Participation of male cytoplasm during gamete fusion in an angiosperm *Plumbago zeylanica*. *Science*, **210**, 200–201.

Salema, R. and Badenhuizen, N.P. (1969) Nucleic acids in plastids and starch formation. *Acta Botanica Neerlandica*, **18** (1), 203–215.

Samson, N., Bausher, M.G., Lee, S.-B. *et al.* (2007) The complete nucleotide sequence of the coffee (*Coffea arabica* L.) chloroplast genome: organization and implications for biotechnology and phylogenetic relationships amongst angiosperms. *Plant Biotechnology Journal*, **5**, 339–353.

Saski, C., Lee, S.-B., Daniell, H. *et al.* (2005) Complete chloroplast genome sequence of *Glycine max* and comparative analyses with other legume genomes. *Plant Molecular Biology,* **59**, 309–322.

Sato, S., Nakamura, Y., Kaneko, T. et al. (1999). Complete structure of the chloroplast genome of *Arabidopsis thaliana. DNA Research,* **6** (5), 283–290.

Siddique, M.A., Grossmann, J., Gruissem, W., and Baginsky, S. (2006). Proteome analysis of bell pepper (*Capsicum annuum* L.) chromoplasts. *Plant and Cell Physiology,* **47**, 1663–1673.

Sperling, U., Franck, F., Van Cleve, B. et al. (1998). Etioplast differentiation in Arabidopsis: both PORA and PORB restore the prolamellar body and photoactive protochlorophyllide-F655 to the *cop1* photomorphogenic mutant. *The Plant Cell,* **10**, 283–296.

Staehelin, L.A. (2006) Chloroplast structure: from chlorophyll granules to supramolecular architecture of thylakoid membranes. In: *Discoveries in Photosynthesis. Advances in Photosynthesis and Respiration,* Vol. **20** (Govindjee, Beatty, J.T., Gest, H. and Allen, J.F., eds), Springer, The Netherlands, pp. 717–728.

Stensballe, A., Haid, S., Bauro, G. et al. (2008). The amyloplast proteome of potato tuber. *FEBS Journal,* **275**, 1723–1741.

Stephenson, P.G., Fankhauser, C. and Terry, M.J. (2009) PIF3 is a repressor of chloroplast development, *Proceedings of the National Academy of Sciences of the United States of America,* **106**, 7654–7659.

Turmel, M., Gagnon, M.-C., O'Kelly, C.J. *et al.* (2009) The chloroplast genomes of the green algae *Pyramimonas, Monomastix,* and *Pycnococcus* shed new light on the evolutionary history of prasinophytes and the origin of the secondary chloroplasts of euglenids. *Molecular Biology and Evolution,* **26**, 631–648.

Verhounig, A., Karcher, D. and Bock, R. (2010) Inducible gene expression from the plastid genome by a synthetic riboswitch. *Proceedings of the National Academy of Sciences of the United States of America,* **107**, 6204–6209.

Vigil, E.L. and Ruddat, M. (1985). Development and enzyme activity of protein bodies in proteinoplasts of tobacco root cells. *Histochemistry,* **83**, 17–27.

von Zychlinski, A., Kleffmann, T., Krishnamurthy, N. *et al.* (2005). Proteome analysis of the rice etioplast: metabolic and regulatory networks and novel protein functions. *Molecular Cell Proteomics,* **4** (8), 1072–1084.

Wani, H., Shabir, H., Nadia, K. *et al.* (2010) Plant plastid engineering. *Current Genomics,* **11**, 500–512.

Wischmann, B., Nielsen, T.H., and Moller, B.L. (1999). *In vitro* biosynthesis of phosphorylated starch in intact potato amyloplasts. *Plant Physiology,* **119**, 455–462.

Wise, R.R. and Hoober, J.K. (2006). *The Structure and Function of Plastids.* Springer, Dordrecht.

Wu, S.S.H., Moreau, R.A., Whitaker, B.D., and Huang, A.H.C. (1999). Steryl esters in the elaioplasts of the tapetum in developing *Brassica* anthers and their recovery on the pollen surface. *Lipids,* **34**, 517–523.

Yang, M., Zhang, X., Liu, G. *et al.* (2010) The complete chloroplast genome sequence of date palm (*Phoenix dactylifera* L.). *PLoS ONE,* **5** (9), e12762.

Yao, C.-W., Hsu, B.-D. and Chen, B.-S. (2011) Constructing gene regulatory networks for long term photosynthetic light acclimation in *Arabidopsis thaliana. BMC Bioinformatics,* **12**, 335. doi:10.1186/1471-2105-12-335

Yoshida, Y., Kuroiwa, H., Misumi, O. *et al.* (2010). Chloroplasts divide by contraction of a bundle of nanofilaments consisting of polyglucan. *Science,* **329**, 949–953.

Yukawa, M. and Sugiura, M. (2013) An additional pathway to translate the downstream *ndhK* cistron in partially overlapping ndhC-ndhK mRNAs in chloroplasts. *Proceedings of the National Academy Sciences of the United States of America,* **110**, 5701–5706.

Zeng, Y., Pan, Z., Ding, Y. *et al.* (2011). A proteomic analysis of the chromoplasts isolated from sweet orange fruits [*Citrus sinensis* (L.) Osbeck]. *Journal of Experimental Botany,* **62**, 5297–5305.

Zhang, D.-P. (2007) Signaling to the nucleus with a loaded GUN. *Science*, **316**, 700–701.

Zhang, T., Fang, Y., Wang, X. *et al.* (2012) The complete chloroplast and mitochondrial genome sequences of *Boea hygrometrica*: Insights into the evolution of plant organellar genomes. *PLoS ONE*, **7**(1), e30531.

Further reading

Bock, R. (2015) Engineering plastid genomes, methods, tools and applications in basic research and biotechnology. *Annual Review of Plant Biology*, **66**, 211–246.

Khan, M.S. (2012) Plastid genome engineering in plants: Present status and future trends. *Molecular Plant Breeding*, **3** (9), 91–102.

Liu, T., Zhang, C., Yan, H., Zhang, L., Ge, X. and Hao, G. (2016) Complete plastid genome sequence of *Primula sinensis* (Primulaceae): structure comparison, sequence variation and evidence for *accD* transfer to nucleus. *PeerJ*, **4**, e2101, DOI: 10.7717/peerj.2101.

Park, J.H. and Lee, J. (2016) The complete plastid genome of *Scopolia parviflora* (Dunn.) Nakai (Solanaceae). *Korean Journal of Plant Taxonomy*, **46** (1), 60–64.

Rogalski, M., Vieira, L.N., Fraga, H.P. and Guerra, M.P. (2015) Plastid genomics in horticultural species: importance and applications for plant population genetics, evolution, and biotechnology. *Frontiers in Plant Science*, **6**, 586, doi: 10.3389/fpls.2015.00586.

Smith, D.R. and Keeling, P.J. (2015) Mitochondrial and plastid genome architecture: reoccurring themes, but significant differences at the extremes. *Proceedings of the National Academy of Sciences of the United States of America*, **112** (33), 10177–10184.

Smith, D.R. and Lee, R.W. (2014) A plastid without a genome: evidence from the nonphotosynthetic green algal genus *Polytomella*. *Plant Physiology*, **164**, 1812–1819.

Theng, S. and Wollman, F.-A. (2014) *Plastid Biology. Advances in Plant Biology*, Springer, New York, NY.

van Wijk, K.J. and Baginsky, S. (2011) Plastid proteomics in higher plants: current state and future goals. *Plant Physiology*, **155** (4), 1578–1588.

Wu, Z., Gu, C., Tembrock, L.R. and Ge, S. (2015) Limited polymorphisms between two whole plastid genomes in the genus *Zizania* (Zizaniinae). *Journal of Proteomics and Bioinformatics*, **8**, 253–259.

CHAPTER 11

Photosynthesis

J. Kenneth Hoober

School of Life Sciences, Center for Photosynthesis, Arizona State University, Tempe, AZ, USA

Introduction

Sunlight provides a steady, abundant supply of energy to a world increasingly concerned about energy. The cause of this concern is a disconnect between inputs and outputs of energy forms. Forms of energy that traditionally have been used to drive our mechanical devices are different from those that support living systems. Nuclear energy, which is fraught with waste problems, and hydroelectric, geothermal, or wind power are all forms that can be used for mechanical devices but provide a minor fraction of the total energy input needed in the world economy. The bulk of the available energy has been, and still is, derived from light energy from the sun. Technology has been developed to convert photons to electrons with photovoltaic cells that use absorption of light energy by silicon semiconductor devices to generate electricity as the end product. A more efficient technology is solar thermal energy systems that heat a fluid to drive a steam-powered turbine. These approaches have great promise for a sustainable energy source and are beginning to compete economically with fossil fuels. Another fuel in the future may be photochemical reduction of protons (H^+) to hydrogen (H_2).

Combustion of fossil fuels (coal, oil, and natural gas) is the major source of energy used to generate electricity and power machines. Fossil fuels are products derived from **ancient photosynthesis** but, with a few exceptions, are not directly usable by living systems. Life is sustained by converting photons to electrons through **contemporary photosynthesis**. Except for organisms that live next to thermal vents in the deep oceans and those that can metabolize hydrocarbons, life is sustained by carbohydrates produced by photosynthesis. The major end product of photosynthesis is **glucose 6-phosphate**, which is the predominant starting material for synthesis of cellular components. In addition, metabolism of glucose 6-phosphate provides the energy required for synthetic reactions and growth. In contrast to animals, **glucose** does not occur in the free

Plant Cells and their Organelles, First Edition. Edited by William V. Dashek and Gurbachan S. Miglani.

form to a significant amount in plants. Of fundamental importance to the biosphere, however, is the fact that plants are the primary source of the glucose unit, which they store in polymeric form.

Aerobic metabolism of glucose in all organisms produces CO_2 and H_2O. Persistence of life on earth would not have been possible without a means to convert these "waste products" back to glucose. Otherwise, living organisms would long ago have run out of food. The conversion of CO_2 and H_2O to the 6-carbon molecule glucose $(C_6H_{12}O_6)$ requires an input of a substantial amount of energy, as illustrated by the following relationship:

$$CO_2 + H_2O + [470 \text{ kJ / mol of } CO_2] \longleftrightarrow 1/6(C_6H_{12}O_6) + O_2. \qquad (11.1)$$

The energy required for complete synthesis of a mole of glucose is six times that shown in Equation 11.1. The energy used to drive this reaction is provided by the absorption of light. The molecular apparatus that performs this process—photosynthesis—is located in **thylakoid** (from the Greek, "sac-like") **membranes**, which are confined within the **chloroplast** in plants. Photosynthesis is driven by absorption of light energy by **chlorophyll** (Chl). Wavelengths of light that contain levels of energy that can be accommodated by the electronic structure of Chl, primarily in the blue and red regions of the visible spectrum (photons with energy levels around 280 and 170 kJ/mol, respectively), are absorbed and generate "excited" states of Chl. The photosynthetic apparatus is designed to collect these packets of light energy and convert them to chemical compounds with sufficiently high levels of energy to drive synthesis of glucose from CO_2 and H_2O. The major metabolically available end products that are stored in plants are **starch**, a polymeric form of glucose, or **sucrose**, a disaccharide. Another polymer of glucose is **cellulose**, which is metabolically available to several microorganisms but not to animals. Cellulose is used as a structural component by plants.

Expanded descriptions of various aspects of photosynthesis can be found in Raghavendra (1998), Blankenship (2002), Wise and Hoober (2006), and Rebeiz *et al.* (2010).

Evolution of photosynthesis

Synthesis of tetrapyrroles

Photosynthesis is nearly as old as life itself. Measurements of the carbon isotope content in early fossils suggest that a reaction similar to that catalyzed by the enzyme **ribulose 1,5-bisphosphate carboxylase/oxygenase** (Rubisco) fixed CO_2 into organic material. Because this reaction involves CO_2 as a reactant, CO_2 molecules containing the heavier isotopes of carbon, i.e., ^{13}C and ^{14}C, react slightly more slowly than $^{12}CO_2$. Thus, direct incorporation of CO_2 into organic molecules, which depends upon diffusion, discriminates against the heavier isotopes. A measurable extent of this type of discrimination, resulting in an enrichment of ^{12}C, is thought to be possible only with enzymatically catalyzed reactions that fix CO_2, and thus isotope ratios have been used as a signature of biological activity. Microfossils with such a signature are present in the oldest rocks on earth, 3.7–3.8 billion years (Hohmann-Marriott and Blankenship, 2011).

Tetrapyrroles, required cofactors in photosynthesis, are among the most ancient of biological molecules (see Figure 11.1 for structures). The availability of cyclic, or rather "macrocyclic," tetrapyrroles that chelate a divalent cation—most commonly Fe^{2+} or Mg^{2+} ion—allowed the development of energy transduction mechanisms, either through electron transport (oxidation–reduction) of the central **Fe** atom in heme or through the generation of high-energy, excited states of **Mg**-tetrapyrroles, as with the Chls. As deduced from a phylogenetic analysis of genes encoding biosynthetic enzymes, bacteriochlorophyll (BChl) (Figure 11.1) was the evolutionary precursor of Chl (Hohmann-Marriott and Blankenship, 2011). A comparison of genomic sequences provided evidence that these genes were subsequently introduced into green sulfur bacteria, green nonsulfur bacteria, and cyanobacteria by lateral gene transfer.

Of the two types of photosynthetic **reaction centers**, the type I, with an iron–sulfur complex as the electron acceptor, emerged among the heliobacterial and green sulfur bacterial lines. Type II, with a quinone as the electron acceptor, emerged among the purple bacteria. The two photosystems were combined for the first time in cyanobacteria, possibly by lateral gene transfer. Cyanobacteria were also the first to use Chl rather than BChl as the primary pigment. Consequently, with the higher energies achieved by excited states of Chl, because Chl absorbs shorter wavelengths of light, it became possible about 3.5 billion years ago (BYA) to span the difference in redox potential from oxidation of water to oxygen ($E'_m = +0.816$ V) to reduction of $NADP^+$ to NADPH ($E'_m = -0.342$ V). The result was the generation of molecular oxygen, which, however, did not appear at a significant level in the atmosphere until about 2.7 BYA (Hohmann-Marriott and Blankenship, 2011). Interestingly, the genes for the reaction center proteins that *bind* (B)Chl have a deeper lineage than those that encode the enzymes that catalyze *synthesis* of (B)Chl. The lack of congruence of the genes that encode the BChl/Chl biosynthetic enzymes and the reaction center proteins that bind these pigments indicate that the photosynthetic apparatus is a composite structure, with components recruited from multiple sources. This concept suggests that the earliest proteins for photosynthetic functions were recruited from those already present. Of particular interest is the proposal that *photosynthetic* reaction center proteins were derived from *respiratory* cytochrome *b* by gene duplication and subsequent divergence of the genes to encode proteins with new functions, from the one that bound Fe-porphyrin cofactors (the heme in cytochromes) to the one that bound Mg-chlorin (Chl) cofactors (Xiong *et al.*, 2000).

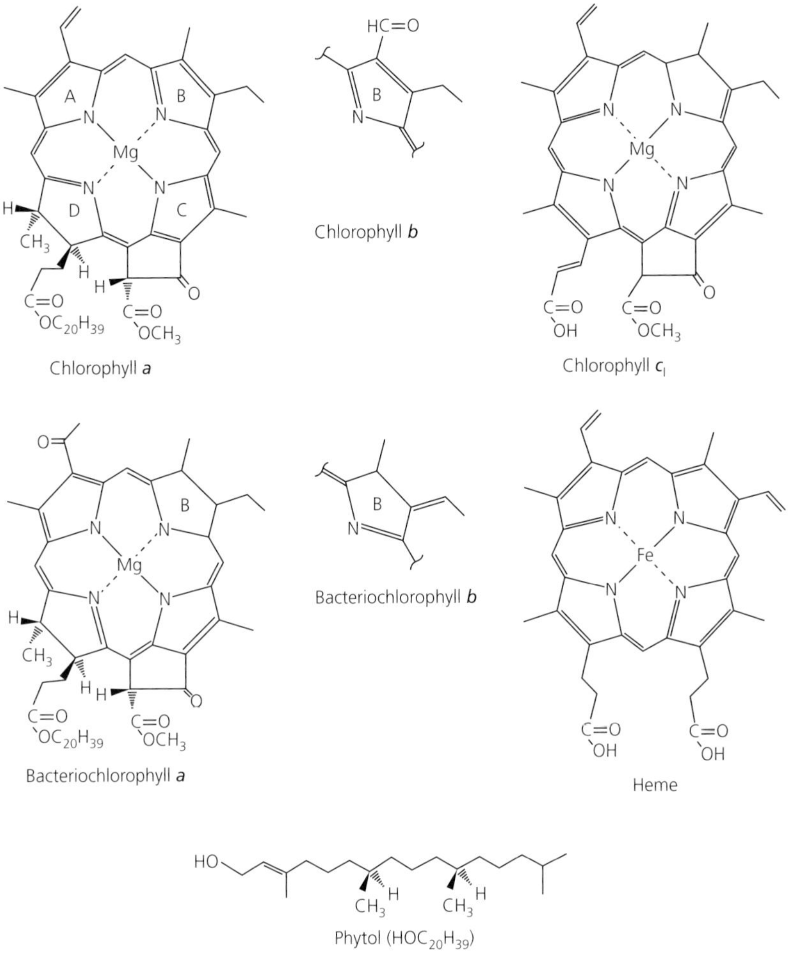

Figure 11.1 The major tetrapyrrole compounds in photosynthetic systems are chlorophylls and hemes. These functional tetrapyrroles are metal complexes, in which either Mg^{2+} or Fe^{2+} is inserted into the molecule during biosynthesis at the stage of protoporphyrin IX. **Chlorophyll *a*** is the major Mg-chlorin in plant cells. The common designation of the pyrrole rings (A–D) is shown in this structure. In addition to the four pyrrole rings, a fifth ring is formed by oxidation and cyclization of the initial methyl propionate side chain on ring C. Chlorophyll *a* is made from its precursor, protochlorophyllide, by the reduction of a double bond in ring D to a carbon–carbon single bond. The propionic acid side chain is then esterified with the 20-carbon, isoprenoid alcohol, **phytol. Chlorophyll *b*** is identical to chlorophyll *a* except that the methyl group on ring B is oxidized to an aldehyde. **Chlorophyll *c*** is similar to protochlorophyllide except that the propionate side chain on ring D is oxidized to the *trans-acrylate* form, which remains unesterified. Addition of the double bond extends the macrocyclic π system to the electronegative carboxyl group. **Bacteriochlorophyll *a*** is the major chlorophyll-type pigment in most anoxygenic photosynthetic bacteria. It is similar to chlorophyll *a* but also has a double bond in ring B reduced to a single bond, and the vinyl group on ring A is oxidized to an acetyl group. **Bacteriochlorophyll *b*** has a vinyl group on ring B in place of the ethyl group. The carbon–carbon single bonds in ring B as well as ring D are the primary characteristics of bacteriochlorophyll. **Heme** is Fe-protoporphyrin IX, which in cytochromes functions by oxidation and reduction of the central iron ion.

Evolution of the photosynthetic membrane and light-harvesting complexes

The essential functional structure for photosynthesis, and indeed universally for biological energy conservation, is a **membrane** that (a) physically separates two different compartments and (b) contains an energy transducing apparatus. The membrane allows the development of an electrochemical gradient, which is an essential intermediate in photosynthesis. Within the membrane reside the reaction centers and a series of electron transfer components. Photochemistry occurs in the reaction centers, initiated by the absorption of light energy by Chl. The flux of photons, even in full sunlight, is sufficient for only a few photons to be absorbed by a Chl molecule in a reaction center per second, a rate that is much too slow for productive photosynthesis. Therefore, photosynthetic organisms also developed structures that contain a large number of accessory pigment molecules that harvest light energy and funnel the energy into reaction centers. As a consequence, sufficient high energy states are generated to efficiently drive synthetic reactions.

Different organisms have developed quite different means to increase absorption of light energy. **Green sulfur bacteria** (e.g., *Chlorobium*), which are obligate *anaerobic* photoautotrophs, contain domains within the cell membrane that include a single reaction center, around which electrons flow, driven by absorption of light energy by BChl *a*. Attached as appendages on the cytoplasmic surface of the photosynthetic domains are specialized antenna complexes, which are nearly pure aggregates of up to 10 000 molecules of BChl *c, d,* or *e,* called **chlorosomes**. Light energy absorbed by chlorosomes is transferred to reaction centers through the FMO complex (after Fenna and Matthews, who first determined its structure, and Olson, who discovered the protein), which is a trimeric arrangement of a protein subunit that binds seven molecules of BChl *a*. The reaction center in green sulfur bacteria has an iron–sulfur complex as the terminal electron acceptor, similar to photosystem (PS) I in plants, and thus these organisms do not produce oxygen.

The facultative photoheterotrophic **purple bacteria** (e.g., *Rhodobacter* and *Rhodopseudomonas*) induce synthesis of BChl and expression of genes encoding photosynthetic proteins, and thus grow photosynthetically, only in response to low oxygen tension (Drews, 1996). The amount of BChl in these cells is several fold higher under low, as compared to high, light intensity, a reflection of the increased need for light-harvesting capacity. Expansion of the cell membrane through the assembly of photosynthetic domains produces extensive invaginations into the interior of the cell. These invaginations form a photosynthetic "intracytoplasmic membrane" system, which usually remains continuous with the respiratory plasma membrane. When cells are broken, the intracytoplasmic membranes fragment into small vesicles called **chromatophores**. Expansion of membrane surface area in these organisms dramatically increases the capacity for photosynthetic activities and is required to accommodate the light-harvesting complexes (LHCs) that are integral pigment–protein complexes within the membrane structure itself. **Light-harvesting complex 1** (LH-I) surrounds the single type of reaction center, which in this case has a quinone as the electron acceptor similar to PS II in plants (Figure 11.2). LH-I is composed of 16 α,β-heterodimer polypeptide subunits, with each polypeptide binding one BChl a molecule, and forms a ring structure that encircles the reaction center. The ring is slightly distorted by a single copy of PufX, the product of the $pufx$ gene, which apparently facilitates access of electron carriers such as quinone molecules to the reaction center. In this environment, the association with proteins causes the absorption maximum of BChl a in LH-I to shift from its maximum of 773 nm in an organic solvent such as diethyl ether to 875 nm. LH-II, an additional, more peripheral LHC, is present at varying amounts. LH-II is also a ring-like structure, composed of 9 α,β-heterodimers of short polypeptides, 53 and 41 amino acid residues long, respectively. The N-terminus of each of the 9 α subunits is modified to the *N-carboxy*-methionine residue, which coordinates with a BChl a molecule

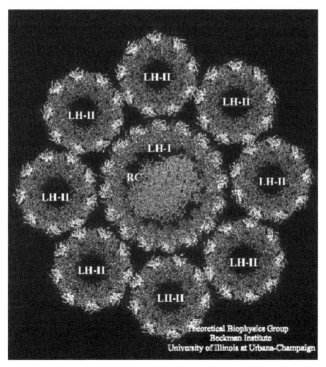

Figure 11.2 Model of the structure of the photosynthetic apparatus in a non-sulfur purple bacterium. The reaction center (RC) is enclosed by the core light-harvesting complex, LH-I. Surrounding the core complex are the peripheral light-harvesting complexes, LH-II, which are present in variable amounts depending upon the growth conditions. Each LH-II complex contains nine bacteriochlorophyll (BChl) a molecules that maximally absorb at 800 nm and 18 BChl a molecules that absorb at 857 nm. Energy is transferred to the LH-I complex, which contains BChl a that absorbs at 875 nm. The reaction center pair of BChl a molecules absorb at 870 nm. Source: Hu *et al.* (2002).

oriented parallel with the membrane near the periplasmic surface. This complex maximally absorbs 800 nm light. The imidazole group of a histidine residue in each α and β peptide coordinates one BChl *a* molecule, which creates a ring of 18 BChl molecules near the cytoplasmic surface of the membrane, oriented perpendicular to the plane of the membrane. These LH-II BChl molecules absorb maximally at 857 nm. Because photons of shorter wavelengths contain more energy than those of longer wavelengths, and excited molecules transfer energy only to those at a lower energy level, this arrangement of the purple bacterial light-harvesting antenna provides efficient "downhill" transfer of energy, from B800 and B857 in LH-II to B875 in LH-I to the reaction center.

Cyanobacteria have an extensive photosynthetic **thylakoid membrane** system that forms concentric layers inside, but separate from, the plasma membrane. Cyanobacteria were the earliest organisms to perform oxygen-producing photosynthesis, with two reaction centers operating in series. One reaction center, **photosystem (PS) II**, is the quinone type (from the purple bacterial lineage), while the second, **PS I**, is the iron–sulfur type (from the green sulfur bacterial lineage). The major light-harvesting function in cyanobacteria is provided by peripheral, exquisitely designed complexes of proteins that contain covalently bound, linear tetrapyrrole chromophores (phycobilins). These complexes, called **phycobilisomes**, are assembled from the subunits phycoerythrin (λ_{max} 565 nm), phycocyanin (λ_{max} 620 nm), and allophycocyanin (λ_{max} 650 nm), named according to their reddish or bluish color, respectively. The arrangement of these subunits, as illustrated in Figure 11.3, also provides the "downhill" flow of excitonic energy from the highest energy chromophore (shortest wavelength λ_{max}) to the reaction center, which is necessary for efficient trapping of light energy (Gantt, 1981).

Origin of the chloroplast in eukaryotic cells

Eukaryotic organisms evolved about 2.7 BYA. An endosymbiotic event, in which an early eukaryotic cell engulfed a prokaryotic cell, occurred about 1.9 BYA and led to the development of mitochondria. Extensive evidence supports the appearance of photosynthesis in eukaryotic plant cells as a second endosymbiotic event in which a cyanobacterium was productively engulfed by an early eukaryotic cell nearly 1.6 BYA (Yoon *et al.*, 2002). This event defined the divergence of animals and plants. It seems that rather quickly thereafter the original endosymbiont evolved into the **chloroplast** and most of the genes in the cyanobacterial cell were transferred to the nucleus or were discarded. Extensive analysis of cyanobacterial genes in the nucleus of the model plant *Arabidopsis thaliana* indicate that about 18% of the total nuclear genes of this plant are derived from the endosymbiont (Martin *et al.*, 2002). Although chloroplasts contain up to 100 copies of the residual **chloroplast genome**, the number of *different* genes that encode proteins in most plant in most chloroplast genomes is in the range of 70–80, only about 5% of that in a modern-day cyanobacterium (Timmis *et al.*, 2004). Analysis of the sequences of the remaining genes, in particular, those encoding the large subunit of ribulose 1,5-bisphosphate carboxylase/oxygenase and ribosomal RNA

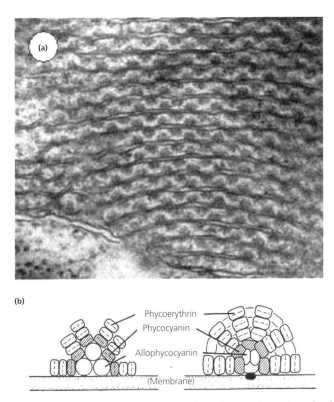

Figure 11.3 Phycobilisomes are light-harvesting complexes in cyanobacteria and red algae. (a) The electron micrograph of a portion of the chloroplast of a red alga shows a parallel array of thylakoid membranes with interdigitating phycobilisomes on the stromal surface. (b) The diagram shows the arrangement of the phycobiliproteins within the structure of phycobilisome from a cyanobacterium (*left*) or a red alga (*right*). Source: Adapted from Gantt (1981).

in algae and plants has shown that the plastids in all photosynthetic eukaryotes have a monophyletic lineage (Bhattacharya and Medlin, 1995; Wise and Hoober, 2006). It is remarkable that the endosymbiotic event that led to the evolution of the plastid seems to have occurred only one time. This is very close to not happening at all, or it happened within a population of cyanobacterial organisms that had substantial similarity and the ability to take advantage of a unique but temporary environment. The rather high uniformity of the chloroplast genome in nearly all plants, with the exception of some rearrangements, supports the single-event hypothesis rather than multiple events. Consequently, chloroplasts have similar features to cyanobacteria, with a separate, extensive, internal thylakoid membrane system. Of course, because genes encoding most of the plastid proteins were now expressed in the nucleus, an elaborate system was required to import these proteins into the plastid after synthesis in the cytosol.

Many algal cells contain a single chloroplast (Figure 11.4), but mature plant cells usually contain numerous—in some cases up to as many as 150—plastids (Figure 11.5). These organelles are surrounded by a two-membrane **envelope**,

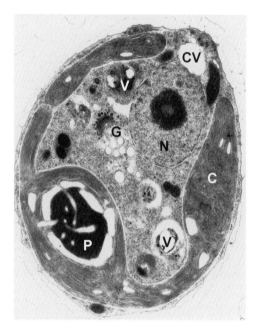

Figure 11.4 Electron micrograph of a cell of the green alga *Chlamydomonas reinhardtii cbnl-113*. The algal cell contains a single, cup-shaped chloroplast (C). Within the base of the chloroplast is a structure called the pyrenoid (P), which is a condensed form of Rubisco, the enzyme that fixes CO_2 into organic compounds. Other organelles, typical of a eukaryotic cell, include the nucleus (N), Golgi apparatus (G), digestive small vacuoles (V), and a contractile vacuole (CV). Source: From Park *et al.* (1999). Micrograph courtesy of Dr. Hyoungshin Park.

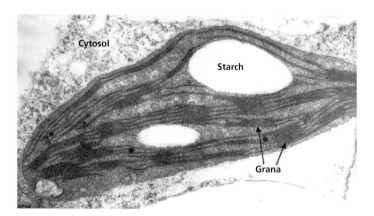

Figure 11.5 Electron micrograph of a chloroplast in a young leaf of tobacco. Each cell contains numerous chloroplasts, perhaps a hundred or more, distributed around the periphery near the cell membrane. Thylakoid membranes within the chloroplast are arranged in stacks, designated grana, connected by stromal lamellae. The chloroplast is surrounded by a double-membrane envelope. Starch grains, the final storage form of the products of photosynthesis, accumulate in the stroma. A small portion of the large vacuole that occupies the center of the cell is visible at the bottom of the micrograph. A portion of the cytosol is shown at the top of the micrograph. Source: Micrograph courtesy of Dr. Hyoungshin Park.

which contains specific transport systems through which proteins and metabolites pass. The outer membrane contains pores that are less discriminatory than the more specific transport systems that reside in the inner membrane. The envelope membranes are tightly appressed, such that transport complexes on the outer and inner membranes provide continuous passageways. Current evidence indicates that the outer membrane of the chloroplast envelope resembles more the outer bacterial membrane than the endocytic membrane that presumably enclosed the engulfed bacterial cell. If the progenitor of the chloroplast was similar to modern cyanobacteria, which are "Gram-negative" organisms that have a cell wall sandwiched between two membranes surrounding the cell, then the evolution of the plastid envelope involved loss of the cell wall. Primitive algae, such as *Cyanophora*, contain chloroplast-like organelles called **cyanelles** that have a prokaryotic-like peptidoglycan layer between the inner and outer membranes of the organelle, which is thought to be a relic of their cyanobacterial ancestor.

Cyanobacteria contain only Chl *a*, whereas plants and their algal ancestors in the genus Charales, and the eukaryotic green algae that belong to the Chlorophyta, contain Chl *b* in addition to Chl *a*. The gene encoding **Chl *a* oxygenase**, the enzyme that catalyzes oxygenation of Chl *a* to make Chl *b*, also has a monophyletic lineage, with the gene in the prokaryotic prochlorophyte organisms such as *Prochloron* and *Prochlorococcus* as the most ancient (Tanaka *et al.*, 2010). Chl *a* oxygenase activity apparently was present in the early endosymbiont, long before the plant-type **light-harvesting Chl *a/b*-binding proteins** (LHCPs) developed. The ancestral prokaryotic endosymbiont may have contained both phycobilisomes and a prochlorophyte-type LHC containing Chl *b*. Subsequently, ancestral photosynthetic eukaryotes would have diverged into either (a) organisms that lost the ability to make phycobiliproteins but gained the source of the abundant family of Chl *a/b*-binding proteins that provide the antenna in green plants or (b) the ancestors of the red algae that retained the ability to make phycobilisomes but lost synthesis of Chl *b*.

Many species of algae are products of "secondary" endosymbiotic events, in which an entire eukaryotic green or red alga was engulfed by another nonphotosynthetic eukaryotic cell (Raven and Allen, 2003). These latter algae contain three or four membranes surrounding the plastid, the two of the original chloroplast envelope, the derivative of the original algal cell membrane, and the endocytic membrane of the secondary host. Most of these species contain Chl *c* instead of Chl *b* (see Figure 11.1). Such secondary endosymbiosis explains the majority of algal biodiversity (Yoon *et al.*, 2002). Even more remarkable are organisms among the dinoflagellates that are products of a third, "tertiary" endosymbiosis (Stoebe and Maier, 2002). Although secondary endosymbiotic events occurred many times, with host organisms with diverse lineages, the plastids themselves all show a single, that is, monophyletic, origin. This chapter will emphasize photosynthesis in the green plants, including algae in the Chlorophyta.

Development of the chloroplast

Maturation of the chloroplast

In seed plants, the plastid begins development during seed germination as a simple, double-membrane vesicle called a **proplastid**. In contrast to most single-celled photosynthetic organisms, which are able to synthesize Chl and the photosynthetic apparatus in darkness, higher plants require light for the conversion of protochlorophyllide (Pchlide) *a* to chlorophyllide (Chlide) *a* (the "-ide" ending indicates the absence of the isoprene alcohol esterified on the side chain of ring D). This reaction involves the stereochemical reduction of a double bond between carbon-17 and carbon-18 in ring D (see Figure 11.1) by reduced **nicotinamide adenine dinucleotide phosphate** (NADPH), catalyzed by **Pchlide-NADPH oxidoreductase** (POR). In plants, this reaction requires the absorption of light energy by the substrate, Pchlide. Three forms of POR occur: PORA preferentially binds to Pchlide *b* and PORB has a higher affinity for Pchlide *a*. The complex of POR, Pchlide, and NADPH is abundant in plastids (**etioplasts**) in plants grown in darkness, but PORA is rapidly degraded upon exposure of plants to light. The level of PORB is sustained during most of chloroplast development. PORC is induced by light and is predominantly present in fully matured light-grown plants.

The process by which the thylakoid system forms depends upon the conditions under which the plant cell is grown. In seedlings that are exposed early to light, initial thylakoid membranes form in the proplastid by invagination of the inner membrane. Thus, photosynthetic domains are expelled as vesicles from the envelope by a series of accessory proteins involved in vesicle traffic. The vesicles then fuse to form and expand the developing thylakoid system. A mutant of *A. thaliana*, deficient in an activity designated vesicle-inducing plastid protein 1 (VIPP1), is unable to induce vesicle formation from the plastid envelope, and consequently does not make thylakoid membranes. The formation of vesicles has been observed during the initial stage of chloroplast development (a process commonly referred to as greening) of the alga *Chlamydomonas reinhardtii* (Figure 11.6) and in plant leaves treated with inhibitors of vesicle fusion. These results provide direct demonstrations of the envelope as the source of material for thylakoid membranes (Wise and Hoober, 2006). This process has also been demonstrated by the application of specific inhibitors of vesicle traffic to isolated chloroplasts (Westphal *et al.*, 2001).

Alternatively, when seedlings are grown from germination in darkness, the resulting **etioplasts** contain a highly organized, tubular membrane structure called the **prolamellar body**. PORA is the predominant protein in the prolamellar body and exists in a complex with its substrates, Pchlide and NADPH. Galactosyl diglycerides are the major lipids of this structure. Exposure of these seedlings to light results in the reduction of Pchlide to Chlide *a* and dispersal of the prolamellar body into "prothylakoid" membranes. These rudimentary membranes expand

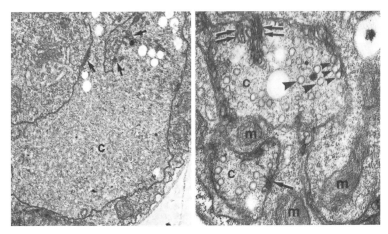

Figure 11.6 *Left panel:* portion of the chloroplast of a dark-grown, degreened cell of the alga *C. reinhardtii y1*. The chloroplast (C) is depleted of thylakoid membranes, which were diluted as the cell continued to grow and divide. *Right panel*: portion of the chloroplast of a dark-grown, degreened cell of alga *C. reinhardtii y1* after 15 min of greening in light. *Arrowheads* point to vesicles that appear to be derived from the chloroplast envelope in response to exposure to light. *Arrows* point to an extensive invagination of the chloroplast envelope. The image suggests that the development of the thylakoid membrane system occurs with membrane vesicles emanating from the envelope. Several profiles of mitochondria (m) are included in the image. Source: Adapted from Hoober *et al.* (1991).

by the addition of proteins. Some proteins, such as those of the core complex of photosynthetic units, are made within the plastid, while others, in particular the proteins of the light-harvesting antennae, are imported after synthesis in the cytosol. Lipids are synthesized predominantly on envelope membranes, and the envelope is also the location of the latter steps in Chl and carotenoid synthesis. Thus, the envelope is an important interface between the chloroplast and cytosol and serves as the platform for biogenesis of the extensive thylakoid membrane. In plant cells exposed to light, membrane formation occurs over several hours to several days, depending on the organism, to achieve the mature chloroplast.

The three-dimensional arrangement of membranes in the chloroplast of plants has been established by electron microscopy. The thylakoid membrane, separate from the inner envelope membrane, is differentiated into cylindrical stacks of "appressed" membranes, designated **grana**, that are interconnected with "unappressed," stromal membranes (Figures 11.5 and 11.7). The highly elaborated, folded membrane system encloses a single, continuous lumen. This arrangement seems to maximize efficiency of the overall process. Thylakoid membranes in algal cells are less differentiated, and in many cases are appressed over much of their surface (see Figure 11.4). Most important for the process of photosynthesis is the formation of the luminal compartment, as will be explained later.

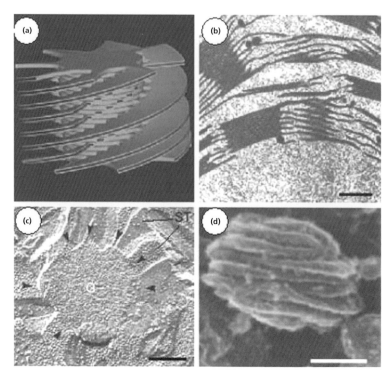

Figure 11.7 Arrangement of stromal membranes around the grana as depicted by (a) a computer model constructed from electron micrographs as shown in panels (b), (c), and (d). Reconstruction suggests that each thylakoid in a granum is connected to its neighbor through the stromal lamellae (ST) and that the thylakoid system is a continuous membrane enclosing a single lumen. Source: Mustárdy and Garab (2003).

Role of Chlorophyll in chloroplast development

Chlorophyll (Chl) is required not only for photosynthesis but also for the assembly of the photosynthetic apparatus and, to a large extent, formation of the thylakoid membrane (Hoober *et al.*, 2007). The two major forms in green plants are Chl *a* and Chl *b* (Figure 11.1). Chls are "macrocyclic," highly conjugated, tetrapyrrole structures with a central Mg^{2+} atom and exist mostly, if not entirely, associated with proteins. Mg^{2+} in Chl usually forms five coordination bonds. Four of the ligands for such bonds are provided by the pyrrole nitrogens within the Chl molecule. The ligand for the fifth, axial coordination bond is provided by solvent (water) or a functional group on a protein molecule. In the reaction centers and LHCs, the preferred ligand to Chl *a* is the electron-rich, neutral imidazole side chain of histidine. Chl *b* differs from Chl *a* only in the oxidation of the methyl group on pyrrole ring B to an aldehyde. Although this reaction results in a spectral shift (Figure 11.8), and consequently expands the spectral range for absorbance of light by the chloroplast, the introduction of the additional oxygen atom also influences the coordination chemistry of the central Mg^{2+}

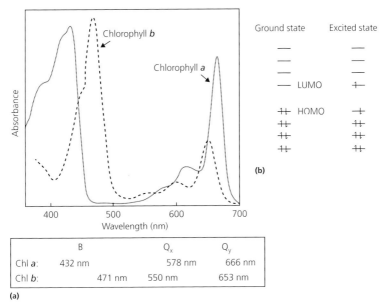

Figure 11.8 (a) Absorption spectra of Chl *a* and Chl *b* in methanol. Indicated below the spectra are the wavelengths of maximal absorption along the major red vector, Q_y, the perpendicular vector Q_x, and the blue absorption band, designated the B or Soret absorption peak. See text for details. (b) A diagram that describes absorption of light energy by elevation of an electron from the highest occupied molecular orbital (HOMO) to the lowest unoccupied molecular orbital (LUMO).

atom. The electronegativity of the oxygen atom results in the withdrawal of electron density away from the pyrrole nitrogens toward the periphery of the molecule. Consequently, the Mg^{2+} atom in Chl *b* expresses a more positive point charge than that in Chl *a* and becomes a stronger Lewis acid (Hoober *et al.*, 2007). As a result, Chl *b* binds more strongly than Chl *a* to the dipolar solvent, water. The imidazole side chain of histidine, which is the preferred ligand for Chl *a*, is normally not a ligand for Chl *b*, probably because the imidazole side chain does not have a sufficiently large dipole to displace the strongly bound water ligand. In fact, most molecules of Chl *b* in LHCs retain water as the axial ligand. The recent high-resolution structural determination of the LHCII from spinach has defined the specific binding sites for Chl *a* and Chl *b* (Liu *et al.*, 2004). Ligands that favor interaction with Chl *b* usually contain oxygen and thus exhibit a strong dipole, which provides an electrostatic character to the coordination bonds (Hoober *et al.*, 2007). Most of these assignments were suggested from results of reconstitution of LHCs with mutant forms of the apoprotein (LHCP) in which an amino acid that provides a ligand is replaced by an amino acid that lacks a functional group in the side chain (Remelli *et al.*, 1999). In mutant strains of plants that lack the ability to make Chl *b*, very few LHCs accumulate. Thus, Chl *a* alone is not sufficient to complete the assembly of most

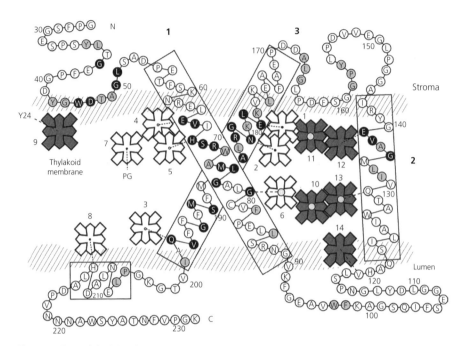

Figure 11.9 Model of the distribution of Chl *a* (*open symbols*) and Chl *b* (*closed symbols*) in the major light-harvesting complex of green plants. Coordination bonds are indicated by a solid line from the center of a Chl symbol to an amino acid. Five Chl molecules have water as a ligand (*central dots*). These Chl molecules form hydrogen bonds between the water molecules and another Chl molecule or an amino acid. Source: Adapted from Green and Durnford (1996), with assignments as described in Liu *et al*. (2004)

of the complexes. It was proposed that the stronger interaction between Chl *b* and the proteins is essential for the assembly of stable Chl–protein complexes (Hoober *et al*., 2007) (Figure 11.9).

About one-third of the Chl *a* and essentially all of the Chl *b* reside in peripheral LHCs. The apoproteins, LHCPs, of the major and minor LHCIIs, associated with PS II, have molecular masses in the range of 25–30 kDa. (A Dalton (Da), or the unified atomic mass unit, is a unit of mass that is 1/12 the mass of a carbon atom (atomic mass, 12.000), or approximately the mass of a hydrogen atom.) LHCPs are encoded by a large family of *Lhcb* genes in the nuclear genome, and bind between 8 and 14 Chl molecules per protein molecule. The major LHCII in plants exists as a trimer. The major LHCPs in LHCII are designated Lhcb1, Lhcb2, and Lhcb3. Each Lhcb1 protein binds about 14 Chl molecules, 8 Chl *a* and 6 Chl *b*, along with 3 xanthophylls, usually 2 lutein and 1 neoxanthin molecules (Liu *et al*., 2004). Minor LHCIIs exist as monomers. Lhcb4 (CP24) binds eight Chls (6 Chl *a* and 2 Chl *b*), Lhcb5 (CP26) binds nine Chls (6 Chl *a* and 3 Chl *b*), and Lhcb6 (CP29) binds ten Chls (5 Chl *a* and 5 Chl *b*). The minor LHCs also contain fewer carotenoids, usually with 1 lutein, 0.5 neoxanthin, and 0.5–1 violaxanthin.

In plants, four apoproteins, Lhca1, Lhca2, Lhca3, and Lhca4, are involved in the antenna for PS I. Because of sequence homology, the structure of LHCI monomers is probably quite similar to that of LHCII. Apoproteins of LHCI coordinate 6–9 Chl a, 3 Chl b, 1 lutein, and 0.5 violaxanthin and bind substoichiometric amounts of β-carotene instead of neoxanthin. LHCI occurs as dimers, generally with heterodimers of Lhca1 with Lhca4, which comprise LHCI-730 that absorbs maximally at 730 nm, and Lhca2 with Lhca3, which provides LHCI-680, with maximal absorption at 680 nm.

Assembly of LHCs during chloroplast development

Experiments with a model organism, the alga *C. reinhardtii*, have been instructive in regard to the pathway for the insertion of the major LHCPs into the thylakoid membrane. The chloroplast of cells grown in the light is filled with thylakoid membranes. When green cells of a mutant strain that is unable to make Chl in darkness were subsequently grown in darkness, the membranes were diluted among the progeny, leaving the chloroplast nearly depleted of membranes (see Figure 11.6). The exposure of the dark-grown cells to light initiates Chl synthesis and membrane assembly. The assembly of LHCs can be monitored by Förster resonance energy transfer from Chl b to Chl a as these molecules are brought sufficiently close (10 Å or less) as a result of incorporation into LHCs (Figure 11.10). With the technique of immunoelectron microscopy, the initially assembled LHCs were detected in the chloroplast envelope and associated invaginations, as shown in Figure 11.11 (White *et al.*, 1996). LHCPs made in excess accumulated outside of the chloroplast in vacuoles. Further, when cells were grown in darkness, conditions under which Chl synthesis did not occur, the proteins were not retained by the chloroplast and instead accumulated in the cytosol and vacuoles (Park and Hoober, 1997). A model that illustrates the current concepts in LHCII assembly is shown in Figure 11.12.

Other than the requirement for Chl, and in particular Chl b, little is known about the mechanism of incorporation of these proteins into membranes. Chl a oxygenase, the enzyme that catalyzes synthesis of Chl b, is localized on chloroplast envelope membranes. Eventual accumulation of these proteins in thylakoid membranes is determined by molecular interactions that occur initially within the envelope, even before the proteins are completely imported into the plastid. In mutants that are unable to make Chl b, few if any of the LHCPs are retained in the chloroplast, although the plants still produce Chl a. Other factors are also likely involved in facilitating assembly of LHCs. For example, mutants that lack the ability to synthesize the major carotenoid in LHCs, the xanthophyll lutein, assemble LHCs much slower than normal. Further, mutants that are deficient in subunits of a stromal complex designated the "chloroplast signal-recognition particle" are deficient in LHCs, which suggests that these proteins may be involved in the assembly of the complex. Plants lacking several proteins, including the membrane protein ALB3, are deficient in LHC assembly. ALB3 seems to be involved in facilitating insertion of LHCPs into the membrane. Kinetic measurements of the

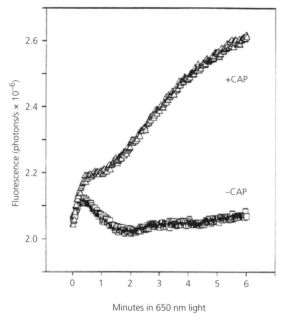

Minutes in 650 nm light

Figure 11.10 Forster resonance energy transfer analysis of assembly of LHCs during greening of the alga *C. reinhardtii y1*. Cells were exposed to light of 650 nm, which was sufficient for photoconversion of Pchlide to Chlide *a* and also for excitation of Chl *b*. Fluorescence of Chl *a* was continuously monitored at 680 nm. After the first minute of illumination in untreated cells, energy absorbed by newly assembled LHCs was trapped as the result of connection of the complexes with reaction centers. When chloramphenicol (+CAP, 50 μg/ml) was added, which inhibited synthesis of reaction center core proteins on chloroplast ribosomes, the energy absorbed by "free" LHCs was reemitted as fluorescence. Source: Adapted from White *et al.* (1996).

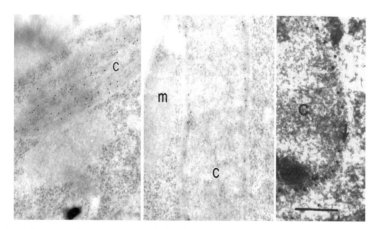

Figure 11.11 Immunoelectron microscopic localization of LHCPs in the developing chloroplast of *C. reinhardtii y1*. *Left panel*: LHCPs in a chloroplast developing normally during 6 h of exposure to light at the normal growth temperature of 25°C were localized to thylakoid membranes in the interior of the chloroplast (c). Thin sections of cells were treated with IgG antibodies raised against LHCP, followed by incubation with protein-A, which binds tightly to IgG molecules. Protein-A was conjugated with gold particles that were visible in the electron microscope. *Middle panel*: Companion cells to those in the *left panel* were exposed to light in the presence of a high concentration of chloramphenicol (200 μg/ml), which suppressed synthesis of Chl. LHCPs were detected in the chloroplast (c) only within the envelope. The micrograph also includes a mitochondrion (m). *Right panel*: Dark-grown cells of the alga were exposed to light at 38°C, which provided conditions that enabled detection of the first LHCPs to be incorporated into membranes during chloroplast development. After 15 min of exposure to light, gold particles were observed in the chloroplast (c) only over the chloroplast envelope, designated by the arrows. Source: Adapted from White *et al.* (1996).

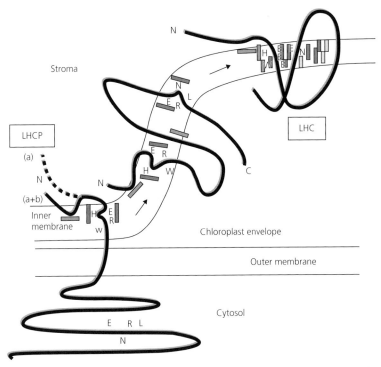

Figure 11.12 Model for assembly of light-harvesting complexes based on experimental results obtained with the model alga *C. reinhardtii*. LHCPs are imported into the chloroplast, N-terminus first, directed by the "transit sequence." The transit sequence enters the stroma and is removed by a specific protease, while most of the protein still extends into the cytosol. Two Chl *a* molecules possibly bind to a motif with the generic sequence—ExxHxR—in the first membrane-spanning region of the protein. Extensive evidence suggests that Chl *b* is required to retain the protein in the chloroplast. Chl *b* possibly binds initially to the backbone peptide bond at tyrosine-24 (see Figure 11.9), and in its absence (dashed N-terminal region (a)) the protein is not held sufficiently strongly to prevent retraction into the cytosol (Park and Hoober, 1997). Binding of the initial Chls to N-terminal sites allows the remainder of the LHC to assemble (a + b) within the envelope inner membrane. Upon assembly, LHCs associate with newly assembled PS II and the resulting expansion of the inner envelope membrane causes formation of vesicles (see Figure 11.6). Source: Adapted from Hoober *et al.* (2007).

development of photosynthetic activities indicate that the connection between photosystems and LHCs occur immediately upon the assembly of the antenna complexes, most likely in the chloroplast envelope.

Absorption of light energy

Excitation of chlorophyll

Chlorophylls, as cyclic tetrapyrroles, are conjugated, organic molecules that efficiently absorb light. Within the electronic structure of the molecules are orbitals that have distinctly different energy levels. Wavelengths of light with energy levels that match the allowed electronic transitions within the molecule are described by the absorption spectrum (Figure 11.8). Photons with more or less energy, that is, an energy packet that does not match an allowed transition, pass through the molecule without being absorbed. Absorption spectra are interpreted in terms of electrons lifted from the energetically highest *occupied* molecular orbitals (HOMOs) to the energetically lowest *unoccupied* molecular orbitals (LUMOs). For these transitions to occur, the molecule must absorb a photon that contains precisely the energy by which the orbitals differ. The ability of the Chl molecule to absorb light is also a function of the direction of the interacting waveform, with stronger absorption occurring when the directional vector is parallel to the *y* axis of the molecule, which lies across pyrrole rings A and C of the macrocycle (see Figure 11.1). Consequently, this generates the Q_y absorption band, which usually has the lowest energy (longest wavelength). The absorption of light with the vector along the shorter axis, across pyrrole rings B and D, generates the Q_x absorption band. The shorter wavelength of absorbed photons indicated by the Q_x band, therefore, reflects a larger energy gap that must be crossed, but these electrons reach the same LUMO when excited as those absorbed in the Q_y band. These transitions occur with photons with wavelengths between 550 and 700 nm. Another set of allowed transitions is described by the higher energies required to reach the LUMO+1 orbital. The absorption of photons with the appropriate energy for this transition is described by the B (or Soret) bands, with wavelengths in the 420–470 nm range.

The energy content of a photon is related to its frequency, as described by the following relationship:

$$E = h\nu = \frac{hc}{\lambda}$$

(11.2)

Here, E is the energy in Joules, h is the Plank's constant, 6.63×10^{-34} J-s, ν is the frequency, λ is the wavelength, and c is the speed of light, 3.0×10^{10} cm/s. The excited state of the molecule can return to the ground state by several means, either by release of vibrational energy as heat, release of a photon of slightly lesser energy (**fluorescence**), or transfer of the packet of energy—an **exciton**—to another molecule. Rapid decay from the LUMO+1 to the LUMO orbital occurs by an internal, vibrational mode. Energy leaves the molecule either by radiationless energy transfer to an acceptor, fluorescence, or heat by decay from the LUMO to the ground state. Most of the energy entering photosynthetic systems is absorbed by LHCs, which contain the bulk of the Chl in thylakoid membranes. Excitons are transferred extremely rapidly, on the order of hundreds of femtoseconds (10^{-15} s), between Chl molecules within an antenna complex. Excitons exit an LHC through a specific Chl *a* molecule and may travel through many Chl–protein complexes before eventually reaching a reaction center. Trapping of energy from LHCI by PS I occurs in about 25 ps (25×10^{-12} s), whereas transfer of energy from LHCII to the reaction center of PS II occurs in about 130 ps (130×10^{-12} s).

The reaction centers

The heart of each **reaction center** is a pair of Chl *a* molecules in close, nearly parallel juxtaposition (in PS I, one of the pair is Chl *a'*, which is a stereoisomer around carbon-13^2). This pair of Chl *a* molecules in PS I has been designated P700 because of its absorption maximum. In PS II, the special pair is designated as P680. When an exciton reaches one of the Chl molecules in the reaction center, an electron is lifted to a higher energy level. The "excited" molecule consequently becomes a strong reducing agent and donates an electron to a nearby electron acceptor. The acceptor thus achieves a negative charge, while the donor has a positive charge. This **charge separation** is the key photochemical event in photosynthesis.

The reaction center of PS II (Figure 11.13) is composed of two very similar proteins, designated D1 (38.0 kDa) and D2 (39.4 kDa), which are encoded by genes *psbA* and *psbD*, respectively, in the chloroplast genome. These two proteins together bind six molecules of Chl *a*. The PS II reaction center also contains two molecules of **pheophytin *a*** (Chl *a* molecules without the central Mg^{2+} atom), a tightly bound **plastoquinone** molecule, and an iron atom. The biochemically purified PS II reaction center also includes the α (9 kDa) and β (4 kDa) subunits of cytochrome b_{559} and a small peptide product (4 kDa) of the *psbI* gene. Each reaction center is associated with two Chl *a*-protein complexes that act as "core" antenna, CP43 (51 kDa) and CP47 (56 kDa), encoded by the *psbC* and *psbB* genes, respectively, which were initially designated according to their molecular masses as estimated by electrophoresis. CP43 and CP47 bind 12 and 14 Chl *a* molecules, respectively. A cluster of Mn^{2+} ions on the electron donor side of the reaction center is involved in oxygen evolution (Barber, 2008). This complex is stabilized by three proteins, 26.5, 20.2, and 16.5 kDa in mass. The complete **core complex** contains approximately 25 different proteins. The core complex was purified by sucrose gradient centrifugation, and cryoelectron microscopy revealed that PS II exists as a dimeric structure that also includes the minor LHCs CP26 and CP29 and a single, tightly bound LHCII trimer per reaction center (Figure 11.14). This overall structure has been referred to as the **supercomplex**. In thylakoid membranes, approximately four to six additional, more loosely bound, LHCII trimers are associated with each PS II reaction center, and thereby increase the absorptive cross-sectional area of the photosynthetic unit. The minor LHCIIs connect the major, peripheral antenna to the reaction center. The slightly more blue-shifted spectra of the major LHCs relative to the absorption maximum of the reaction center ensures that energy will flow toward, and be trapped by, the lower energy levels of P680. Together with the functional antenna, a PS II unit contains a total of about 280 Chl molecules per reaction center.

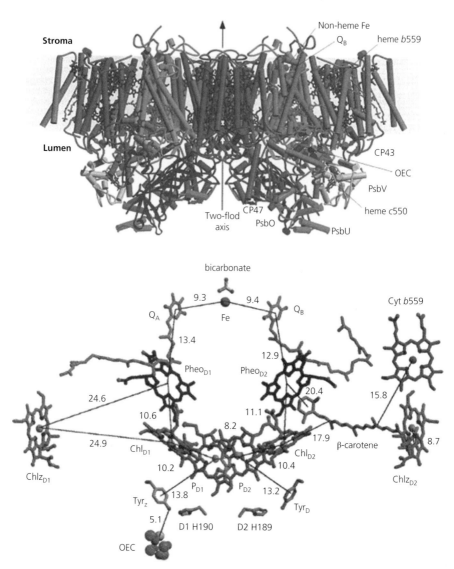

Figure 11.13 Structure of PS II from the cyanobacterium *Thermosynechococcus elongatus*. *Upper*: View of the PS II dimer as viewed from the side. Extending from the luminal side of the membrane is the oxygen evolving complex (OEC). PsbO on the luminal side of the complex is the 33 kDa protein that stabilizes the Mn cluster. *Lower*: Arrangement and distances of the cofactors within the reaction center shown with the proteins removed. P_{D1} and P_{D2} are the Chl *a* molecules of the special pair, P680. Electrons flow from the Mn cluster (oxygen-evolving complex, OEC) on the luminal surface to P680 through a tyrosine radical, Tyr_z, and then in sequence to Chl_{D1}, $Pheo_{D1}$, Q_A, and Q_B on the stromal surface (see text for details). Source: Ferreira *et al.* (2004).

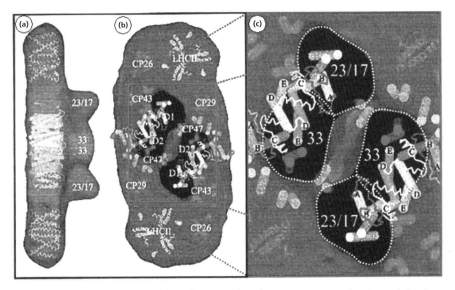

Figure 11.14 Structural model of the subunits within the PS II supercomplex. (a) and (b) show representations from the side and luminal surface, respectively. Proteins of the Mn cluster involved in oxygen evolution protrude from the luminal surface, as shown in (a). The reaction centers, composed of proteins Dl and D2, occur as a dimer, each with its associated Chl–protein complexes. Higher magnification of the luminal surface is shown in (c), with the positions of the 17, 23, and 33 kDa proteins of the Mn cluster indicated in outline over the reaction center subunits. Source: Nield *et al.* (2000).

In the PS II reaction center, electrons are transferred from P680 to a Chl *a* and then to a pheophytin *a* molecule on protein D1. Experimental data suggest that the initial charge separation occurs predominantly between the Chl *a* and pheophytin *a* to form the radical pair state Chl$^+$ Phe$^-$, followed by electron transfer from the P680 Chl *a* on D1 to produce the terminal radical pair state P680$^+$ Phe$^-$. The electron is then transferred to a tightly bound plastoquinone molecule (Q_A) on subunit D2 and then on to a loosely held plastoquinone (Q_B) on D1 to generate the semiquinone form. After a second electron is transferred to the latter acceptor, along with two protons (H$^+$), the reduced quinol (PQH$_2$) leaves the reaction center and is replaced by another plastoquinone molecule from the pool in the membrane. The reduction of plastoquinone to the quinol is a two-electron reaction, as shown in Equation 11.3, and involves the uptake of two protons from the stroma. In oxygenic photosynthesis, the electron hole in the special pair is then filled by an electron abstracted from water by the oxygen-evolving complex (see later, "Generation of end products").

$$\text{(structure)} + 2\,e^- + 2\,H^+ \underset{\text{Oxidation}}{\overset{\text{Reduction}}{\rightleftharpoons}} \text{(structure)} \qquad (11.3)$$

R = C$_{45}$ Isoprenoid sidechain

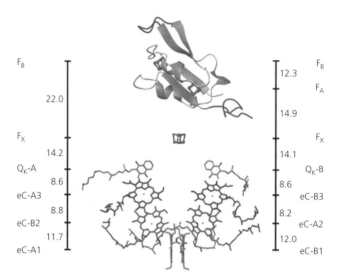

F_B	22.0	12.3	F_B
			F_A
		14.9	
F_X	14.2	14.1	F_X
Q_K-A	8.6	8.6	Q_K-B
eC-A3	8.8	8.2	eC-B3
eC-B2	11.7	12.0	eC-A2
eC-A1			eC-B1

Figure 11.15 Positions of the cofactors in the reaction center of PSI from the *Synechococcus elongatus*, with the center-to-center distances between them in Angstroms. Pairs of Chls and phylloquinones are arranged in two branches, A and B, from the P700 Chls (eC-Al and eC-Bl) to the iron–sulfur center, F_X. Iron–sulfur clusters F_A and F_B are shown within protein PsaC. Phylloquinones are indicated by Q_K, and one or both may be the electron acceptor A_1 (see Figure 11.18). Source: Jordan *et al.* (2001).

The reaction center of PS I is also composed of two principle proteins, PsaA (84 kDa) and PsaB (83 kDa), encoded by genes *psaA* and *psaB* in the chloroplast genome (Jordan *et al.*, 2001). These proteins bind 6 Chl *a* molecules, 2 **phylloquinone** molecules, and an **iron–sulfur center** (Figure 11.15). The structure of the PS I core complex was resolved to 2.5 Å, which revealed that the complete complex contains a total of 12 protein subunits, 90 additional Chl *a* molecules, a total of three Fe_4S_4 centers (designated F_A, F_B, and F_X), and 22 various isomers of β-carotene. Seventy-nine of the Chls are bound to the large proteins of the PS I reaction center, PsaA and PsaB, mostly to the N-terminal domains that serve as the core antenna. In the excited state, P700 donates an electron to an adjacent Chl *a* molecule (A_0). The electron is then transferred to a phylloquinone molecule (A_1) and subsequently to the series of iron–sulfur centers. From F_X, electrons are transferred to the Fe_4S_4 cluster of the electron-carrying protein, **ferredoxin**. Whereas the reaction center of PS II exists as a dimeric structure, PS I in plants is monomeric. (However, in cyanobacteria and another photosynthetic prokaryote *Prochlorococcus*, PS I is trimeric.) Estimates indicate that about five dimeric LHCIs are associated with each PS I complex. Along with the light-harvesting antenna, LHCI, PS I contains a total of about 215 molecules of Chl.

Both reaction centers have essentially a symmetrical structure. The two subunits in each reaction center differ slightly but create two potential branches for electron transfer. Interestingly, electron flow occurs preferentially across one branch. A substantial amount of work has been devoted to determine whether any electrons flow across the alternate branch. Although evidence exists for functional ability of both branches, electrons normally flow over only one branch in PS II. In PS I, the rate of electron transfer is about 10-fold faster (~13 ns) on the PsaB side than on the PsaA side (~140 ns). With one branch being kinetically much more rapid, the activity of the alternate branch can only be observed when the primary pathway is impaired by mutation.

Fluorescence induction curves

A useful technique has been refined over the past two decades to noninvasively monitor energy flow through PS II. When the quinone acceptors of electrons from the reaction center are oxidized, light energy is nearly quantitatively trapped by energy transduction to produce electrons that reduce the quinones first to a semiquinone when one electron is added and then to the fully reduced quinol when the second electron is added. However, when photons are absorbed by the LHCs more rapidly than the rate at which the reduced quinol at the Q_B site can exchange with an oxidized plastoquinone molecule, Q_A remains reduced and electron flow through PS II is blocked. Subsequent excitons produced from absorption of photons by LHCII are deactivated by nonradiative decay with release of heat or, in a minor fraction of events, by the emission of photons of slightly lower energy, the process of **fluorescence**. Consequently, transients of fluorescence intensity provide information on the state of the reaction center. Figure 11.16 describes the fluorescence transient when light-grown cells of the alga *C. reinhardtii* were exposed first to a low-intensity red light, which yields a low level of fluorescence of Chl *a* molecules when the reaction center remains oxidized (F_o). Subsequent more intense irradiation causes a rise first to a plateau level, indicative of PS II reaction centers that are not directly connected to electron transport chains. Without a means for electrons to exit these reaction centers, they are rapidly reduced. Fluorescence then continues to increase, but more slowly, to a peak value (F_p) as Q_A becomes reduced in the bulk of the PS II centers. The fluorescence transient, therefore, indicates that electrons initially arrive at Q_A, and then Q_B, at a higher rate than reduced Q_B (quinol) can be oxidized. As PS I and the carbon fixation pathway become activated, the kinetically more rapid PS I drains electrons from PS II through the electron transport chain, Q_B is oxidized, the backup in PS II is relieved, and the fluorescence decreases to a steady-state level (F_s). However, when electron transport from Q_A is blocked by a herbicide that binds to the Q_B site, the efflux of energy from the reaction center is blocked and fluorescence rises to a maximal level (F_m). The same maximal level can be achieved with a brief, intense flash of white light that floods the reaction center with excitons. The quantum yield, Φ, an indication of the integrity of the reaction center, is described by the simple relationship, $\Phi = (F_m - F_o)/F_m = F_v/F_m$. This value for chloroplasts in most organisms grown under normal conditions is 0.6–0.8, as shown in Figure 11.16.

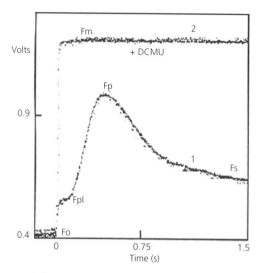

Figure 11.16 Fluorescence induction curve obtained with green cells of the alga *C. reinhardtii*. Cell suspensions were dark adapted before exposure to modulated measuring light of low intensity (2.5 μmol photons/m²/s) for determination of F_o and then to actinic light (80 μmol photons/m²/s) to obtain fluorescence transients. Fluorescence was measured at room temperature in the absence (curve 1) or presence (curve 2) of 10 μM 3-(3,4-dichlorophenyl)-L,L-dimethylurea (DCMU), a herbicide that binds to the Q_B site and blocks electron transfer to plastoquinone. Source: Adapted from White and Hoober (1994).

Generation of end products

Evolution of oxygen

Oxygen is produced only by photosynthetic systems that use Chl in coupled reaction centers. In these organisms, photosystems PS I and PS II operate in series and thus span a wider redox range than possible with the single reaction center in photosynthetic bacteria. The oxidation of water requires an oxidant with a redox potential more positive than $E_m = +0.816$ V, the mid-point potential of the O_2/H_2O couple. BChl in reaction centers of green and purple bacteria achieves redox potentials of only $E_m = \sim +0.5$ V, because the energy of a photon at 870 nm, the absorption maximum of the purple bacterial special-pair BChl, is considerably less than that at 680 nm, the absorption maximum of P680 in PS II (from Eq. 11.2, 138 kJ vs. 180 kJ/mol photons). After donating an electron to pheophytin, the oxidized reaction center of PS II, P680⁺, with a redox potential $E_m = +1.2$ V, abstracts electrons from a tyrosine residue in D1. The tyrosine radical, designated Z, has a redox potential of about +0.93 V, which is sufficient to abstract electrons from water. Z connects P680 with the oxygen-evolving complex located on the luminal surface of the membrane (see Figures 11.13 and 11.14). The core of this complex is a structure formed by four manganese atoms coordinated with several oxygen atoms and one each of Cl⁻ and Ca²⁺ ions (Ferreira *et al.*, 2004; Barber, 2008). The

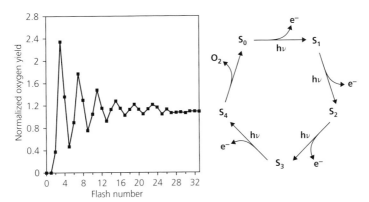

Figure 11.17 Oxygen evolution as a function of light flashes. *Left:* Pierre Joliot and Bessel Kok discovered this periodic pattern when chloroplasts were exposed to a series of flashes sufficiently short for only a single turnover of P680 with each flash. This pattern indicated that four photons were required to abstract four electrons from water to generate one molecule of oxygen. Studies of this process led to the scheme shown *on the right*, which suggests the Mn cluster of the oxygen-evolving complex is progressively driven through four states. Because oxygen is released spontaneously from S_4, and the first burst of oxygen occurs after the third flash, these results indicate that the resting state is S_1. Because not every center was hit with each flash, the periodicity was gradually damped as the experiment proceeded.

Mn cluster is stabilized by three proteins, 26.5, 20.2, and 16.5 kDa in mass. P680$^+$ abstracts one electron at a time from tyrosine Z, which in turn pulls an electron from the Mn cluster. Consequently, for an oxygen molecule to be released by the complex, four electrons from two water molecules must be consecutively collected by the Mn cluster and transferred to P680. Electron collection from water and delivery to P680 occurs by changes in the valence states of Mn. When chloroplasts are exposed to a series of very brief flashes of light, such that only a single turnover of P680 occurs per flash, the result shown in Figure 11.17 is obtained, with a burst of oxygen produced every fourth flash. For each oxygen molecule produced, four protons from water are released into the *luminal* environment.

The proton motive force and adenosine triphosphate

The electron acceptor of PS II, Q_B, is reduced on the stromal surface of the membrane. The reduction of each plastoquinone molecule to the quinol (PQH_2) involves the uptake of two protons from the *stromal* environment (Eq. 11.3). PQH_2 is oxidized by a complex of membrane proteins, intermediate between PS II and PS I, that contains cytochrome b_6, cytochrome f, and a Rieske iron–sulfur (Fe_2S_2) center. The iron ions in the **cytochrome b_6/f complex** accepts only electrons and, as electrons are transferred to a site on the luminal side of cytochrome b_6, the protons from PQH_2 are released into the *luminal* compartment (Figure 11.18). If this complex operated as a simple, forward transfer of electrons, one H$^+$ would be transported across the membrane per each electron

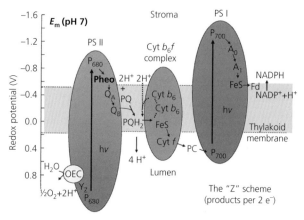

Figure 11.18 The "Z-scheme" of linear electron transport in photosynthesis is generated when the sequence of electron carriers is plotted according to their mid-point redox potentials. Thick arrows indicate the changes in redox potential of the electron donors in each reaction center, the Chl pairs P680 and P700, when they receive excitation energy from the antenna. The physical positions of the electron carriers Pheo, Q_A, and Q_B in PS II and A_0, A_1, and the Fe_4S_4 clusters in PS I are shown in Figures 11.13 and 11.15. Plastoquinone (PQ) is reduced by PS II on the stromal side and reoxidized by the cytochrome b_6/f complex on the luminal side of the membrane, which contains a Rieske Fe_2S_2 center. Operation of the "Q cycle," which transfers electrons through cytochrome b_6, results in transfer of 4 H^+ from the stroma to the lumen. Plastocyanin (PC) shuttles electrons from cytochrome f to PS I. Ferredoxin (Fd), a small protein containing a Fe_4S_4 cluster, is the first soluble, stromal electron carrier to accept electrons from PS I. Ferredoxin reduces $NADP^+$ to NADPH in a reaction catalyzed by the flavin-containing enzyme, ferredoxin-$NADP^+$ reductase. Arrows trace the path of electrons through the complexes.

transferred from PS II to the cytochrome b_6/f complex. However, experimental measurements indicate that *two* H^+ are transported per each electron. To explain this observation, a mechanism called the **Q-cycle** has been proposed. In this cycle, one electron from the quinol, PQH_2, in the "Q_o" site of the cytochrome b_6/f complex is transferred to one of the two heme groups, b_L, in cytochrome b_6 and the second to the Rieske iron–sulfur center, which can only accept one electron. The electron in cytochrome b_6 is then transferred from the b_L to the b_H heme, which reduces a plastoquinone tightly bound in the "Q_i" site to the semiquinone form. An electron from oxidation of a second PQH_2 in the Q_o site is also transferred through b_L and b_H to fully reduce the semiquinone (PQ^-) in Q_i to the quinol ($PQ^=$), while the other electron is transferred to the Rieske center. In the process, two additional protons (H^+) are pulled from the stroma during reduction of the plastoquinone to produce PQH_2 in the Q_i site. The PQH_2 is then released from the Q_i site and binds to the Q_o site. The cycling of electrons through cytochrome b_6, and the oxidation of the second PQH_2 by the complex, results in transfer of a second H^+ per each electron, or two H^+ per each electron passing through the cytochrome b_6/f complex, from the stroma to the lumen. Each electron that reaches the Rieske center is transferred to cytochrome f and then to a small luminal, copper-containing protein called **plastocyanin**.

The oxidation of water releases one H^+ per electron. When the uptake of protons accompanying reduction of plastoquinone (Q_B) on the acceptor side of PS II is combined with the subsequent release of protons into the lumen by the Q-cycle, a net gain of three H^+ is achieved in the thylakoid lumen, relative to the stroma, during transfer of one electron from water to NADPH. Therefore, the reduction of two $NADP^+$ by four electrons from two water molecules results in a total gain of $12H^+$ in the thylakoid lumen. The resulting proton gradient across the membrane generates a **proton motive force**, which is roughly equally divided between the proton concentration gradient and the accompanying voltage potential ($\Delta\psi$) caused by the imbalance of positive charges across the membrane. This force drives synthesis of **adenosine triphosphate** (ATP). The experimentally measured ratio of H^+/ATP is approximately 4, which indicates that three ATP molecules are synthesized per two NADPH molecules, the stoichiometry that is required for CO_2 fixation by the **reductive pentose phosphate cycle** (see discussion later). Whereas the stromal pH is 7.5–8.0 when the chloroplast is actively engaged in photosynthesis, the luminal pH is maintained between 5.8 and 6.5, which provides a transmembrane gradient of approximately a hundred-fold difference in proton concentration.

The ability to develop the proton motive force requires a physical boundary between these compartments, which is provided by the thylakoid membrane. Maintaining a small volume of the lumen is also required to achieve a relatively high proton concentration. The activity of ATP synthase is regulated by the concentration of inorganic phosphate (Pi) in the stroma (Eq. 11.4); only when the phosphate level is reduced to a very low level, such that the consequently low activity of ATP synthase slows H^+ flow across the membrane does the luminal pH drop below 5.8.

ATP synthase is a bipartite protein, with a membrane component, CF_0, attached to a large, peripheral domain, CF_1, which extends from the membrane on the stromal surface (Figure 11.19). This enzyme was designated the "chloroplast coupling factor (CF)" because of its function in coupling electron transfer during photosynthesis to ATP synthesis. (The designation "CF" distinguished the chloroplast ATP synthase from the similar enzyme in mitochondria, which was designated "F.") The energy in the proton motive force causes CF_0, an integral, annular structure of 12–14 "c" subunits, to rotate within the plane of the membrane. The stromal domain, CF_1, is a "segmented sphere" of six subunits, three α subunits, and three β subunits in alternating order, and is held stationary by the "a," "b," and "δ" subunits that extend out from the membrane and contact the CF_1 domain. The "γ" subunit, which is attached to CF_0, extends up through the central core of CF_1. As CF_0 rotates, it drives rotation of the γ subunit, which causes conformational changes in the catalytic sites on the β subunits. The reaction

$$\text{ADP} + \text{inorganic phosphate (Pi)} \longleftrightarrow \text{ATP} + H_2O \qquad (11.4)$$

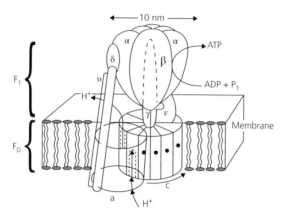

Figure 11.19 Structure of the CF_0CF_1 ATP synthase. The membrane-spanning complex, CF_0, includes approximately 12 copies of the "c" subunit, which are associated with the γ and ε subunits. This complex rotates within the membrane, driven by the flux of protons through the membrane. The "a," "b," and δ subunits extend from the membrane and attach to, and serve as a "stator" for, the CF_1 complex composed of three each of the α and β subunits. Rotation of the γ subunit causes consecutive conformational changes in CF_1. Source: Elston *et al.* (1998).

is thermodynamically reversible, although the hydrolysis (reverse) reaction is often considered practically irreversible in an aqueous environment. (Many biosynthetic pathways are pulled forward by coupling the energy released by ATP hydrolysis to reactions that are of themselves thermodynamically unfavorable.) In one conformation, one of the β subunits binds ADP and Pi extremely tightly and the reversible reaction (Eq. 11.4) occurs. When the γ subunit rotates 120°, the next conformation causes the affinity for ATP in this site to be essentially lost and the product ATP diffuses from the enzyme. At the same time, the catalytic site in the adjacent β subunit changes from one that loosely binds ADP and Pi to the tight-binding form that makes ATP. Again, when rotation of the γ subunit continues, ATP dissociates from this site. Thus, ATP synthase operates as a molecular rotary motor by undergoing conformational changes in which each of the three catalytic sites is changed, in sequence, from one that loosely binds ADP and Pi to the tight-binding site that generates ATP to the low affinity site from which ATP dissociates. When the CF_1 domain is disconnected from CF_0, the enzyme operates in reverse as an ATP hydrolase.

Production of NADPH

Electrons are transferred from the copper ion in **plastocyanin** to PS I on the luminal surface. A specific subunit of PS I, the protein PsaF, is required in addition to a surface loop of PsaB, one of the PS I reaction center proteins, to provide a docking site for plastocyanin that brings the reduced copper atom (Cu^+) near P700. As described before, the transfer of excitons from LHCI to P700 powers the transfer of electrons from P700 across the membrane, through a Chl *a* molecule (A_0) and the phylloquinone (A_1) to the iron–sulfur centers on the stromal surface of PS I. The soluble iron–sulfur protein, **ferredoxin**, is then reduced on the stromal surface. Ferredoxin is a one-electron carrier, whereas the ultimate electron acceptor NADP$^+$ is reduced to NADPH in a two-electron reaction. This reaction is catalyzed by **ferredoxin-NADP$^+$ reductase**, an enzyme that carries a **flavin adenine dinucleotide** (FAD) prosthetic group. The FAD cofactor of the reductase accepts two electrons, one at a time, to fully reduce the flavin, which then reduces NADP$^+$ in a two-electron reaction to generate NADPH (Eq. 11.5), the major reductant for CO$_2$ fixation in the chloroplast. Reduced plastocyanin fills the electron hole in P700 as electrons flow into NADPH.

$$FADH_2 + NADP^+ \longleftrightarrow FAD + NADPH + H^+ \qquad (11.5)$$

Distribution of the photosystems in thylakoid membranes

The operation of the photosystems in oxygenic photosynthesis is usually diagrammed as shown in Figure 11.18. The complete system, with two reaction centers acting in series, provides a linear flow of electrons from water to NADP$^+$. This process has been described as the Z-scheme, based on the flow of electrons through components that are arranged according to the midpoint of their redox potentials. The system can be short-circuited, depending upon environmental conditions, when reduced ferredoxin adds electrons back into the chain at the level of plastoquinone or the cytochrome b_6/f complex. The net effect of this cyclic process is a transfer of protons from the stroma to the thylakoid lumen, resulting in a proton gradient from which ATP is formed but NADPH is not produced. This **cyclic electron flow** is required to maintain balance in the energy requirements with the chloroplast.

Juxtaposition of the photosystems, which is implied by Figure 11.18, exists in only a small part of the thylakoid membrane. In reality, the photosystems are mostly quite distant from each other. In a typical chloroplast of a higher plant, such as spinach, tobacco, or barley, about 80% of the membrane lies within the grana, which are cylindrical stacks 10 ± 5 lamellae high and about 0.4–0.5 μm in diameter (see Figure 11.7). The remaining 20% of the membrane is as "stromal lamellae" that interconnect the grana. PS I, which requires access to ferredoxin and NADP$^+$ in the stroma, and ATP synthase are localized predominantly on stromal lamellae and at the edges and surfaces of grana. About 85% of PS II, with its full antenna (designated PS IIα), is within the grana, while a minor portion, with a smaller antenna (PS IIβ) is in the stromal lamellae. This distribution of photosystems within the membrane system has been established biochemically, with purified fragments of granal membranes enriched in PS II and fragments of stromal lamellae containing most of the PS I.

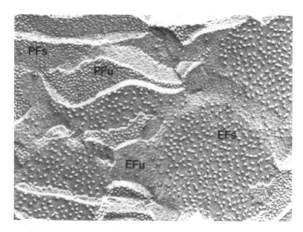

Figure 11.20 Freeze-fracture image of thylakoid membranes from spinach. An ice block containing the membranes was fractured, which revealed the inner faces of the lipid bilayer membrane. Protein complexes that spanned the membrane were retained on one leaflet or the other depending on which side of the membrane the proteins were most strongly anchored in the surrounding ice. Interior faces of the stacked, granal membranes provided the surfaces designated EF_s for the luminal leaflet and PF_s for the stromal leaflet. Corresponding faces for stromal thylakoid are designated EF_U and PF_U. Large particles on the EF surfaces represent PS II core complexes, which are more abundant in granal membranes. A high density of smaller particles on the PF_U face represent PS I complexes. LHCII complexes were retained by the PF_s leaflet. This image shows that PS II and PS I particles do not occur to any significant extent in each other's territory. Source: Adapted from Staehelin (2003). Micrograph courtesy of Dr. Andrew Staehelin.

A direct demonstration of the density of PS II units in granal membranes was provided by freeze-fracture images of the membrane interior (Figure 11.20). In this technique, membranes are embedded in ice. When the ice block is fractured, the fracture plane travels along the internal bilayer interface of the membrane, the path of least resistance. Consequently, the interior of the membrane is exposed. A high density of 14–20-nm diameter particles, similar in size to the PS II dimeric "supercomplex" (about 16 nm without LHCII antenna), was revealed on the interior face of the luminal half of stacked, granal membranes (EF_s face), whereas the facture face of the luminal half of unstacked, stromal membranes (EF_U face) contained particles about half this size and at a much reduced density. The complementary interior face of the stromal half of the unstacked membrane, the PF_U face, contained a very high density of 10–12-nm diameter particles, a size expected for monomeric PS I. The abundant LHCII particles were retained by the outer half of the granal membrane (PF_s face) when the membrane was fractured. The PS II particles on the EF_s face interdigitate between these PF_s particles, which reflects the close contact of the antenna with the core complexes.

Long-range segregation of most of the PS II and PS I units presents an interesting problem. It is generally accepted that PS II and PS I function in series, although cyclic electron flow can occur around PS I, mediated by return of electrons from ferredoxin back to plastoquinone and cytochrome b_6/f. Mutants that are deficient in electron transfer from plastocyanin to P700 are light sensitive, which indicates that the major dissipation of energy absorbed by PS II is via PS I. Thus, the flow of electrons from water to NADP$^+$ must occur rapidly, mediated presumably by highly mobile electron carriers. The rapid flow of electrons is reflected in the fluorescence transient illustrated in Figure 11.16. The major intermediate between the two photosystems, the cytochrome b_6/f complex, is distributed on granal and stromal membranes. This complex has been isolated and crystallized as a large, dimeric structure that is expected to have limited mobility because of its size. PQH$_2$, the electron carrier from PS II to the cytochrome b_6/f complex, is a mobile membrane lipid, but the rate of long-range molecular diffusion in thylakoid membranes is two orders of magnitude slower than expected for a fluid lipid membrane, possibly the result of the high density of protein complexes in the membrane. At this rate, the $t_{1/2}$ of electron transfer from PS II would be on the order of many minutes instead of the measured rate of several hundred milliseconds. Studies of electron flow in thylakoid membranes indicated that only a few plastoquinone molecules (about 6) are associated with each PS II reaction center, which shuttle electrons over short distances to nearby cytochrome b_6/f complexes. Reduction of the cytochrome b_6/f complex is kinetically the slowest redox step in electron flow.

The short-range diffusion limit of plastoquinone in thylakoid membranes leaves the task of electron transport from PS II-cytochrome b_6/f microdomains to PS I up to plastocyanin. Diffusion of plastocyanin, the small protein that transfers electrons from the cytochrome b_6/f complex to PS I, is localized in the thylakoid lumen, which is a narrow compartment with many proteins extending from the membrane into this space. Diffusion may also be restricted by the lack of bulk water within the lumen. Thus, it is clear that the rate of transfer of electrons from PS II to ferredoxin is much slower than the rate of conversion of excitons to electrons in the reaction centers, which occur on a time scale of picoseconds to nanoseconds. The conclusion emerges that the rate of photosynthesis is limited, at saturating intensities of light, by the rate of diffusion of these electron carriers, or the equilibrium transfer of electrons along a series of plastocyanin molecules. Although the activation of the CO$_2$ fixation cycle and the transfer of electrons out of PS I maintain the PS II reaction center in a relatively oxidized state, at high light intensities an energy "backup" in PS II is readily detected by the release of some of the absorbed energy as fluorescence.

The elaborately folded granal structure raises an interesting question about the function of this arrangement, because structure and function are usually complementary. One possible explanation for membrane differentiation, with PS II in grana and PS I in stromal lamellae, follows from their kinetic differences. PS I has much faster trapping kinetics than PS II, and were PS I sufficiently near to the major antenna, LHCII, most of the absorbed energy would be drained off by PS I. Because the proton gradient required for ATP synthesis is generated by electron flow from PS II to the cytochrome b_6/f complex, the segregation of the photosystems is essential to maintain a high level of ATP synthesis, via the proton gradient, relative to NADPH production. An active PS II is also required for oxygen evolution. To some extent, this problem has been solved in cyanobacteria, which do not form grana, by the connection of the light-harvesting phycobilisomes only to PS II. In some eukaryotic algae, thylakoid membranes are also not appressed, and this problem is partially solved by the number of PS II units being several fold higher than that of PS I. Many eukaryotic algae have thylakoid membranes that are appressed over most of their surface, without discrete distinction between granal and stromal membranes.

Photoinhibition: damage and repair of the PS II reaction center

When LHCII absorbs light energy at levels in excess of the rate at which electron flow can dissipate energy in the PS II reaction center, photooxidative damage to the D1 subunit (PsbA) can occur. Such conditions are caused when light absorption exceeds the chloroplast's capacity for CO_2 fixation and is referred to as **photoinhibition**. The resultant inactive PS II reaction center can only be repaired by the replacement of the damaged D1 subunit with a newly synthesized protein, which occurs at a surprisingly rapid rate. The repair requires partial dissociation of the reaction center, removal of the damaged D1, and its replacement. A compensatory, or protective, mechanism in most plants and algae is the ability of LHCII monomers to dissociate from PS II at high light intensities and move toward PS I, when PS II is running faster than PS I, and the plastoquinone pool consequently becomes highly reduced. Dissociation of LHCII from PS II requires phosphorylation of the protein, a reaction that is regulated by binding of PQH_2 to cytochrome b_6. The phosphorylation of LHCII is catalyzed by a specific protein kinase, Stt7 (*s*tate *t*ransition, *t*hylakoid). As a consequence, more energy is transferred from LHCII to PS I, a condition referred to as **state 2**. When light intensities return to lower levels, the kinase activity is reduced, and phosphatases remove the phosphate group from LHCII. The reconnection of LHCII with—and the transfer of energy predominantly to—PS II restores the default condition, referred to as **state 1** (Allen, 2003).

Protection of PS II by carotenoids

Although repair of photodamaged PS II reaction centers can occur over a period of minutes, plants have developed another adaptive mechanism to protect D1 by releasing absorbed energy as heat in the antenna, before it is transferred to the reaction center. This process is called **nonphotochemical quenching** of

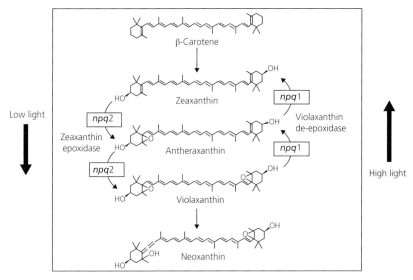

Figure 11.21 The xanthophyll cycle. Xanthophylls are oxygenation products of α- and β-carotene. The major product from α-carotene is lutein, a dihydroxy carotenoid in light-harvesting complexes. Oxidation of β-carotene produces zeaxanthin, an isomer of lutein, which is normally converted through antheraxanthin to violaxanthin, a major xanthophyll in chloroplasts. Violaxanthin is the precursor of neoxanthin, another major xanthophyll. In plants exposed to high light intensity, which results in extensive reduction of the plastoquinone pool and generation of a large pH gradient across the thylakoid membrane, violaxanthin is converted back to zeaxanthin by de-epoxidation. Zeaxanthin is able to quench excited states of Chl and prevents buildup of excitation pressure on PS II. Mutant strains defective in the ability to "non-photochemically quench" excited states are designated *npq1* and *npq2* and lack the ability to interconvert violaxanthin and zeaxanthin. Source: Adapted from Niyogi (1999).

Chl fluorescence (NPQ) and is triggered by two factors, an excessive proton gradient across the thylakoid membrane and synthesis of **zeaxanthin**. (Zeaxanthin is a member of a class of carotenoids, the "xanthophylls," that contain oxygen.) At high light intensities, the de-epoxidation reaction, by which the xanthophyll violaxanthin is enzymatically converted to zeaxanthin, is stimulated (Figure 11.21). The energy level of the lowest excited singlet state (denoted S_1) of zeaxanthin is $14\,550\,cm^{-1}$, which is lower than the S_1 state of Chl a, about $14\,700\,cm^{-1}$. The S_1 energy level of violaxanthin, at $14\,880\,cm^{-1}$, is too high for energy transfer from Chl. Elevated amounts of zeaxanthin will quench excited states of Chl and allow dissipation of absorbed light energy by thermal decay within the antenna. The interconversion of violaxanthin and zeaxanthin in response to light intensity has been described as the **xanthophyll cycle** (Demmig-Adams and Adams, 1992).

High levels of zeaxanthin alone are not sufficient for nonphotochemical quenching, as shown by *npq2* mutants that lack the enzyme that catalyzes epoxidation of zeaxanthin to violaxanthin and, therefore, accumulate zeaxanthin.

A large pH gradient across the thylakoid membrane, which is a factor in feedback regulation of electron transport, is also required. The effects of both factors are mediated by the PsbS protein, which is structurally similar to apoproteins of LHCII and also binds Chl. Mutants lacking PsbS are unable to perform the quenching required for the protection of the reaction center. When two pH-responsive glutamate residues in loops of the protein exposed on the luminal side of the membrane were changed to glutamine residues, or when a pH-responsive aspartate residue at the luminal end of the first membrane-spanning domain was changed to a glycine, the resultant PsbS protein was unable to perform efficient quenching of energy in the antenna.

Additional carotenoids, including the xanthophyll **lutein** in the core of LHCII, and the α- and β-carotenes and xanthophylls within the membrane environment, provide general protection against photooxidative damage. When the energy in excited Chl molecules is not transferred to the reaction center or quenched by protective mechanisms, the initial singlet state, which is short-lived, can convert to the triplet state. Because this **intersystem crossing** involves reversal of the orientation of the spin of an electron, the Chl molecule now has two unpaired electrons with parallel spin, which restrains its decay to the ground state. Consequently, triplet states are relatively long-lived and decay only when an electron returns to the antiparallel orientation. Triplet states, however, can react with oxygen, and the transfer of energy to oxygen allows one of the unpaired, parallel-spin electrons in the *ground-state, triplet* oxygen molecule to flip. The resulting antiparallel orientation of the two electrons allows pairing to produce *singlet* oxygen, which has an unfilled orbital. Singlet oxygen is extremely reactive, and particularly in a membrane environment containing fatty acids with many unsaturated double bonds, will cause damage of the membrane by two-electron oxidation reactions. The abundance of oxygen molecules produced by photosynthesis, therefore, requires a mechanism to prevent this outcome. Carotenoids effectively quench the triplet state of Chls and thus provide a major protective "barrier" to damage of the photosynthetic apparatus (Wise and Hoober, 2006).

Incorporation of carbon as CO$_2$ into carbohydrate

The reductive pentose phosphate or "C$_3$" cycle

The energy-laden products, ATP and NADPH, of *photo*-synthesis, referred to as the "light reactions," are produced in very small quantities. These compounds are immediately consumed in **anabolic** (synthetic) reactions and recycled. As such, they are not designed as energy-storage products. Photosynthesis absorbs much more energy over a day than the plant needs to sustain itself moment to moment. The excess is converted to storage forms such as starch or sucrose for use during the night. From many years of selection of desired traits by plant

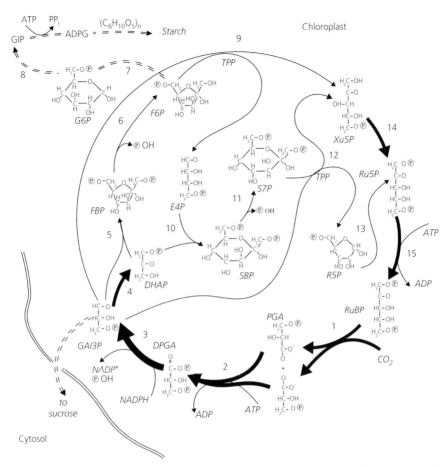

Figure 11.22 The reductive pentose phosphate (C_3) cycle. The heavy lines indicate reactions of the cycle. Numbers next to each reaction refer to the reactions listed in Table 11.1.

breeders, the vegetables we currently find in the grocery store have an ability to accumulate storage products in extraordinarily large quantities.

Three molecules of ATP and two molecules of NADPH, the stoichiometry produced by the light reactions, drive the **reductive pentose phosphate cycle**, with each turn of the cycle resulting in incorporation of one molecule of CO_2 into carbohydrate. Thus, the cycle must operate six times for synthesis of one molecule of glucose (Figure 11.22). ATP is used to synthesize **ribulose 1,5-bisphosphate** from ribulose 5-phosphate in the reaction catalyzed by **ribulose 5-phosphate kinase**. Ribulose 1,5-bisphosphate then reacts with CO_2 in a reaction catalyzed by the most abundant protein in the biosphere, **ribulose 1,5-phosphate carboxylase/ oxygenase**, or "Rubisco." The enzyme is so named because it cannot completely discriminate between CO_2 and O_2 (Figure 11.23). The products of the carboxylation reaction are two molecules of **3-phosphoglycerate**. The reaction with oxygen results in one molecule of 3-phosphoglycerate and one of **2-phosphoglycolate**

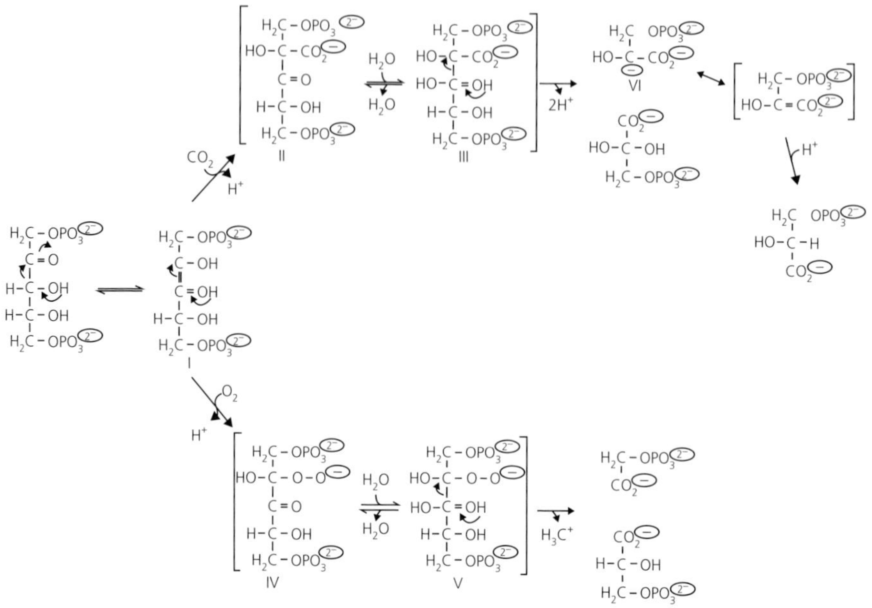

Figure 11.23 Reactions catalyzed by Rubisco. The reaction of ribulose 1,5-bisphosphate with CO_2 results in two molecules of 3-phosphoglycerate (*upper pathway*). The reaction with O_2 results in one molecule of 3-phosphoglycerate and one of 2-phosphoglycolate (*lower pathway*). Source: Hartman and Harpel (1994).

(Hartman and Harpel, 1994). The latter reaction is not anabolic, and subsequent metabolism of 2-phosphoglycolate results in production of CO_2 in the pathway known as **photorespiration**.

Rubisco is a large enzyme, occurring in eukaryotic photosynthetic organisms as a protein with eight large subunits of about 52 kDa in mass and eight small subunits between 12 and 16 kDa in mass. The total mass of the enzyme is about 520 kDa. Active sites on the enzyme are formed at the interface of the large subunits of alternating orientation, which provides eight active sites for each protein molecule. The enzyme requires Mg^{2+} for activity, which can only be bound to the enzyme after a side-chain ε-amino group of lysine-201 reacts with a molecule of CO_2 to form a carbamate group

$$
\begin{array}{cc}
H & O \\
| & || \\
\end{array}
$$
(-lysine–N–C–O⁻).

This interesting reaction, in which CO_2 is required not as a substrate but to *activate* the enzyme, is catalyzed by **Rubisco activase**, a protein that undergoes conformational changes driven by hydrolysis of ATP. The carbamate negative charge, along with side-chain carboxyl groups of glutamate and aspartate residues, coordinates the Mg^{2+} in the active site. The Mg^{2+} ion consequently positions the "substrate" CO_2 to achieve reaction with the other substrate, ribulose 1,5-bisphosphate. The enzyme is very sluggish, catalyzing only about three reactions per second per protein molecule. The concentration of CO_2 that provides half-maximal activity of the enzyme, designated the K_M, is 10µM, or approximately the concentration of CO_2 in water at 30°C and ambient atmospheric concentrations of CO_2. Thus the enzyme normally operates at only half its maximal rate, which suggests that plants should respond well to an elevated concentration of CO_2 in the atmosphere. The enzyme evolved in the early biosphere under conditions in which CO_2 was 10 fold or more higher than that in the current atmosphere and O_2 levels were very low. Nature has not been able to improve much on this reaction over the eons, and has apparently compensated for the seemingly inefficient catalytic activity by making large quantities of the enzyme. Nevertheless, the activity of Rubisco remains one of the limiting factors in photosynthesis.

The product of the carboxylation reaction, **3-phosphoglycerate**, is a three-carbon compound, which led to designation of this pathway as the C_3 cycle. As described in the following text, this is one of the three major pathways for assimilation of carbon from the atmosphere. (In the other two, a four-carbon compound is the initial product of carbon assimilation. (See later text.) 3-Phosphoglycerate is converted to **1,3-bisphosphoglycerate** in a reaction with ATP. Because two molecules of 3-phosphoglycerate are now in play, this step in the cycle uses two molecules of ATP. These two, along with the ATP molecule involved in synthesis of ribulose 1,5-bisphosphate, account for the three required for each CO_2 molecule fixed. The product, 1,3-bisphosphoglycerate, has an "activated" carboxyl group, which can be reduced by NADPH to **glyceraldehyde 3-phosphate**. This reaction uses the two NADPH molecules produced in the light reactions. The enzyme is named for the reverse reaction, **glyceraldehydes 3-phosphate dehydrogenase**. The remaining steps of the cycle are rearrangements of the molecules to arrive at a net molecule of glucose for each of six turns of the cycle. The reactions involved in assimilation of CO_2 are listed in Table 11.1.

The "C_4" pathway

Whereas approximately 90% of land plants contain *only* the reductive pentose phosphate cycle, the remainder developed additional pathways for assimilation of carbon (Ku *et al.*, 1996). These alternate pathways are adaptations that provide greater efficiencies in more hostile environments, such as areas with higher temperatures and more arid climates. In these pathways, HCO_3^- is the initial carbon source instead of CO_2. Bicarbonate is formed by the dissociation of carbonic acid, which is formed when CO_2 in solution is hydrated (Eq. 11.6).

$$CO_2 + H_2O \leftrightarrow H_2CO_3 \leftrightarrow HCO_3^- + H^+ \qquad (11.6)$$

Hydration of CO_2 is catalyzed very rapidly by the enzyme **carbonic anhydrase**. Because the solubility of CO_2 in water decreases with increasing temperature, the concentration of bicarbonate can therefore reach much higher levels than that of CO_2 under these conditions. Consequently, incorporation of carbon into organic

Table 11.1 Reactions in the reductive pentose phosphate cycle.

A. Fixation of CO_2

(1) Ribulose-1,5-bisphosphate + CO_2 $\xrightarrow{\text{Rubisco}}$ 2,3-phosphoglycerate

(2) 3-Phosphoglycerate + ATP $\xrightarrow{\text{Phosphoglycerate kinase}}$ 1,3-bisphosphoglycerate + ADP

(3) 1,3-Bisphosphoglycerate + NADPH + H^+ $\xrightarrow{\text{Glyceraldehyde 3-phosphate dehydrogenase}}$ glyceraldehyde 3-phosphatae + $NADP^+$ + Pi

B. Synthesis of glucose

(4) Glyceraldehyde 3-phosphate $\xleftrightarrow{\text{Triose-phosphate isomerase}}$ dihydroxyacetone phosphate

(5) Glyceraldehyde 3-phosphate + dihydroxyacetone phosphate $\xrightarrow{\text{Fructose bisphosphate aldolase}}$ fructose-1, 6-bisphosphate

(6) Fructose-1,6-bisphosphate $\xrightarrow{\text{Fructose bisphosphatase}}$ fructose 6-phosphate + Pi

(7) Fructose 6-phosphate $\xrightarrow{\text{Glucose phosphate isomerase}}$ glucose 6-phosphate

(glucose 6-phosphate $\xrightarrow{\text{Glucose 6-phosphatase}}$ glucose + Pi)

(8) Glucose 6-phosphate $\xrightarrow{\text{Phosphoglucomutase}}$ glucose 1-phosphate(substrate for starch synthesis)

C. Regeneration of ribulose 1,5-bisphosphate

(9) Fructose 6-phosphate + glyceraldehyde 3-phosphate $\xrightarrow{\text{Transketolase}}$ xylulose 5-phosphate + erythrose 4-phosphate

(10) Erythrose 4-phosphate + dihydroxyacetone phosphate $\xrightarrow{\text{fructose bisphosphate aldolase}}$ sedoheptulose 1,7-bisphosphate

(11) Sedoheptulose 1,7-bisphosphate $\xrightarrow{\text{sedoheptulose bisphosphatase}}$ sedoheptulose 7-phosphate + Pi

(12) Sedoheptulose 7-phosphate + glyceraldehyde 3-phosphate $\xrightarrow{\text{transketolase}}$ ribose 5-phosphate + xylulose 5-phosphate

(13) Ribose 5-phosphate $\xrightarrow{\text{ribose phosphate isomerase}}$ ribulose 5-phosphate

(14) Xylulose 5-phosphate $\xrightarrow{\text{ribulose phosphate 3-epimerase}}$ ribulose 5-phosphate

(15) Ribulose 5-phosphate + ATP $\xrightarrow{\text{ribulose 5-phosphate kinase}}$ ribulose 1,5-bisphosphate + ADP

molecules is more efficient. (Because of the minimal difference in diffusion rates between bicarbonate ions with different isotopes of carbon than with CO_2, products of the C_4 pathway can be identified by the increased content of ^{13}C.) The initial reaction in this pathway is catalyzed by **phospho*enol*pyruvate carboxylase**, which adds the one-carbon unit to **phospho*enol*pyruvate** (Eq. 11.7). In contrast to the reaction catalyzed by Rubisco, O_2 is not a substrate or inhibitor of this reaction. The product, oxaloacetate, is unstable and is rapidly reduced to **malate** or converted to the amino acid **aspartate** by a transamination reaction (Figure 11.24). The initial product of carbon assimilation is a four-carbon intermediate, and thus the pathway has become known as the **C_4 pathway**.

$$H_2C{=}\overset{\overset{\displaystyle PO_3^=}{|}}{C}{-}\overset{\overset{\displaystyle O}{\|}}{C}{-}O^- + HCO_3^- \rightarrow {}^-O{-}\overset{\overset{\displaystyle O}{\|}}{C}{-}CH_2{-}\overset{\overset{\displaystyle O}{\|}}{C}{-}\overset{\overset{\displaystyle O}{\|}}{C}{-}O^- + Pi \qquad (11.7)$$

Phospho*enol*pyruvate   Oxaloacetate

The presence of the C_4 pathway is usually accompanied by a leaf morphology referred to as **Kranz** (crown) anatomy (Figure 11.25). The vascular tissue

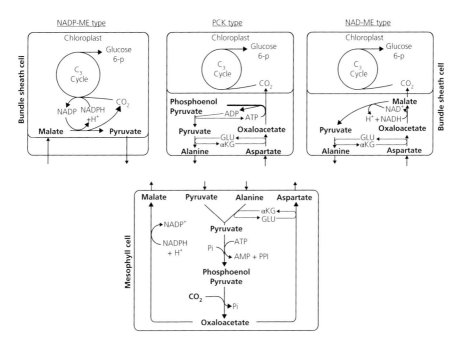

Figure 11.24 Fluxes of carbon in the C_4 pathways of photosynthesis. In mesophyll cells, CO_2 is rapidly hydrated to carbonic acid, which dissociates to form the bicarbonate ion, HCO_3^- (not shown). Bicarbonate reacts with phospho*enol*pyruvate to form oxaloacetate, catalyzed by phospho*enol*pyruvate carboxylase. Oxaloacetate is either reduced to malate or transaminated to aspartate, depending on the specific plant. These intermediates are transported to adjacent bundle sheath cells, where a reversal of these reactions provides a source of CO_2. This CO_2 is fixed in the bundle sheath cells by the reductive pentose phosphate (C_3) pathway. The differing pathways in the variety of C_4 plants are illustrated by the upper three schemes.

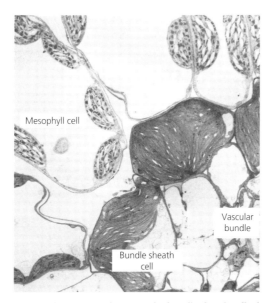

Figure 11.25 Morphological differentiation between the bundle sheath cells that surround the vascular tissue and the outer layer of mesophyll cells in leaves of a C_4 plant. Thylakoid membranes in the chloroplasts of bundle sheath cells are not organized into grana, in contrast to the extensive granal structures in mesophyll cells. Rubisco and the C_3 pathway are localized almost entirely in bundle sheath cells, while phospho*enol*pyruvate carboxylase and the C_4 pathway are localized in the mesophyll cells. Source: Micrograph courtesy of Dr. C. C. Black, Jr.

in these plants is surrounded by a layer of **bundle sheath** cells that perform the C_3 pathway. Chloroplasts in bundle sheath cells contain thylakoid membranes that are not differentiated into grana or tightly appressed. Surrounding the bundle sheath cells are layers of **mesophyll** cells that initiate the C_4 pathway. Thylakoid membranes in the chloroplasts in these cells are differentiated into the typical granal and stromal membranes. Interestingly, some species, such as in the family *Chenopodiaceae*, contain both pathways, with chloroplasts differentiated into the two types that are typical of bundle sheath and mesophyll cells *within the same cell*.

Phospho*enol*pyruvate carboxylase is located predominantly in mesophyll cells. The initial products of carbon assimilation, malate or aspartate, are transported to bundle sheath cells, where they are oxidized or deaminated and then decarboxylated to generate CO_2. The CO_2 thus produced is used by Rubisco in chloroplasts of the bundle sheath cells to initiate the typical reductive pentose phosphate cycle (see Figure 11.22). The other product of the decarboxylation reaction is pyruvate, which in plants that use malate as the "carbon carrier" is returned to the mesophyll cells. Pyruvate is transaminated to alanine in those plants that use aspartate, which is also returned to the mesophyll cells. In these plants, the carrier must transport nitrogen and carbon. The advantage of the interplay between these cells is a significantly greater concentration of CO_2 in the chloroplast of bundle sheath cells than could be achieved under normal atmospheric conditions, which results from its production *in situ*. Moreover, bundle sheath cells have a lower PS II to PS I ratio than mesophyll cells, which reduces the amount of oxygen produced by PS II during the light reactions. Bundle sheath cells thus seem to perform more cyclic photosynthetic electron transport than mesophyll cells. The higher concentration of CO_2, combined with a lower oxygen level, increases the efficiency of the "carboxylation" reaction of Rubisco over the "oxygenation" reaction, which allows these plants to thrive under less favorable conditions. However, this adaptation has an energy cost of two additional ATP molecules consumed per CO_2 assimilated. The additional energy is required to convert pyruvate to phospho*enol*pyruvate in mesophyll cells, which expends two equivalents of ATP to convert adenosine monophosphate (AMP), the product of this reaction catalyzed by **pyruvate orthophosphate dikinase**, to ADP and then to ATP (see Figure 11.24).

The development of the C_4 pathway probably occurred just over 7 MYA, when the carbon isotope ratio in fossil organic material abruptly changed from one that showed strong discrimination against ^{13}C, a characteristic of CO_2 fixation by Rubisco, to a less discriminatory ratio characteristic of bicarbonate incorporation by phospho*enol*pyruvate carboxylase. The change in ratio indicated a rapid expansion of C_4 plants across the earth. C_4 plants have high rates of photosynthesis and growth, and 11 of the 12 most productive species are C_4 (Raghavendra, 1998). Although only about 3% of the approximately 250 000 current species

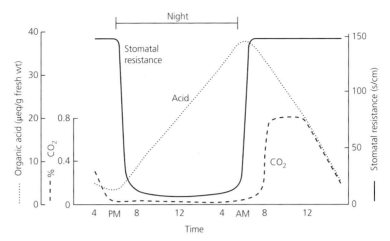

Figure 11.26 Changes in internal gas phase CO_2 concentration (-----), acid content of the vacuole, primarily malate (⋯), and the stomatal resistance to diffusion of water vapor (—) in a typical CAM plant. The drop in stomatal resistance at nightfall marked the opening of these leaf structures. CO_2 entering through the stomata at night was incorporated into malate by the CAM pathway (Figure 11.27). At dawn, the stomata close and subsequent decarboxylation of malate markedly increased the CO_2 levels within the plant during the day. Over the course of the day, the released CO_2 was fixed into carbohydrate (starch) by the reductive pentose phosphate cycle.

of plants contain the C_4 pathway, the agricultural importance of several of these species (e.g., maize, sugarcane, and sorghum) results in about 30% of the primary productivity of plants.

An interesting variation on the C_4 pathway occurs in plants that live in unusually hot, dry climates. These plants keep their **stomates**—openings in the leaves through which O_2, CO_2, and H_2O exchange with the atmosphere—closed during the day but open during the night (Figure 11.26). Photosynthesis during the day generates starch, which is stored in chloroplasts. At night, starch is metabolized through glycolysis to phospho*enol*pyruvate, which is carboxylated to oxaloacetate by **phospho*enol*pyruvate carboxylase**. Oxaloacetate is reduced to malate and stored in the large vacuole in the cells. During the day, malate is decarboxylated and the resulting CO_2 is fixed by Rubisco and the C_3 cycle, driven by photosynthetic electron transport (Figure 11.27). This pathway, in which the light reactions and the CO_2 assimilation reactions are separated *temporally*, is a remarkable adaptation to a hostile environment and is referred to as **crassulacean acid metabolism**. Plants containing this pathway are commonly known as "CAM plants." Leaves of these plants are generally thick and fleshy and store large quantities of water. The most common examples of such plants are the cacti and agave plants that survive surprisingly well in the hot, arid deserts of south-western United States and Mexico.

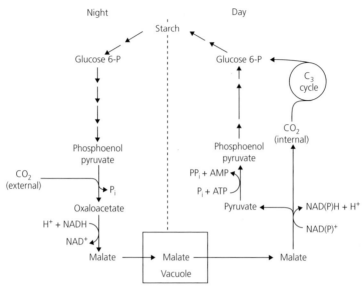

Figure 11.27 Pathway for the flux of carbon in crassulacean acid metabolism in CAM plants. During the night, starch is degraded through a glycolytic pathway to phospho*enol*pyruvate. CO_2 entering through the stomata from the atmosphere at night is hydrated to carbonic acid (not shown) and fixed into oxaloacetate by phospho*enol*pyruvate carboxylase (see Figure 11.24). Oxaloacetate is reduced by cytosolic NAD^+-malate dehydrogenase to malate, which is transported into and accumulates in the vacuole of the cell. During the day, malate is released from the vacuole and is converted to pyruvate through a decarboxylation reaction catalyzed by $NADP^+$-dependent malate dehydrogenase. Pyruvate is converted to phospho*enol*pyruvate by an unusual reaction catalyzed by pyruvate phosphate dikinase, which provides the substrate for glucose and starch synthesis by a reversal of glycolysis (gluconeogenesis). The CO_2 released by the decarboxylation reaction of malate is refixed photosynthetically by the reductive pentose phosphate cycle. Loss of CO_2 produced in this reaction is prevented by closure of the stomata during the day (see Figure 11.26).

Regulation of CO_2 assimilation by light

The activity of a large number of soluble enzymes in the chloroplast stroma is regulated to prevent "futile" reactions that would dissipate the energy captured by the reactions on the thylakoid membrane. A general factor controlling enzyme activity is the change in environment in the stroma between the day and night. During light-driven photophosphorylation, transfer of protons from the stroma to the thylakoid lumen for generation of the proton gradient results in the pH of the stroma rising to nearly 8, whereas the pH is near 7 during the night. For an enzyme with a pH optimum near 8, this change in pH is sufficient to dramatically affect activity. The lower concentration of H^+ in the stroma during the day is compensated for by an increase in the concentration of Mg^{2+}. Several enzymes, including Rubisco, require Mg^{2+} for activity. The higher pH and concentration of Mg^{2+} support significantly higher activity of these enzymes in the light.

Light energy is required to drive the photosynthetic reactions, and these reactions in turn regulate the activity of several key enzymes in CO_2 assimilation. Chloroplasts regulate many of their enzymes by covalent modification. A common mechanism is disulfide-sulfhydryl interchange ($-S—S– \leftrightarrow 2 –SH$) between cysteine residues in the protein, in which light-driven reduction of the disulfide bond to sulfhydryl groups occurs during the day and reoxidation to the disulfide form in darkness. For example, at the end of the day, the photosystems stop driving the electron transport required to generate the proton gradient. Left to its own, ATP synthase would then run in reverse, as an ATP hydrolase, using the remaining ATP made during the day for a retrograde transfer of protons from stroma to the lumen. This hydrolysis of ATP is inhibited by oxidation of a pair of sulfhydryl groups in the γ subunit, cysteine-199 and cysteine-205, to a disulfide bond, which locks the rotary motion of the γ subunit of ATP synthase (see Figure 11.19). Upon activation of electron transport at dawn, ferredoxin is reduced by PS I, which in turn reduces a small (12 kDa) protein called **thioredoxin**. Thioredoxin contains cysteine residues that undergo reversible oxidation and reduction. Production of reduced thioredoxin leads to reduction of the disulfide bond in the γ subunit, and the activity of ATP synthase is restored.

Regulation of the reductive pentose phosphate cycle is critical because the cycle must run in the forward direction during the day, using ATP and NADPH to drive synthesis of glucose 6-phosphate from CO_2. Glucose 6-phosphate is the substrate for storage of carbohydrate as starch. In darkness, the cycle operates in reverse and is then called the *oxidative* pentose phosphate cycle. Glucose 6-phosphate, produced by breakdown of starch, is metabolized in the oxidative pathway to 6-phosphogluconate and then decarboxylated to ribulose 5-phosphate, with $NADP^+$ reduced to NADPH in each step. These reactions serve as the primary source of NADPH and ribose 5-phosphate in darkness. To prevent "futile" cycling, or to ensure that the system does not work against itself, the cycle is regulated. Glucose 6-phosphate dehydrogenase, which oxidizes glucose 6-phosphate to 6-phosphogluconate, is inhibited by the product of the reaction, NADPH. At pH 8, the pH of the stroma during active photosynthesis, the ratio of concentrations of NADPH over $NADP^+$ is sufficiently high to completely inhibit the enzyme. In addition, the enzyme is also inhibited by ribulose 1,5-bisphosphate, which is the substrate for Rubisco. Thus, by reducing $NADP^+$ to NADPH and synthesizing ATP from ADP and phosphate, the light reactions of photosynthesis ensure that the oxidative pentose phosphate pathway is blocked during the day. Furthermore, just to make sure, glucose 6-phosphate dehydrogenase is inactivated in the light through reduction by thioredoxin.

Several enzymes in the cycle, including fructose 1,6-bisphosphatase, sedoheptulose 1,7-bisphosphatase, ribulose 5-phosphate kinase, and NADP-dependent glyceraldehyde dehydrogenase are activated in the light by reduced thioredoxin. Each of these enzymes of the carbon assimilation cycle is inactivated when two cysteine-SH groups in the enzyme are oxidized to a

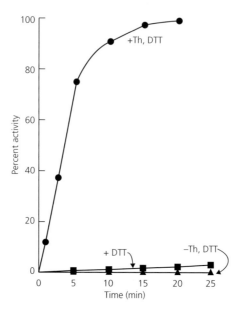

Figure 11.28 Activation of the chloroplastic NADP⁺-dependent malate dehydrogenase. The oxidized form of the enzyme, in which two cysteine residues are linked with a disulfide (–S—S–) bond, is inactive. Dithiothreitol, a 4-carbon molecule containing a sulfhydryl group (–SH) on each terminal carbon, was ineffective in activating the enzyme but was capable of reducing thioredoxin to its sulfhydryl form. Addition of reduced thioredoxin rapidly activated the enzyme by reducing the disulfide bond to sulfhyryl groups. Source: Adapted from Ferte *et al.* (1982).

disulfide (–S—S–). Restoration of activity requires light-activated PS I to reduce ferredoxin, which then reduces thioredoxin. Although the overall redox state of the chloroplast remains highly reducing during the night, the redox potentials of sulfhydryl groups on these regulated enzymes are unusually low and are oxidized to the disulfide form when the status of the stroma becomes only slightly more oxidative.

A particularly well-studied enzyme that is regulated by sulfhydryl–disulfide interchange is the **NADP-dependent malate dehydrogenase**, which is involved in the C$_4$ pathway of carbon fixation, but it is also present in C$_3$ chloroplasts. As shown in Figure 11.28, the enzyme is inactive in the oxidized form and is fully activated by treatment with reduced thioredoxin. In the oxidized form, a disulfide bond occurs near the C-terminus of the protein, between cysteine-361 and cysteine-373 (Ashton *et al.*, 2000). This bond causes the otherwise flexible C-terminal extension to be constrained into a sharp turn and fold into the active site (Figure 11.29). Consequently, binding of malate to the active site is blocked and no conversion to oxaloacetate is measured. Addition of a sulfhydryl reducing agent such as dithiothreitol is relatively ineffective in activating the enzyme as compared with reduced thioredoxin, which suggests that thioredoxin has an additional activity, perhaps through protein–protein interactions.

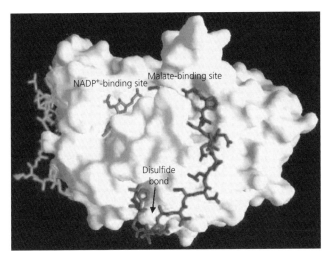

Figure 11.29 Model of the structure of the oxidized, inactive NADP$^+$-dependent malate dehydrogenase. The figure shows a surface view of the enzyme monomer with a stick view of the C-terminal extension (*dark grey*) that extends into the active site near the bound NADP$^+$ (*light grey*), which is partially obscured by the active site loops. The presence of the negatively charged C-terminus in the active site stabilizes the binding of the positively charged nicotinamide ring of NADP$^+$ but not NADPH. Thioredoxin reduces the disulfide bond between cysteine-361 and cysteine-373 (the enzyme contains 385 amino acids) and promotes dissociation of the C-terminal extension from the active site. Source: Adapted from Ashton *et al.* (2000).

Rubisco is also a highly regulated enzyme. In darkness, when the carbamate group of Rubisco dissociates, the enzyme binds one of its substrates, **ribulose 1,5-bisphosphate**, very tightly (K_d of 20 nM) in a nonproductive complex. Without the carbamate group, the enzyme cannot bind the Mg^{2+} ion that is required for binding the other substrate, CO_2. The formation of the carbamate group at the active site is thus effectively blocked. Rubisco activase, which requires ATP for activity, facilitates the removal of the ribulose 1,5-bisphosphate and thus allows the activation of Rubisco (Hartman and Harpel, 1994). To further ensure inactivity of Rubisco in darkness, any of the enzyme remaining in the carbamylated form binds very tightly (K_d of 32 nM) to a reaction-intermediate analog, 2-carboxyarabinitol 1-phosphate, which inhibits enzymatic activity. When reductants are generated in the light, a phosphatase is activated that hydrolyzes the phosphoester bond, which causes the concentration of free 2-carboxyarabinitol 1-phosphate to drop below the K_d and allows dissociation of the inhibitor from the active site. Moreover, the kinase reaction that phosphorylates ribulose 5-phosphate to produce ribulose 1,5-bisphosphate, the substrate for Rubisco, is activated by thioredoxin. Consequently, the reaction catalyzed by Rubisco can only proceed when conditions for CO_2 fixation are suitable. Rubisco activase is also inactive in darkness and is activated by thioredoxin. The remarkable regulation of these enzymes ensures that ATP is not dissipated in darkness needlessly, and reactions that use ATP as a substrate occur only in the light when it can be replenished.

End products of carbon assimilation

Starch

Similar to the storage of excess calories by animals in the form of lipids and a polymer of glucose, plants also use these materials to store energy. Lipids, the long-term storage form of carbon with highest caloric content per unit weight, accumulate primarily as fat in adipose tissue in animals and in oil bodies in the seeds of plants. In animals, glucose is polymerized into a highly branched molecule, glycogen. In plants, glucose is polymerized into long, mostly linear polymers, **starch** or **amylose**, which has roughly 2000 glucose units per polymer. Starch is slightly branched, much less so than another, more abundant, highly branched polymer, amylopectin. Up to 10 000 glucose units occur in amylopectin, which has an average chain length between branches of about 21. Chains are linked by α-1,4 glycosidic bonds (Figure 11.30), with branches extending from the carbon-6 hydroxymethyl group in α-1,6 bonds. Starch accumulates as large aggregates as "grains" in the stroma of chloroplasts (see Figures 11.4 and 11.5).

In the stroma of the chloroplast, glucose 6-phosphate, the ultimate product of the reductive pentose phosphate cycle, is converted to glucose 1-phosphate by **phosphoglucomutase. ADP-glucose pyrophosphorylase** catalyzes the reaction between ATP and glucose 1-phosphate to generate **ADP-glucose** and pyrophosphate (Eq. 11.8).

$$ATP + glucose\ 1 - P \leftrightarrow ADP\text{-glucose} + PPi \tag{11.8}$$

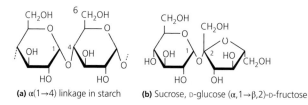

(a) α(1→4) linkage in starch   **(b)** Sucrose, D-glucose (α,1→β,2)-D-fructose

Figure 11.30 Structures of the linkages in starch and sucrose. The linear chain of starch consists of glucose molecules linked α(1,4). Branches occur by additional α(1,6) linkages, in which a glucose molecule at a branch point is linked to other molecules through the 6-hydroxyl as well as the 4-hydroxyl groups. In sucrose, the glycosidic bond links the anomeric carbon-1 of glucose, with the hydroxyl group in the α configuration, with the anomeric carbon-2 of fructose, with the hydroxyl in the β configuration to form an αl, β2 linkage. The net result of formation of a glycosidic bond is the elimination of a molecule of water.

ADP serves as the "leaving group" when the oxygen in the 4-OH of the glucose residue at the end of a polymer attacks carbon-1 of ADP-glucose (Eq. 11.9), the reaction catalyzed by **starch synthase**. Thus the polymer grows at the "nonreducing" end, which is also the end from which degradation occurs to release glucose 1-phosphate by the activity of **starch phosphorylase** (Eq. 11.10).

$$\text{HO-Glucose-}\left(\text{glucose}\right)_n + \text{ADP-glucose} \leftrightarrow \text{Glucose}\left(\alpha\text{-1,4}\right)$$
$$\text{-}O\text{-Glucose-}\left(\text{glucose}\right)_n + \text{ADP} \tag{11.9}$$

$$\text{Glucose-}\left(\text{glucose}\right)_n + \text{Pi} \leftrightarrow \text{glucose 1-phosphate} + \left(\text{glucose}\right)_n \tag{11.10}$$

ADP-glucose pyrophosphorylase is the rate-limiting activity in starch synthesis and is allosterically activated by 3-phosphoglyceric acid, the initial product of CO_2 fixation. Therefore, the enzyme is most active during periods of maximal rates of photosynthesis. The enzyme is inhibited by phosphate, whose concentration increases in darkness. The higher concentration of Pi in the chloroplast promotes the breakdown of starch in darkness, catalyzed by **starch phosphorylase**.

Sucrose

Whereas starch forms insoluble granules, **sucrose** is a highly soluble end product of photosynthesis. Sucrose is the main form of carbon transport in plants and plays a key role in reproduction and propagation, such as in flower nectar. As a readily available source of energy, sucrose sustains the initial stages of growth after dormant periods in temperate plants. Sucrose is found in all cells of a plant and is enzymatically hydrolyzed to fructose and glucose, the ultimate building blocks for all other organic compounds in plants.

Unlike starch, sucrose is synthesized in the cytosol. Carbon leaves the chloroplast primarily as the triose-phosphates, glyceraldehyde 3-phosphate, or dihydroxyacetone-phosphate, in exchange for inorganic phosphate. This exchange is facilitated by a specific transporter on the chloroplast envelope. In the cytosol, the reversal of the glycolytic pathway converts triose-phosphates to fructose 1,6-bisphosphate, fructose 6-phosphate, and then glucose 6-phosphate. Glucose 1-phosphate is made from glucose 6-phosphate in the reaction catalyzed by **phosphoglucomutase** and is then used to produce uridine diphosphate or **UDP-glucose** by **UDP-glucose pyrophosphorylase**. This reaction is mechanistically similar to the reaction in the plastid (Eq. 11.8) but uses uridine triphosphate (UTP) rather than ATP. UDP-glucose reacts with fructose 6-phosphate to form sucrose 6-phosphate in the reaction catalyzed by **sucrose-phosphate synthase**. This enzyme is inhibited by high concentrations of Pi, as would occur in darkness, and also by phosphorylation of a serine residue. Sucrose 6-phosphate is hydrolyzed to sucrose by **sucrose-phosphate phosphatase**.

Glucose and fructose are linked in sucrose between the "reducing" groups of each (i.e., $\alpha 1,\beta 2$) (Figure 11.30), which produces a much more stable disaccharide than maltose or lactose. Excess sucrose is transported from the cytosol into the vacuole by a specific transport system on the tonoplast—the membrane that surrounds the vacuole.

Conclusions for the reactions of photosynthesis

ATP and NADPH are the chemical currencies that photosynthetic organisms use for synthesis of glucose from CO_2 and H_2O. Glucose is stored as starch in the chloroplast and used to support metabolism of the plant in the dark or is consumed by animal cells to provide energy for their metabolism and growth. Intermediates in the pathway to glucose 6-phosphate, at the level of triose-phosphates, are also exported from the chloroplast and used for synthesis of sucrose in the cytosol, the other major storage form of energy as carbohydrates in plants. Although less in bulk, sucrose is highly desired as a source of energy by animals. Metabolism of starch and sucrose generates CO_2 and H_2O, and the cycle gets repeated.

References

Allen, J.F. (2003) State transitions—a question of balance. *Science*, **299**, 1530–1532.

Ashton, A.R., Trevanion, S.J., Carr, P.D. *et al.* (2000) Structural basis for the light regulation of chloroplast NADP malate dehydrogenase. *Physiologia Plantarum*, **110**, 314–321.

Barber, J. (2008) Crystal structure of the oxygen-evolving complex of photosystem II. *Inorganic Chemistry*, **47**, 1700–1710.

Bhattacharya, D., and Medlin, L. (1995) The phylogeny of plastids: a review based on comparisons of small-subunit ribosomal RNA coding regions. *Journal of Phycology*, **31**, 489–498.

Blankenship, R.E. (2002) *Molecular Mechanisms of Photosynthesis*. Blackwell Science, Ltd., Oxford, UK.

Demmig-Adams, B., and Adams, W.W. III. (1992) Photoprotection and other responses of plants to high light stress. *Annual Reviews of Plant Physiology and Plant Molecular Biology*, **43**, 599–626.

Drews, G. (1996) Forty-five years of developmental biology of photosynthetic bacteria. *Photosynthesis Research*, **48**, 325–352.

Elston, T., Wang, H., and Oster, G. (1998) Energy transduction in ATP synthase. *Nature*, **391**, 510–513.

Ferreira, K.N., Iverson, T.M., Maghlaoui, K. *et al.* (2004) Architecture of the photosynthetic oxygen-evolving center. *Science*, **303**, 1831–1838.

Ferte, N., Meunier, J.-C., Ricard, J. *et al.* (1982) Molecular properties and thioredoxin-mediated activation of spinach chloroplastic NADP-malate dehydrogenase. *FEBS Letters*, **146**, 133–138.

Gantt, E. (1981) Phycobilosomes. *Annual Reviews of Plant Physiology*, **32**, 327–347.

Green, B.R., and Durnford, D.G. (1996) The chlorophyll-carotenoid proteins of oxygenic photosynthesis. *Annual Review of Plant Physiology and Plant Molecular Biology*, **47**, 685–714.

Hartman, F.C., and Harpel, M.R. (1994) Structure, function, regulation, and assembly of D-ribulose-1,5-bisphosphate carboxylase/oxygenase. *Annual Review of Biochemistry* **63**, 197–234.

Hohmann-Marriott, M.F., and Blankenship, R.E. (2011) Evolution of photosynthesis. *Annual Review of Plant Biology*, **62**, 515–548.

Hoober, J.K., Boyd, C.O., and Paavola, L.G. (1991) Origin of thylakoid membranes in *Chlamydomonas reinhardtii* y-1 at 38°C. *Plant Physiology*, **96**, 1321–1328.

Hoober, J.K., Eggink, L.L., and Chen, M. (2007) Chlorophylls, ligands and assembly of light-harvesting complexes in chloroplasts. *Photosynthesis Research*, **94**, 387–400.

Hu, X., Ritz, T., Damjanović, A., Autenrieth, F., and Schulten, K. (2002) Photosynthetic apparatus of purple bacteria. *Quarterly Reviews of Biophysics*, **35**, 1–62.

Jordan, P., Fromme, P., Witt, H.T. *et al.* (2001) Three-dimensional structure of cyanobacterial photosystem I at 2.5 Å resolution. *Nature*, **411**, 909–917.

Ku, M.S.B., Kano-Murakami, Y., and Matsuoka, M. (1996) Evolution and expression of C_4 photosynthesis genes. *Plant Physiology*, **111**, 949–957.

Liu, Z., Yan, H., Wang, K. *et al.* (2004) Crystal structure of spinach major light-harvesting complex at 2.72 Å resolution. *Nature*, **428**, 287–292.

Martin, W., Rujan, T., Richly, E. *et al.* (2002) Evolutionary analysis of *Arabidopsis*, cyanobacterial, and chloroplast genomes reveals plastid phylogeny and thousands of cyanobacterial genes in the nucleus. *Proceedings of the National Academy of Sciences of the United States of America*, **99**, 12246–12251.

Mustárdy, L., and Garab, G. (2003) Granum revisited. A three-dimensional model—where things fall into place. *Trends in Plant Science*, **8**, 117–122.

Nield, J., Orlova, E.V., Morris, E.P. *et al.* (2000) 3D map of the plant photosystem II supercomplex obtained by cryoelectron microscopy and single particle analysis. *Nature Structural Biology*, **7**, 44–47.

Niyogi, K.K. (1999) Photoprotection revisited: genetic and molecular approaches. *Annual Review of Plant Physiology and Plant Molecular Biology*, **50**, 333–359.

Park, H., and Hoober, J.K. (1997) Chlorophyll synthesis modulates retention of apoproteins of light-harvesting complex II by the chloroplast in *Chlamydomonas reinhardtii*. *Physiologia Plantarum*, **101**, 135–142.

Park, H., Eggink, L.L., Roberson, R.W., and Hoober, J.K. (1999) Transfer of proteins from the chloroplast to vacuoles in *Chlamydomonas reinhardtii* (Chlorophyta): a pathway for degradation. *Journal of Phycology*, **35**, 528–538.

Raghavendra, A.S. (1998) *Photosynthesis: A Comprehensive Treatise*. Cambridge University Press, Cambridge, UK.

Raven, J.A., and Allen, J.F. (2003) Genomics and chloroplast evolution: what did cyanobacteria do for plants? *Genome Biology*, **4**, 209.

Rebeiz, C.A., Benning, C., Bohnert, H.J. *et al.* (2010) *The Chloroplast: Basics and Applications. Advances in Photosynthesis and Respiration*, Vol. **31**. Springer, Dordrecht.

Remelli, R., Varotto, C., Sandonà, D. *et al.* (1999) Chlorophyll binding to monomeric light-harvesting complex: a mutational analysis of chromophore-binding residues. *The Journal of Biological Chemistry*, **274**, 33510–33521.

Staehelin, L.A. (2003) Chloroplast structure: from chlorophyll granules to supra-molecular architecture of thylakoid membranes. *Photosynthesis Research*, **76**, 185–196.

Stoebe, B., and Maier, U.G. (2002) One, two, three: nature's tool box for building plastids. *Protoplasma*, **219**, 123–130.

Tanaka, R., Ito, H., and Tanaka, A. (2010) Regulation and functions of the chlorophyll cycle. In: *The Chloroplast: Basics and Applications, Advances in Photosynthesis and Respiration*, Vol. **31**. (Rebeiz, C.A., Benning, C., Bohnert, H., *et al.*, eds) Springer, Dordrecht, pp. 55–77.

Timmis, J.N., Ayliffe, M.A., Huang, C.Y., and Martin, W. (2004) Endosymbiotic gene transfer: organelle genomes forge eukaryotic chromosomes. *Nature Review Genetics*, **5**, 123–135.

Westphal, S., Soll, J., and Vothknecht, U.C. (2001) A vesicle transport system inside of chloroplasts. *FEBS Letters*, **506**, 257–261.

White, R.A., and Hoober, J.K. (1994) Biogenesis of thylakoid membranes in *Chlamydomonas reinhardtii* y1: a kinetic study of initial greening. *Plant Physiology*, **106**, 583–590.

White, R.A., Wolfe, G.R., Komine, Y., and Hoober, J.K. (1996) Localization of light-harvesting complex apoproteins in the chloroplast and cytoplasm during greening of *Chlamydomonas reinhardtii* at 38°C. *Photosynthesis Research*, **47**, 267–280.

Wise, R.R., and Hoober, J.K. (2006) *The Structure and Function of Plastids. Advances in Photosynthesis and Respiration*, Vol. **23**. Springer, Dordrecht.

Xiong, J., Fischer, W.M., Inoue, K. *et al.* (2000) Molecular evidence for the early evolution of photosynthesis. *Science*, **289**, 1724–1730.

Yoon, H.S., Hackett, J.D., Pinto, G., and Bhattacharya, D. (2002) The single, ancient origin of chromist plastids. *Proceedings of the National Academy of Sciences of the United States of America*, **99**, 15507–15512.

Further reading

Chi, W., He, B.Y., Mao, J., Jiang, J.J. and Zhang, L.X. (2015) Plastid sigma factors: their individual functions and regulation in transcription. *Biochimica et Biophysica Acta-Bioenergetics*, **1847** (9), 770–778.

Goldbeck, J. and VanderEst, A. (2014) *The Biophysics of Photosynthesis. Biophysics for the Life Sciences*, Springer, New York, NY.

Jensen, P.E. and Leister, D. (2014) Chloroplast evolution, structure and functions. *F1000 Prime Reports*, **6**, 40, doi:10.12703/P6-40.

Nawrocki, W.J., Tourasse, N.J., Taly, A., Rappaport, F. and Wollman, F. (2015) The plastid terminal oxidase: its elusive function points to multiple contributions to plastid physiology. *Annual Review of Plant Biology*, **66**, 49–74.

Theng, S. and Wollman, F.-A. (2014) *Plastid Biology. Advances in Plant Biology*, Springer, New York, NY.

van Vijk, K. (2015) Protein maturation and proteolysis in plant production, mitochondria and peroxisomes. *Annual Review of Plant Biology*, **66**, 75–111.

CHAPTER 12

Vacuoles and protein bodies

William V. Dashek[1] and Amy M. Clore[2]

[1] Retired Faculty, Adult Degree Program, Mary Baldwin College, Staunton, VA, USA
[2] Division of Natural Sciences, New College of Florida, Sarasota, FL, USA

Vacuoles

Structure

There are at least two types of vacuoles in plants, lytic (L) and protein storage vacuoles (PSVs) (Taiz, 1992; Wink, 1993; Paris *et al.*, 1996; Marty, 1999; De, 2000; Robinson and Rogers, 2000). Frigerio *et al.* (2008) hypothesized that there are multiple vacuoles in certain plant cells. The central vacuoles can be categorized as PSVs in seeds (Jiang *et al.*, 2001) and as the vegetative vacuoles in non-seeds. The lytic vacuole occupies the central portion of a mature plant cell and can constitute 90% of the cell's volume (Figure 12.1). This vacuole is surrounded by a tonoplast membrane containing soluble *N*-ethylmaleimide-sensitive factor attachment protein receptors or SNARES (Carter *et al.*, 2004). SNARES constitute two categories of integral proteins: *v*-SNAREs and *t*-SNARES. The *v*-SNAREs reside on vesicle membranes and play roles in membrane fusion, while the *t*-SNARES are located on target compartment membranes (Bassham and Blatt, 2008; Kim and Brandizzi, 2012). SNAREs can also be classified into four structural types: Qa, Qb, Qc, and R (Lipka *et al.*, 2007). The tonoplast appears to contain SNAREs from each of four groups necessary for membrane fusion (Carter *et al.*, 2004). Lytic vacuoles occur in the cells of young seedlings, following seed germination and are ubiquitous in cells during vegetative growth (Zheng and Staehelin, 2011).

Chemical composition

The protein compositions in diverse PSVs are presented in Table 12.1. During seed germination, the proteins are degraded to yield nutrients for the developing embryo. This degradation occurs through the action of hydrolytic enzymes (Jiang and Rogers, 2001).

The central vacuole in non-seeds possesses a pH of 5.5–6.0 and can be quite complex in composition (Leigh, 1997). It can contain alkaloids (Figure 12.2;

Plant Cells and their Organelles, First Edition. Edited by William V. Dashek and Gurbachan S. Miglani.
© 2017 John Wiley & Sons, Ltd. Published 2017 by John Wiley & Sons, Ltd.

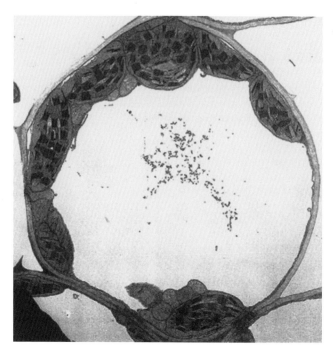

Figure 12.1 TEM of plant cell illustrating the large, lytic central vacuole. Source: From http://bioweb.wku.edu/courses/biol22000/11organelles/Fig.html.

Table 12.1 Examples of proteins present in protein storage vacuoles.*

Protein	Plant system	Reference(s)
Glutenins (glutelins), gliadins	Wheat endosperm	Hurkman *et al.* (2013)
Vicilin and legumin	*Pisum sativum* cotyledons	Robinson *et al.* (1997)
Glycinin and β-conglycinin	Soybean seeds	Mori *et al.* (2004)
Zeins (maize prolamins)	Maize endosperm (PSVs in aleurone only)	Reyes *et al.* (2011)
Phaseolin	Common bean	Park *et al.* (2004)

*Eudicot seeds can contain primarily 2S albumins and 7S and 11S globulins (Otegui *et al.*, 2006).

Kutchan *et al.*, 1986; Facchini, 2001), amino acids (Echeverria, 2000), hydrolytic enzymes (Hara-Nishimura and Hatsugai, 2011), ions (Barkla and Pantoja, 1996; Andreev, 2001; Surpin and Raikhel, 2004), organic acids, pigments (Evert, 2006), tannins (see Figure 12.3; Khanbabaee and van Ree, 2001), shikimate pathway intermediates, vitamins, alkaloids, phenylpropanoids, flavonoids (Tohge *et al.*, 2011), and water. Proteome analysis of tonoplasts from cell suspension cultures of *Arabidopsis thaliana* revealed 163 proteins, most notably V-type and H^+ ATPases (Shimaoka *et al.*, 2004). The ATPases acidify the central vacuoles (Taiz, 1992).

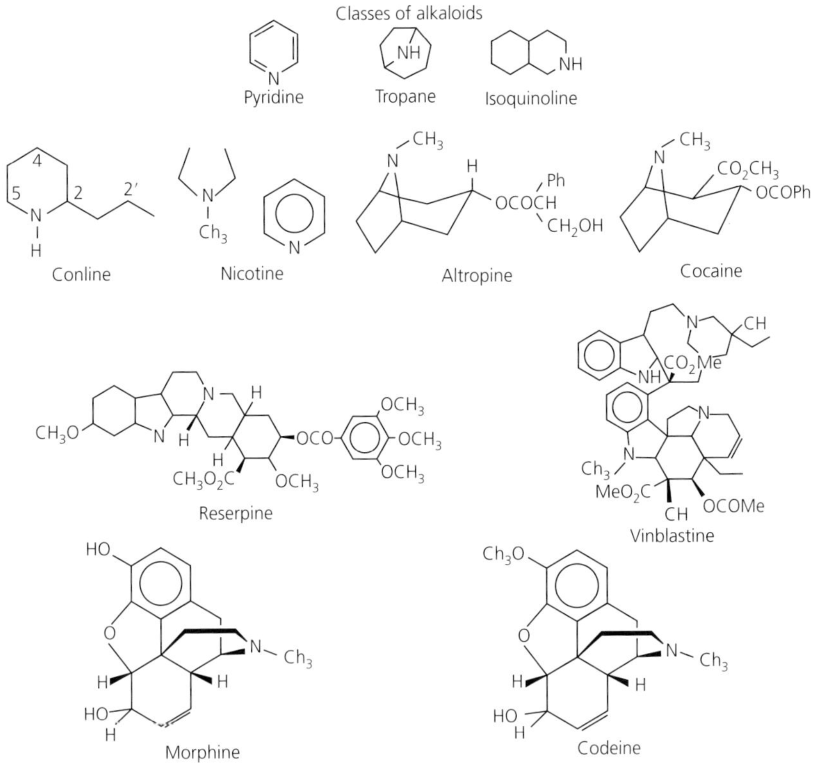

Figure 12.2 Structures of alkaloids. Source: Dashek and Harrison (2006).

Whereas the tonoplast of lytic vacuoles in pea was reported to contain gamma tonoplast intrinsic protein (TIP 1;1, Jauh *et al.*, 1998), the membrane of pea PSV was found to house alpha- and delta-TIPs (TIP 3;1 and TIP 2;1, respectively; for designations, see Gattolin *et al.*, 2011), all of which are aquaporins (Jauh *et al.*, 1999). More recent work by Gattolin *et al.* (2011) has indicated that TIP 3;1 and the lesser characterized 3;2 are localized to the PSV tonoplast during seed development and germination in *Arabidopsis*. However, TIP types and distributions can change during development, and multiple TIPs tend to coexist on given tonoplasts, making the situation complex (Hunter *et al.*, 2007; Gattolin *et al.*, 2011). Nonetheless, TIP 3;1 appears to be a consistent marker for the PSV tonoplast based on the aforementioned studies. In addition to TIPs, the tonoplast contains phospholipids, free sterols, ceramide monohexoside, and digalactosyldiacylglycerol (Yoshida and Uemura, 1986).

The PSV of certain plants can house a crystalloid structure that may be composed of phytic acid (Jiang *et al.*, 2001). The appearance of proteins in the PSV appears to be a consequence of peptide signal sequences (Bassham and Raikhel, 1997).

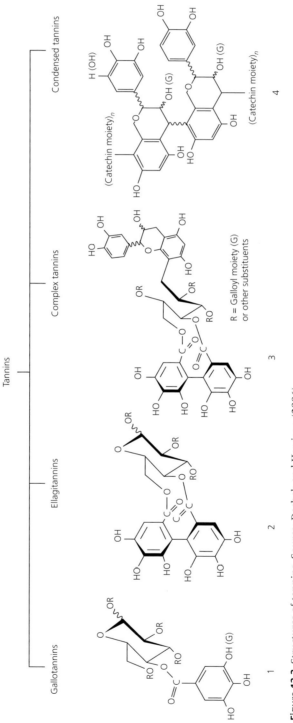

Figure 12.3 Structures of tannins. Source: Dashek and Harrison (2006).

Biogenesis

PSVs appear to arise *de novo*, at least in pea cotyledons (Hoh *et al.*, 1995). With regard to transport of storage proteins to PSVs, there appears to be more than one route in seeds (Figure 12.4). Robinson *et al.* (1998) reported that multivesicular bodies (MVBs-late endosomes, Jiang *et al.*, 2002a), which may be formed by the aggregation of Golgi-derived vesicles, fused with the forming PSVs. In 2011, Reyes *et al.* (2011) noted that PSVs in maize aleurone contain glycoproteins that are "trafficked" via a Golgi–MVB pathway. They also stated that MVBs are likely to fuse with multilayered compartments (Robinson *et al.*, 1998) prior to fusing with the PSV. In the pea cotyledon, storage proteins are

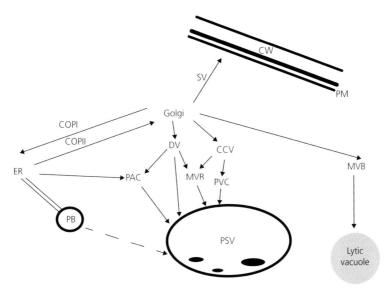

Figure 12.4 Schematic diagram of transport through the endomembrane system with an emphasis on the various pathways taken by storage proteins. CCVs, clathrin-coated vesicles; COPI, coat protein I-coated vesicles; COPII, coat protein II-coated vesicles; CW, cell wall; DVs, dense vesicles; ER, endoplasmic reticulum; MVBs, multivesicular bodies; PAC, precursor-accumulating vesicles; PB, protein body; PM, plasma membrane; PSV, protein storage vacuole; PVC, prevacuolar compartment; and SVs, secretory vesicles. Generally, cargo is taken from the ER to the Golgi (for processing and sorting) by COPII-coated vesicles. Some cargo is ultimately sent to the cell periphery for exocytosis via secretory vesicles. Cargo destined for the lytic vacuole is sent from the Golgi via MVBs. COPI-coated vesicles transport retrieved (escaped) ER-resident cargo back to the ER. Some storage proteins remain resident as PBs in the lumen of the ER, although it appears that in certain cases these bodies may be released and fuse with PSV in a process involving autophagy (dashed line). As shown in the diagram and discussed in the text, other storage proteins can leave the Golgi and move to the PSV by several different pathways depending on the type of storage protein and plant tissue. Transport via the PAC vesicles appears to allow bypass of the Golgi. Endocytosis (omitted for the sake of clarity) can also be the source of lytic vacuole contents and involves CCVs and MVBs (also known as the late endosomes). A diagram with additional detail pertaining to storage protein trafficking can be found in Ibl and Stoger (2012).

sorted in the cisternae of the Golgi apparatus (Hillmer *et al.*, 2001). A Golgi-independent route of transport of certain storage proteins to PSVs involves precursor-accumulating vesicles (PACs). These vesicles contain aggregates of endoplasmic reticulum (ER)-derived precursor proteins (Hara-Nishimura *et al.*, 1998, 2004).

Historically, the origin of the central lytic vacuole has involved two possible models. In one model, the vacuole arises from the *trans*-Golgi network (Boller and Wiemken, 1986; Marty, 1999). In another model, the vacuole originates via dilation of a portion of the smooth ER or SER (Hilling and Amelunxen, 1985). Recent evidence indeed points to the ER as the primary source for lytic vacuole biogenesis in *A. thaliana* root meristem cells (Viotti *et al.*, 2013; Viotti, 2014). Whatever the case may be, the formation of the large central vacuole involves the fusion of smaller vacuoles (Boller and Wiemken, 1986). A possible role for microtubules in the organization of *Physcomitrella patens* vacuoles has been proposed (Oda *et al.*, 2009). In this organism, microtubule-induced tubular protrusions and cytoplasmic strands are important for vacuolar distribution. Hicks *et al.*, 2004 have reported tubular vacuolar formation and highly mobile cytoplasmic invaginations for germinating pollen. However, they stated that the mechanism of vacuole biogenesis in pollen may differ from other plant tissues. For example, Zheng and Staehelin (2011) reported that PSVs can be transformed into lytic vacuoles in root meristematic cells of tobacco.

At least some proteins of the lytic vacuole appear to be transported from the Golgi apparatus by clathrin-coated vesicles or CCVs (see literature in Okita and Rogers, 1996). The CCVs seem to arise from the *trans*-Golgi apparatus and the early endosome (Figure 12.4). Proteins of the PSVs can arise from protein bodies or PBs (Figure 12.4).

Transport of solutes across the tonoplast

Transport across the tonoplast involves two types of proton pumps as well as channels, carriers, ABC transporters and proton antiport mechanisms (Martinoia *et al.*, 2000; Shimaoka *et al.*, 2004).

Proton pumps

The proton pumps are H^+ or V-type adenosine triphosphatase (V-ATPases) and vacuolar pyrophosphatase or V-PPases (Sze *et al.*, 1992; Maeshima, 2001; Martinoia *et al.*, 2007). These ATPases are responsible for acidifying the central vacuole. The pumps generate electrochemical gradients, which can be employed to transport cations by a proton antiport mechanism or anions (Martinoia *et al.*, 2000; Geisler and Venema, 2011). The tertiary structure of the V-ATPase is analogous to the F-type ATPases. This ATPase is the largest complex in the tonoplast and possesses a molecular weight (MW) of 400–650 kDa. The complex is composed of two functional subcomplexes: V_1, which contains the catalytic site for

ATP hydrolysis and V_o, the transmembrane sector. See Maeshima (2001); this author can be consulted for greater molecular detail.

The V-PPase is a prominent tonoplast protein (Schmidt *et al.*, 2007). It is an alternative vacuolar proton pump where PP_i is employed as the main energy source. This protein, with a subunit MW of approximately 64.5–75 kDa, is essential for maintaining the acidity of the large central vacuole (Terrier and Romieu, 2001).

Aquaporins

In addition to those discussed earlier, the major intrinsic membrane protein, α-TIP (3;1), is an aquaporin found in PSVs of *Phaseolus vulgaris* cotyledons (Daniels *et al.*, 1999; Maurel *et al.*, 2008). Maurel *et al.* (2008) reported that the water channel activity of α-TIP can be regulated by phosphorylation. Furthermore, Daniels *et al.* (1999) state that α-TIP may "function in seed desiccation, cytoplasmic osmoregulation, and/or seed rehydration."

ABC transporters

The ATP-binding cassette (ABC) transporters belong to a superfamily of active transport proteins that are energized by ATP binding and/or hydrolysis (Theodoulou, 2000). These transmembrane proteins (Martinoia *et al.*, 2007) can function to transport significant numbers of secondary metabolites, such as alkaloids and terpenoids (Yazaki, 2006). These transporters are directly energized by Mg ATP and are not dependent on the electrochemical force (Martinoia *et al.*, 2000). In 2007, Martinoia *et al.* discussed vacuolar transporters in depth.

Channels

Central lytic vacuolar channels include ones that transport Ca^{++}, K^+, and likely anions including Cl^- (Johannes *et al.*, 1992; Malmström, 2006). The Ca^{++} channel (Ward *et al.*, 2009) is a voltage-gated type (open or closed pore) in the vacuolar membrane of broad bean guard cells (Allen and Sanders, 1994). In addition, there are malate channels in tonoplast membranes (Cheffings *et al.*, 1997). Less is known regarding the molecular biology of anion channels/transporters, but it is clear the members of given families are diverse in terms of substrate specificity, localization, and regulation (Barbier-Brygoo *et al.*, 2011). Thorough discussions of ion channels in membranes occur in Hedrich and Schroeder (1989) and Malmström (2006). Malmström (2006) presented an in-depth biophysical review of membrane transport. Vacuolar ion channels as signaling mechanisms in plant nutrition have been reviewed by Isayenkov *et al.* (2010).

Carriers

Carriers are proteins that facilitate the diffusion of solutes and some are phosphorylated (Tanner and Caspari, 1996). Malmström (2006) presented evidence that carriers behave analogous to enzymes. Perhaps the best-known plant tonoplast carriers are those that transport sugars (Neuhaus, 2007). Kaiser and Heber

(1984) reported that sucrose transport is carrier-mediated in barley mesophyll protoplasts. Indeed, Endler *et al.* (2006) more recently identified a vacuolar sucrose transporter in *A. thaliana* and barley mesophyll using proteomics.

Antiporters

Plant vacuolar tonoplasts contain Ca^{++}/H^+ and $Na^+(K^+)/H^+$ antiporters (Maeshima, 2001; Rodriguez-Rosales *et al.*, 2009), among others. For antiporters, the direction of H^+ transport is opposite to the direction of the other ion. Antiporters are to be contrasted with symporters where the direction of H^+ and the other ion are the same. The $Na^+(K^+)/H^+$ antiporters are believed to help maintain cellular ion concentrations by moderating the transport of Na^+ (and K^+) from the cytosol to the vacuole (Xu *et al.*, 2010). The Na^+/H^+ and K^+/H^+ antiport activities may also contribute to pH regulation (Rodriguez-Rosales *et al.*, 2009). A summary of some of the major transport mechanisms associated with the vacuole is presented in Table 12.2. Etxeberria *et al.* (2012) discuss these mechanisms in depth.

Functions of the vacuoles

Lytic vacuole functions are diverse and appear in Table 12.3. The lytic vacuole functions are to be contrasted with those of the PSVs (Table 12.4). Sometimes, the distinction between lytic vacuoles and PSVs is not absolute. For example, Jiang *et al.* (2001) reported that a PSV can have both compartmentalized lytic and storage functions.

Table 12.2 Summary of some major transport mechanisms in through the tonoplast.

Ion pumps	Energy source	Ion channels	Energy source	Ion carriers	Energy source
H^+	ATP	Cl^- (anion)	Membrane potential	$Na^+ (K^+)$	H^+ antiport
H^+	PPi			Ca^{++}	H^+ antiport
				Mg^{++}	H^+ antiport

Table 12.3 Functions of the central lytic vacuole.

Function	Reference(s)
Accumulation of ions	Barkla and Pantoja (1996), Andreev (2001), and Surpin and Raikhel (2004)
Detoxification and homeostasis	Surpin and Raikhel (2004) and Taiz (1992)
Generation of defense responses	Andreev (2001), Surpin and Raikhel (2004), and Taiz (1992)
Maintenance of turgor pressure	Evert (2006) and Neuhaus and Martinoia (2011)
Regulation of cytoplasmic pH	Barkla and Pantoja (1996)
Sequestering of CO_2, malate and secondary metabolites	Barkla and Pantoja (1996) and Andreev (2001)
Programmed cell death	Hara-Nishimura and Hatsugai (2011)

Table 12.4 Functions of the protein storage vacuole.

Function	Reference(s)
Storage protein reserve accumulation	Ibl and Stoger (2012)
Storage of phosphorus and cations	Otegui *et al*. (2002)
Storage of protective compounds, for example, lectins, chitinases, and proteolytic enzymes	DeHoff *et al*. (2009) and Gruis *et al*. (2004)
Autophagy	Levanony *et al*. (1992) and Reyes *et al*. (2011)
Proteins with antimicrobial activity	de Souza Cândido *et al*. (2011)

Of recent interest is the role of the vacuole in programmed plant cell death (Lam *et al.*, 2000; Gray, 2004). This death has been defined as "a sequence of (potentially interruptible) events that lead to the controlled and organized destruction of the cell" (Reape *et al.*, 2008). Hara-Nishimura and Hatsugai (2011) stated that plant vacuoles participate in both destructive and nondestructive mechanisms. The former results from vacuolar membrane collapse causing release of vacuolar hydrolytic enzymes into the cytosol. The nondestructive mechanism centers about fusion of the plasmalemma and the vacuolar membrane permitting discharge of defense proteins.

PBs and other protein storage compartments

Terminology, locations, and relationships to particular storage proteins

As alluded to previously, there are many storage proteins in plant tissues that accumulate into protein storage compartments. Depending on species and tissue types, these can include PSVs and/or spherical ER-based accretions known as PBs. PBs have rather recently been defined as being distinct from the PSVs introduced earlier (Herman and Larkins, 1999; Ibl and Stoger, 2012). Storage protein compartments (including PBs and PSVs) are largely found in seeds but are also sometimes found in roots, leaves, stems, fruits, and flowers (Evert, 2006). Although they primarily contain protein, some are also known to harbor such compounds as phytic acid (Bewley and Black, 1985) and calcium oxalate crystals (Evert, 2006). Nonproteinaceous globoid regions can occur, as is the case within the large protein accumulations found in procambial and storage parenchyma cells in the cotyledons of *A. thaliana* (Figure 12.5).

Major classes of storage proteins include the albumins, globulins, glutelins, and prolamins (prolamines) as referred to previously (Table 12.1). The albumins are water-soluble proteins that are important for storage in eudicots, while globulins are soluble in certain salt solutions and are found in eudicots (Jones *et al.*, 2013) and in most, if not all, monocots (Ibl and Stoger, 2012). Glutelins are

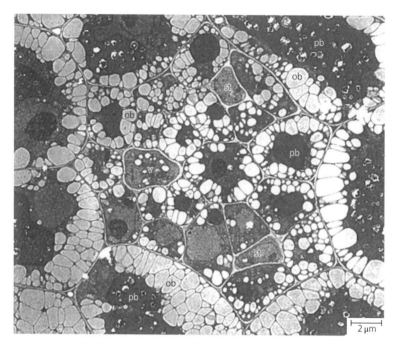

Figure 12.5 Micrograph of an immature vascular bundle in the cotyledon of an *Arabidopsis thaliana* embryo. Globoid-containing protein accumulations occur in the procambial cells as well as in the surrounding storage parenchyma. ob, oil bodies; pb, globoid-containing protein "bodies"; st, immature sieve tube; v, immature vessel. Source: Busse and Evert (1999). Reproduced with permission of University of Chicago Press.

soluble in alkali solutions and prevalent in monocot seeds, while the prolamins are unique to grass endosperm and are soluble in alcohol (Buchanan *et al.*, 2000; Jones *et al.*, 2013). Some of the most heavily investigated storage proteins are those in legumes and grains because of their nutritional and economic significance. Generally speaking, legumes accumulate their storage proteins (mainly globulins and albumins) in PSVs as do some grains, such as wheat and barley. Other grains, including maize and sorghum, store their storage proteins (primarily prolamins) in PBs within the rough ER, or RER (Buchanan *et al.*, 2000).

Storage protein "trafficking" and related controversies

Investigations of the synthesis and subcellular transport of storage proteins have helped to elucidate details about the secretory pathway, not only in plant cells but also in eukaryotes in general (Vitale and Galili, 2001). The exact steps of storage protein "trafficking" are still being elucidated, and some details remain controversial in both legumes and grains. One point of consensus is that all seed storage proteins originate at the RER into which they are co-translationally inserted (as reviewed in Ibl and Stoger, 2012). Although it is widely held that storage proteins in legumes and other eudicots are *generally* transported to the Golgi on their way to storage vacuoles (Buchanan *et al.*, 2000; Wang *et al.*, 2012; Jones *et al.*, 2013), recent work has highlighted the existence of possible variations. In the case of the clover-like *Medicago truncatula*, there are indications that the ER-to-vacuole pathway plays a role in mature protein aggregate formation as does the fusion of smaller aggregations into larger ones (Abirached-Darmency *et al.*, 2012). Evidence for this more direct form of transport (bypassing the Golgi) includes the enduring presence of ribosome-bound membranes surrounding the bodies within the vacuolar compartment. As discussed by Abirached-Darmency *et al.* (2012), this observation is consistent with similar findings in pumpkin and soybean, in which the route taken (either through the Golgi or bypassing it) depends on protein composition (Mori *et al.*, 2004).

Certain monocots (including some grains) also store proteins in storage vacuoles. In wheat, many of the prolamins form aggregates known as aleurone grains. The proteins are initially inserted into the RER during translation but are then transported via vesicles to the vacuoles by multiple mechanisms that appear to be both Golgi-dependent and -independent (Levanony *et al.*, 1992; Tosi, 2012). A Golgi-independent process also occurs in the peripherally located aleurone layer of maize endosperm, in which the alcohol-soluble zeins (the prolamin storage proteins of maize) are stored in PSVs (Reyes *et al.*, 2011).

In contrast to the latter examples, in the more internal starchy endosperm tissue of maize kernels, the zeins are co-translationally inserted into the RER where they form into PBs that remain in the ER lumen (see Figure 12.6, comparing zein distributions in the starchy endosperm versus aleurone regions). The prolamins in rice also remain in the ER (Crofts *et al.*, 2005).

It is somewhat controversial whether these prolamins are inserted at random sites on the ER and self-aggregate, or whether they are targeted to specific sites on the ER. Evidence exists for both models. Research in maize revealed fine meshworks of actin around regions of the ER (PB-ER) in which PBs form (Abe *et al.*, 1991; Clore *et al.*, 1996), with microtubules aligned adjacent to these regions (Clore *et al.*, 1996). Cytoskeleton-bound polysomes have been found in endosperm homogenates containing PBs (Davies *et al.*, 1993). Such observations led multiple groups to hypothesize that the cytoskeleton is involved in anchoring and/or transporting the zein mRNAs specifically to sites of zein PB assembly (Abe *et al.*, 1991; Clore *et al.*, 1996). On the other hand, early high resolution *in situ* hybridization results seemed to indicate that the zein mRNAs may be associated with both cisternal as well as PB-ER (Kim *et al.*, 2002). In conjunction with experiments in a heterologous expression system (yeast), in which the zeins were found to be able to self-assemble into reasonably normal PBs, this finding led to the alternative conjecture that zeins may simply be inserted into the ER at random locations coalescing into more organized bodies based on their chemical properties (Kim *et al.*, 2002).

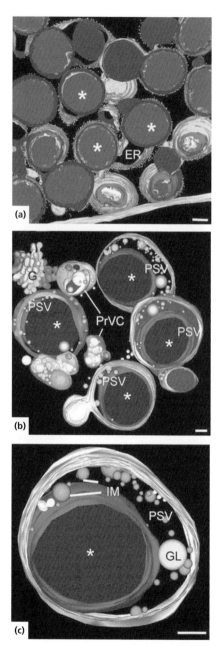

Figure 12.6 Prolamins reside in distinct subcellular compartments in the different layers of maize (*Zea mays* L.) endosperm as shown by these electron tomographic models. In the starchy endosperm (a), they form protein bodies that remain in the endoplasmic reticulum (ER), whereas in the aleurone layer (b and c), they are found in inclusions (*) in PSVs. A single PSV is depicted in c, showing also the presence of a globoid (GL) and intravacuolar membranes (IM). G, Golgi. PrVC; prevacuolar compartment. Bars – 100 nm each. Source: Reyes *et al.* (2011). Reproduced with permission of The American Society of Plant Biologists.

Researchers in the Okita laboratory, studying rice, similarly reported the presence of cytoskeletal elements around PB-ER, and subsequently published a series of papers indicating that rice prolamin RNAs contain so-called zipcodes (cis-acting RNA sequences first described in animal systems) that target them for PB-ER (as reviewed in Crofts *et al.*, 2005). One may be tempted to speculate that prolamins are targeted/organized by different mechanisms in the two grains, but the Okita group rather recently identified similar zipcodes on the mRNA encoding one of the zeins, namely the 10 kDa δ-zein (Washida *et al.*, 2009). Furthermore, a different fixation procedure revealed greater heterogeneity of zein mRNA distribution at the RER surface than seen previously (Washida *et al.*, 2004), hinting that perhaps the earlier hypothesis about prolamin mRNA targeting in maize is correct. Alternatively, the two hypotheses may not be mutually exclusive, that is, perhaps some degree of mRNA targeting and zein diffusion and interaction both play roles in PB formation.

Rice and maize accumulate non-prolamin storage proteins as well. Rice has long been known to contain a second type of "protein body" (a PSV body termed PB-II), containing glutelins. The glutelins are co-translationally inserted predominantly into the cisternal ER (and not the PB-ER) during translation (Li *et al.*, 1993). More recently, maize has also been found to accumulate additional proteins besides zeins, including uncleaved legumin-1, which may accumulate in compartments resembling storage vacuoles (Yamagata *et al.*, 2003). Nonetheless, zeins retained in the ER are the dominant form of protein storage in the maize starchy endosperm.

Prolamin protein body structure in maize

Regardless of exactly how the zein proteins "find" one another, the organization of the resulting PBs during their ontogeny has been well established. The zeins lack canonical ER retention/retrieval signals, and aggregation into PBs may largely prevent their escape from the ER. However, interactions with chaperones such as binding luminal protein (BiP) and potential membrane interactions have also been hypothesized to contribute to retention (as reviewed by Vitale and Ceriotti, 2004). The process of aggregation was elucidated by immunocytochemistry (Lending and Larkins, 1989) and begins with deposition of β- and γ-zein. γ-zein has since been shown to sequester α-zein in the endosperm of transgenic tobacco and is hypothesized to be important for PB nucleation (Coleman *et al.*, 1996). In the course of normal PB formation in maize, small interior compartments of α- and δ-zein form after these proteins somehow penetrate the initial aggregates of β- and γ-zein. These compartments ultimately coalesce and expand to fill the interior of the bodies that are ultimately surrounded by a "shell" of γ- and β-zein (Figure 12.7; see also Holding and Larkins, 2006).

Recombinant protein accumulation and practical applications

Some new and exciting work is being conducted to capitalize on the PB system in order to produce and stabilize recombinant proteins netting high yields. A new technology, named Zera® (ERA Biotech), has been pioneered to make use of a domain of γ-zein for PB induction in a wide range of hosts (including tobacco, *Medicago*, insect, fungal, and even primate cells). Such medically relevant proteins as human growth hormone and epidermal growth factor have

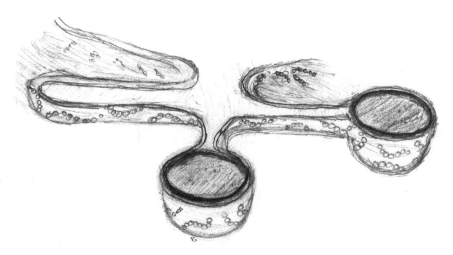

Figure 12.7 Depiction of the structure of mature protein bodies (PBs) residing within the rough endoplasmic reticulum of maize (*Zea mays* L.) starchy endosperm cells. Two protein bodies are depicted in cross section. Just inside of the ribosome-studded ER membrane exists the thin outer (darker) layer of each protein body, consisting of β- and γ-zeins, while the interior is filled with α- and δ-zeins. The relative levels of shading coincide with differences in electron density in these two PB regions, with the outer layer being more electron dense than the inner core. Source: Drawn by Ms. Laney Clore under the supervision of A. Clore.

been produced using Zera® in these hosts (see Schmidt, 2013 for a review). Other PB-inducing technologies have been devised as well, such as the use of zeolin, a chimeric protein consisting of phaseolin, and 89 amino acids of γ-zein. Zeolin has the benefit of being more nutritionally balanced (having both sulfur-rich amino acids characteristic of zeins and lysines characteristic of phaseolin), making it potentially attractive for nutritional quality improvement efforts (Mainieri *et al.*, 2004). Another example employs synthetic elastin-like polypeptides (ELPs). ELPs have been used to make PB-inducing fusions for the accumulation of a variety of proteins, ranging from those in spider silk to types of anti-HIV antibodies; however, accumulation efficacy may vary with protein types (Conley *et al.*, 2011).

Conclusions and perspectives for protein body/protein storage compartment research

On the surface, storage protein synthesis and deposition may seem simplistic, but these are actually quite complex processes. Recent research has both increased our understanding and also brought to light new complexities. We have seen that within a particular tissue, similar storage proteins may be stored in different organelles in adjacent cell layers (as is the case for the prolamins within the starchy versus aleurone layers of maize endosperm). In other cases, even within the same individual cells, different organelles may contain mature storage protein aggregations. Added to this complexity are the various pathways storage proteins take through the endomembrane system, sometimes involving the Golgi and other times not. Despite a steady stream of new research studies, controversies still exist, for example, with respect to mRNA targeting to particular regions of the ER. Nonetheless, scientists are now able to manipulate nature's system of storing these proteins in fascinating ways to produce a variety of recombinant proteins for potential medical and nutritional uses. Continued basic research will be needed to further our understanding of these vital pathways.

References

Abe, S., You, W. and Davies, E. (1991) Protein bodies in corn endosperm are enclosed by and enmeshed in F-actin. *Protoplasma*, **165**, 139–149.

Abirached-Darmency, M., Dessaint, F., Benlicha, E. and Schneider, C. (2012) Biogenesis of protein bodies during vicilin accumulation in *Medicago truncatula* immature seeds. *BMC Research Notes*, **5**, 409.

Allen, G.J. and Sanders, D. (1994) Two voltage-gated, calcium release channels coreside in the vacuolar membrane of broad bean guard cells. *The Plant Cell*, **6**, 685–694.

Andreev, I.M. (2001) Functions of the vacuole in higher plant cells. *Russian Journal of Plant Physiology*, **48**, 672–680.

Barbier-Brygoo, H., De Angeli, A., Filleur, S. *et al.* (2011) Anion channels/transporters in plants: from molecular bases to regulatory networks. *Annual Review of Plant Biology*, **62**, 25–51.

Barkla, B.J. and Pantoja, O. (1996) Physiology of ion transport across the tonoplast of higher plants. *Annual Review of Plant Physiology and Plant Molecular Biology*, **47**, 159–184.

Bassham, D.C. and Blatt, M.R. (2008) SNAREs: cogs and coordinators in signaling and development. *Plant Physiology*, **147**, 1504–1515.

Bassham, D.C. and Raikhel, N.V. (1997) Molecular aspects of vacuole biogenesis. *Advances in Botanical Research*, **25**, 43–58.

Bewley J.D. and Black, M. (1985) Seeds: germination, structure and composition. In: *Seeds: Physiology of Development and Germination*, (Bewley, J.D. and Black, M., eds), Plenum Press, New York, NY, pp. 1–25.

Boller, T. and Wiemken, A. (1986) Dynamics of vacuolar compartmentation. *Annual Review of Plant Physiology*, **37**, 137–164.

Buchanan, B.B., Gruissem, W. and Jones, R.L. (2000) *Biochemistry and Molecular Biology of Plants*, American Society of Plant Physiologists, Rockville, MD.

Busse, J.S. and Evert, R.F. (1999) Pattern of differentiation of the first vascular elements in the embryo and seedling of *Arabidopsis thaliana*. *International Journal of Plant Sciences*, **160**, 1–13.

Carter, C.J., Bednarek, S.Y. and Raikhel, N.V. (2004) Membrane trafficking in plants: new discoveries and approaches. *Current Opinion in Plant Biology*, **7**, 701–707.

Cheffings, C.M., Pantoja, O., Ashcroft, F.M. and Smith, J.A.C. (1997) Malate transport and vacuolar ion channels in CAM plants. *Journal of Experimental Botany*, **48**, 623–631.

Clore, A.M., Dannenhoffer, J.M. and Larkins, B.A. (1996) EF-1α is associated with a cytoskeletal network surrounding protein bodies in maize endosperm cells. *The Plant Cell*, **8** (11), 2003–2014.

Coleman, C.E., Herman, E.M., Takasaki, K. and Larkins B.A. (1996) The maize gamma-zein sequesters alpha-zein and stabilizes its accumulation in protein bodies of transgenic tobacco endosperm. *The Plant Cell*, **8** (12), 2335–2345.

Conley, A.J., Joensuu, J.J., Richman, A. and Menassa, R. (2011) Protein body-inducing fusions for high-level production and purification of recombinant proteins in plants. *Plant Biotechnology Journal*, **9**, 419–433.

Crofts, A.J., Washida, H., Okita, T.W. *et al.* (2005) The role of mRNA and protein sorting in seed storage protein synthesis, transport, and deposition. *Biochemistry and Cell Biology*, **83**, 728–737.

Daniels, M.J., Chrispeels, M.J. and Yeager, M. (1999) Projection structure of a plant vacuole membrane aquaporin by electron cryo-crystallography. *Journal of Molecular Biology*, **294**, 1337–1349.

Dashek, W.V. and Harrison, M. (eds.) (2006) *Plant Cell Biology*, Science Publishers, Enfield, NH.

Davies, E., Comer, E.C., Lionberger, J.M., Stankovic, B. and Abe, S. (1993) Cytoskeleton-bound polysomes in plants. III. Polysome–cytoskeleton–membrane interactions in corn endosperm. *Cell Biology International*, **17**, 331–340.

De, D.N. (2000) *Plant Cell Vacuoles: An Introduction*, CSIRO Publishing, Collingwood, VIC.

De Hoff, P.L., Brill, L.M. and Hirsch, A.M. (2009) Plant lectins: the ties that bind in root symbiosis and plant defense. *Molecular Genetics and Genomics*, **282**, 1–15.

de Souza Cândido, E., Pinto, M.F.S., Pelegrini, P.B. *et al.* (2011) Plant storage proteins with antimicrobial activity: novel insights into plant defense mechanisms. *FASEB Journal*, **25**, 3290–3305.

Echeverria, E. (2000) Vesicle-mediated solute transport between the vacuole and the plasma membrane. *Plant Physiology*, **123**, 1217–1226.

Endler, A., Meyer, S., Schelbert, S. *et al.* (2006) Identification of a vacuolar sucrose transporter in barley and *Arabidopsis* mesophyll cells by a tonoplast proteomic approach. *Plant Physiology*, **141**, 196–207.

Etxeberria, E., Pozueta-Romero, J. and Gonzalez, P. (2012) In and out of the plant storage vacuole. *Plant Science*, **190**, 52–61.

Evert, R.F. (2006) *Esau's Plant Anatomy*, John Wiley & Sons, Inc., New York, NY.

Facchini, P.J. (2001) Alkaloid biosynthesis in plants: biochemistry, cell biology, molecular regulation, and metabolic engineering applications. *Annual Review of Plant Physiology and Plant Molecular Biology*, **52**, 29–66.

Frigerio, L., Hinz, G. and Robinson, D.G. (2008) Multiple vacuoles in plant cells: rule or exception? *Traffic*, **9**, 1564–1570.

Gattolin, S., Sorieul, M. and Frigerio, L. (2011) Mapping of tonoplast intrinsic proteins in maturing and germinating *Arabidopsis* seeds reveals dual localization of embryonic TIPs to the tonoplast and plasma membrane. *Molecular Plant*, **4**, 180–189.

Geisler, M. and Venema, K. (eds.) (2011) *Transporters and Pumps in Plant Signaling*. Signaling and Communication in Plants, vol. **7**, Springer, New York, NY.

Gray, J. (ed.) (2004) *Programmed Cell Death in Plants*, Blackwell Publishing, Oxford.

Gruis, T., Schulze, J. and Jung, R. (2004) Storage protein in the absence of vacuolar processing enzyme family of cysteine proteases. *The Plant Cell*, **16**, 270–290.

Hara-Nishimura, I. and Hatsugai, N. (2011) The role of vacuole in plant cell death. *Cell Death and Differentiation*, **18**, 1298–1304.

Hara-Nishimura, I., Shimada, T., Hatano, K., Takeuchi, Y. and Nishimura, M. (1998) Transport of storage proteins to protein storage vacuoles is mediated by large precursor-accumulating vesicles. *The Plant Cell*, **10**, 825–836.

Hara-Nishimura, I., Matsushima, R., Shimada, T. and Nishimura, M. (2004) Diversity and formation of endoplasmic reticulum-derived compartments in plants. Are these compartments specific to plant cells? *Plant Physiology*, **136**, 3435–3439.

Hedrich, R. and Schroeder, J.L. (1989) The physiology of ion channels and electrogenic pumps in higher plants. *Annual Review of Plant Physiology*, **40**, 539–569.

Herman, E.M. and Larkins, B.A. (1999) Protein storage bodies and vacuoles. *The Plant Cell*, **11**, 601–613.

Hicks, G.R., Rojo, E., Hong, S. *et al.* (2004) Germinating pollen has tubular vacuoles, displays highly dynamic vacuole biogenesis, and requires vacuoles for proper function. *Plant Physiology,* **134**, 1227–1239.

Hilling, B. and Amelunxen, F. (1985) On the development of the vacuole. II. Further evidence for endoplasmic reticulum origin. *European Journal of Cell Biology,* **38**, 195–200.

Hillmer, S., Movafeghi, A., Robinson, D.G. and Hinza, G. (2001) Vacuolar storage proteins are sorted in the cis-cisternae of the pea cotyledon Golgi apparatus. *Journal of Cell Biology,* **152**, 41–50.

Hoh, B., Hinz, G., Jeong, B.-K. and Robinson, D.G. (1995) Protein storage vacuoles form *de novo* during pea cotyledon development. *Journal of Cell Science,* **108**, 299–310.

Holding, D.R. and Larkins, B.A. (2006) The development and importance of zein protein bodies in maize endosperm. *Maydica,* **51**, 243–254.

Hunter, P.R., Craddock, C.P., Di Benedetto, S. *et al.* (2007) Fluorescent reporter proteins for the tonoplast and the vacuolar lumen identify a single vacuolar compartment in *Arabidopsis* cells. *Plant Physiology,* **145**, 1371–1382.

Hurkman, W.J., Tanaka, C.K., Vensel, W.H. *et al.* (2013) Comparative proteomic analysis of the effect of temperature and fertilizer on gliadin and glutenin accumulation in the developing endosperm and flour from *Triticum aestivum* L. cv. Butte 86. *Proteome Science,* **11**, 8.

Ibl, V. and Stoger, E. (2012) The formation, function and fate of protein storage compartments in seeds. *Protoplasma,* **249**, 379–392.

Isayenkov, S., Isner, J.-C. and Maathuis, F.J.M. (2010) Vacuolar ion channels: roles in plant nutrition and signaling. *FEBS Letters,* **584**, 1982–1988.

Jauh, G.Y., Fischer, A.M., Grimes, H.D. *et al.* (1998) δ-Tonoplast intrinsic protein defines unique plant vacuole functions. *Proceedings of the National Academy of Sciences of the United States of America,* **95**, 12995–12999.

Jauh, G.Y., Phillips, T.E. and Rogers, J.C. (1999) Tonoplast intrinsic isoforms as markers for vacuolar functions. *The Plant Cell,* **11**, 1867–1882.

Jiang, L.W. and Rogers, J.C. (2001) Compartmentation of proteins in the protein storage vacuole: a compound organelle in plant cells. *Advances in Botanical Research,* **35**, 139–170.

Jiang, L., Phillips, T.E., Hamm, C.A. *et al.* (2001) The protein storage vacuole. *Journal of Cell Biology,* **155**, 991–1002.

Jiang, L., Erickson, A. and Rogers, J. (2002) Multivesicular bodies: a mechanism to package lytic and storage functions in one organelle? *Trends in Cell Biology,* **12**, 362–367.

Johannes, E., Brosnan, J.M. and Sanders, D. (1992) Parallel pathways for intracellular Ca²⁺ release from the vacuole of higher plants. *The Plant Journal,* **2**, 97–102.

Jones, R., Ougham, H., Thomas, H. and Waaland, S. (2013) *The Molecular Life of Plants,* Wiley-Blackwell/American Society of Plant Physiologists, New York, NY/Rockville, MD.

Kaiser, G. and Heber, U. (1984) Sucrose transport into vacuoles isolated from barley mesophyll protoplasts. *Planta,* **161**, 562–568.

Khanbabaee, K. and van Ree, T. (2001) Tannins: classification and definition. *Natural Product Reports,* **18**, 641–649.

Kim, S.J. and Brandizzi, F. (2012) News and views into the SNARE complexity in *Arabidopsis. Frontiers in Plant Sciences,* **3**, 28.

Kim, C.S., Woo, Y.-M., Clore, A.M. *et al.* (2002) Zein protein interactions, rather than the asymmetric distribution of zein mRNAs on endoplasmic reticulum membranes, influence protein body formation in maize endosperm. *The Plant Cell,* **14**, 655–672.

Kutchan, T.M., Rush, M. and Coscia, C.J. (1986) Subcellular localization of alkaloids and dopamine in different vacuolar compartments of *Papaver bracteatum. Plant Physiology,* **81**, 161–166.

Lam, E., Fukuda, H. and Greenberg, J. (eds.) (2000) *Programmed Cell Death in Higher Plants,* Springer, New York, NY.

Leigh, R.A. (1997) Solute composition of vacuoles. *Advances in Botanical Research*, **25**, 171–194.

Lending, C.R. and Larkins, B.A. (1989) Changes in the zein composition of protein bodies during maize endosperm development. *The Plant Cell*, **1**, 1011–1023.

Levanony, H., Rubin R., Altschuler Y., Galili G. (1992) Evidence for a novel route of wheat storage proteins to vacuoles. *Journal of Cell Biology*, **119**, 1117–1128.

Li, X., Franceschi, V.R. and Okita, T.W. (1993) Segregation of storage protein mRNAs on the rough endoplasmic reticulum membranes of rice endosperm cells. *Cell*, **72**, 869–879.

Lipka, V., Kwon, C. and Panstruga, R. (2007) SNARE-ware: the role of SNARE-domain proteins in plant biology. *Annual Review of Cell and Developmental Biology*, **23**, 147–174.

Maeshima, M. (2001) Chloroplast transporters: organization and function. *Annual Review of Plant Physiology and Plant Molecular Biology*, **52**, 469–497.

Mainieri, D., Rossi, M. and Archinti, M. *et al.* (2004) Zeolin. A new recombinant storage protein constructed using maize gamma-zein and bean phaseolin. *Plant Physiology*, **136**, 3447–3456.

Malmström, S. (2006) Movement of molecules across membranes. In: *Plant Cell Biology*, (Dashek, W.V. and Harrison, M., eds), Science Publishers, Enfield, NH, pp. 131–196.

Martinoia, E., Massonneau, A. and Frangne, N. (2000) Transport processes of solutes across the vacuolar membrane of higher plants. *Plant and Cell Physiology*, **41**, 1175–1186.

Martinoia, E., Maeshima, M. and Neuhaus, H.E. (2007) Vacuolar transporters and their essential role in plant metabolism. *Journal of Experimental Botany*, **58**, 83–102.

Marty, F. (1999) Plant vacuoles. *The Plant Cell*, **11**, 587–599.

Maurel, C., Verdoucq, L. Luu, D.-T. and Santoni, V. (2008) Plant aquaporins: membrane channels with multiple integrated functions. *Annual Review of Plant Biology*, **59**, 595–624.

Mori, T., Maruyama, N., Nishizawa, K. *et al.* (2004) The composition of newly synthesized proteins in the endoplasmic reticulum determines the transport pathways of soybean seed storage proteins. *The Plant Journal*, **40**, 238–249.

Neuhaus, H.E. (2007) Transport of primary metabolites across the plant vacuolar membrane. *FEBS Letters*, **581**, 2223–2226.

Neuhaus, J.-M. and Martinoia, E. (2011) Plant vacuoles. *eLS*, 10.1002/9780470015902. a0001675.pub2

Oda, Y., Higaki, T., Hasezawa, S. and Kutsuna, N. (2009) New insights into plant vacuolar structure and dynamics. *International Review of Cell and Molecular Biology*, **277**, 103–135.

Okita, T.W. and Rogers, J.C. (1996) Compartmentation of proteins in the endomembrane system of plant cells. *Annual Review of Plant Physiology and Plant Molecular Biology*, **47**, 327–350.

Otegui, M.S., Capp, R. and Staehelin, L.A. (2002) Developing seeds of *Arabidopsis* store different minerals in two types of vacuoles and in the endoplasmic reticulum. *The Plant Cell*, **14**, 1311–1327.

Otegui, M.S., Herder, R., Schulze, J. *et al.* (2006) The proteolytic processing of seed storage proteins in *Arabidopsis* embryo cells starts in the multivesicular bodies. *The Plant Cell*, **18**, 2567–2581.

Paris, N., Stanley, C.M., Jones, R.L. and Rogers, J.C. (1996) Plant cells contain two functionally distinct vacuolar compartments. *Cell*, **85**, 563–572.

Park, M., Kim, S.J., Vitale, A. and Hwang, I. (2004) Identification of the protein storage vacuole and protein targeting to the vacuole in leaf cells of three plant species. *Plant Physiology*, **134**, 625–639.

Reape, T.J., Molony, E.M. and McCabe, P.F. (2008) Programmed cell death in plants: distinguishing between different modes. *Journal of Experimental Botany*, **59**, 435–444.

Reyes, F.C., Chung, T., Holding, D., *et al.* (2011) Delivery of prolamins to the protein storage vacuole in maize aleurone cells. *The Plant Cell*, **23**, 769–784.

Robinson, D.G. and Rogers, J.C. (eds.) (2000) *Vacuolar Compartments*, Annual Plant Reviews, Blackwell Publishing, Ames, IA.

Robinson, D.G., Bäumer, M., Hinz, G. and Hohl, I. (1997) Ultrastructure of the pea cotyledon Golgi apparatus: origin of dense vesicles and the action of brefeldin A. *Protoplasma*, **200**, 198–209.

Robinson, D.G., Bäumer, M., Hinz, G. and Hohl, I. (1998) Vesicle transfer of storage proteins to the vacuole: the role of the Golgi apparatus and multivesicular bodies. *Journal of Plant Physiology*, **152**, 659–667.

Rodriguez-Rosales, M.P., Gálvez, F.J., Huertas, R. *et al.* (2009) Plant NHX cation/proton antiporters. *Plant Signaling and Behavior*, **4**, 265–276.

Schmidt, S.R. (2013) Protein bodies in nature and biotechnology. *Molecular Biotechnology*, **54**, 257–268.

Schmidt, U.G., Endler, A., Schelbert, S. *et al.* (2007) Novel tonoplast transporters identified using a proteomic approach with vacuoles isolated from cauliflower buds. *Plant Physiology*, **145**, 216–229.

Shimaoka, T., Ohnishi, M., Sazuka, T. *et al.* (2004) Isolation of intact vacuoles and proteomic analysis of tonoplast from suspension-cultured cells of *Arabidopsis thaliana*. *Plant and Cell Physiology*, **45**, 672–683.

Surpin, M. and Raikhel, N. (2004). Traffic jams affect plant development and signal transduction. *Nature Reviews Molecular Cell Biology*, **5**, 100–109.

Sze, H., Ward, J.M. and Lai, S. (1992) Vacuolar-H^+-ATPases from plants. *Journal of Bioenergetics and Biomembranes*, **24**, 371–381.

Taiz, L. (1992) The plant vacuole. *The Journal of Experimental Biology*, **172**, 113–122.

Tanner, W. and Caspari, T. (1996) Membrane transport carriers. *Annual Review of Plant Physiology and Plant Molecular Biology*, **47**, 595–626.

Terrier, N. and Romieu, C. (2001) Grape berry acidity. In *Molecular Biology and Biotechnology of the Grapevine*, Roubelakis-Angelakis, K.A. (ed.), Kluwer, Dordrecht, pp. 35–58.

Theodoulou, F.L. (2000) Plant ABC transporters. *Biochimica et Biophysica Acta*, **1465**, 79–103.

Tohge, T., Ramos, M.S., Nunes-Nesi, A. *et al.* (2011) Toward the storage metabolome: profiling the barley vacuole. *Plant Physiology*, **157**, 1469–1482.

Tosi, P. (2012) Trafficking and deposition of prolamins in wheat. *Journal of Cereal Science*, **56**, 81–90.

Viotti, C. (2014) ER and vacuoles: never been closer. *Frontiers in Plant Science*, **5**, 20.

Viotti, C., Krüger, F., Krebs, M. *et al.* (2013) The endoplasmic reticulum is the main membrane source for biogenesis of the lytic vacuole in *Arabidopsis*. *The Plant Cell*, **25**, 3434–3449.

Vitale, A. and Ceriotti, A. (2004) Protein quality control mechanisms and protein storage in the endoplasmic reticulum. A conflict of interests? *Plant Physiology*, **136**, 3420–3426.

Vitale, A. and Galili, G. (2001) The endomembrane system and the problem of protein sorting. *Plant Physiology*, **125**, 115–118.

Wang, J., Tse, Y.C., Hinz, G. *et al.* (2012) Storage globulins pass through the Golgi apparatus and multivesicular bodies in the absence of dense vesicle formation during early stages of cotyledon development in mung bean. *Journal of Experimental Botany*, **63**, 1367–1380.

Ward, J.M., Maser, P. and Schroeder, J.I. (2009) Plant ion channels: gene families, physiology and functional genomics analyses. *Annual Review of Physiology*, **71**, 59–82.

Washida, H., Sugino, A., Messing, J. *et al.* (2004) Localization of seed storage protein RNAs to distinct subdomains of the endoplasmic reticulum in maize endosperm cells. *Plant Cell Physiology*, **45** (12), 1830–1837.

Washida, H., Sugino, A., Kaneko, S. *et al.* (2009) Identification of cis-localization elements of the maize10-kDa δ-zein and their use in targeting RNAs to specific cortical endoplasmic reticulum subdomains. *The Plant Journal*, **60**, 146–155.

Wink, M. (1993) The plant vacuole: a multifunctional compartment. *Journal of Experimental Botany*, **44** (Suppl), 231–246.

Xu, K., Zhang, H., Blumwald, E. and Xia, T. (2010) A novel plant vacuolar Na⁺/H⁺ antiporter gene evolved by DNA shuffling confers improved salt tolerance in yeast. *The Journal of Biological Chemistry* **285**, 22999–23006.

Yamagata, T., Kato, H., Kuroda, S. *et al.* (2003) Uncleaved legumin in developing maize endosperm: identification, accumulation and putative subcellular localization. *Journal of Experimental Botany*, **54**, 913–922.

Yazaki, K. (2006) ABC transporters involved in the transport of plant secondary metabolites. *FEBS Letters*, **580**, 1183–1191.

Yoshida, S. and Uemura, M. (1986) Lipid composition of plasma membranes and tonoplasts isolated from etiolated seedlings of mung bean (*Vigna radiata* L.). *Plant Physiology*, **82**, 807–812.

Zheng, K. and Staehelin, L.A. (2011) Protein storage vacuoles are transformed into lytic vacuoles in root meristematic cells of germinating seedlings by multiple, cell type-specific membranes. *Plant Physiology*, **155**, 2023–2035.

Further reading

Abirached-Darmency, M., Dessaint, F., Benlicha, E. and Schneider, C. (2012) Biogenesis of protein bodies during vicilin accumulation in *Medicago truncatula* immature seeds. *BMC Research Notes*, **2012** (5), 409, DOI:10.1186/1756-0500-5-409.

Elbaz-Alon, Y., Rosenfeld-Gur, E., Shinder, V., Futerman, A.H., Geiger, T. and Schuldiner, M. (2014) A dynamic interface between vacuoles and mitochondria in yeast. *Developmental Cell*, **30** (1), 95–102.

Gujratia, V., Leeb, M., Kob, Y., Leeb, S. and Kima, D. (2016) Bioengineered yeast-derived vacuoles with enhanced tissue-penetrating ability for targeted cancer therapy. *Proceedings of the National Academy of Sciences of the United States of America*, **113** (3), 710–715.

Mainieri, D., Morandini, F., Maîtrejean, M., Saccani, A., Pedrazzini, E. and Vitale, A. (2014) Protein body formation in the endoplasmic reticulum as an evolution of storage protein sorting to vacuoles: insights from maize γ-zein. *Frontiers in Plant Science*, **5**, 331, http://dx.doi.org/ 10.3389/fpls.2014.00331.

Saberianfar, R., Sattarzadeh, A., Joensuu, J.J., Kohalmi, S.E. and Menassa, R. (2016) Protein bodies in leaves exchange contents through the endoplasmic reticulum. *Frontiers in Plant Science*, http://dx.doi.org/10.3389/fpls.2016.00693.

Zhang, C., Hicks, G.R. and Raikhel, N.V. (2015) Molecular composition of plant vacuoles: important but less understood regulations and roles of tonoplast lipids. *Plants*, **4**, 320–333.

CHAPTER 13

Systems biology in plant cells and their organelles

Rajdeep Kaur Grewal, Saptarshi Sinha, and Soumen Roy
Department of Physics, Bose Institute, Kolkata, India

Systems biology ushered as an interdisciplinary field of sciences with its aim to gain insights into biology. The name "systems biology" itself describes its role as understanding biology from "systems" level unlike the most prevalent reductionist approach undertaken by biologists so far. Systems biology emphasizes onto the so-called emergent properties of a system constituting genes, metabolites, proteins, or any other molecules. The reductionist approach is based upon investigating a single molecule or component and its mechanism and functions. On the other hand, systems biology delves into the various interactions present among molecules of the system to address the underlying mechanism or functioning of a subcellular or cellular component as a "whole." Many of the cellular processes, rather almost all of them, are results of these interactions. For example, signal transduction results from interactions among various proteins; metabolism—interactions among metabolites and enzymes; and transcription—interactions between genes, RNA, and proteins, etc. Systems biology approach employs new high-throughput experiments and technologies in order to analyze large number of molecules or their interactions. Modeling a system to gain insights and to draw biological relevance is another prime input of this field. Thus, to put forward newer predictions based on experiments as well as theoretical knowledge, systems biology calls for contributions from diverse fields of sciences to understand the complex organismal behavior.

Systems biology—"omics"

The future of cell biology of organelles will definitely benefit with systems biology approaches. The omics of systems biology are genomics, lipidomics, proteomics, metabolomics, transcriptomics, etc. (Kitano, 2002); these omics are

Rajdeep Kaur Grewal and Saptarshi Sinha have contributed equally to this work.

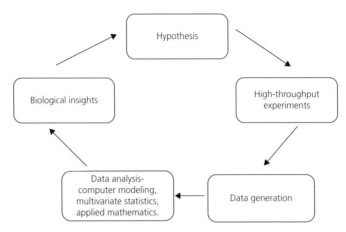

Figure 13.1 Systems biology approach.

Table 13.1 Details of techniques used in various "omics."

Omics	Technique	Importance
Genomics	1. DNA sequencing	1. Determines the sequence of a DNA molecule
	2. Microarray technology	2. Measures the expression of large number of genes
Lipidomics	1. Gas chromatography	1. Separates lipid molecules of different length
	2. NMR spectroscopy	2. Predicts the structure of different fatty acid chains
Metabolomics	1. Mass spectroscopy	1. Predicts the structure of differ metabolites
	2. Proton NMR	2. Protein profiling
Proteomics	1. Gel electrophoresis	1. Separates different polypeptides in a solution depending on the molecular weight
	2. Protein blotting	2. Identifies specific polypeptides
Transcriptomics	1. RNA sequencing	1. Determines sequence of RNA molecules

concerned with the global views of the genome, lipidome, proteome, metabolome and transcriptome, respectively.

Systems biology is different from the reductionist view of biology, for example, isolation and characterization only. As depicted in Figure 13.1, systems biology is a seamless and logical integration of "wet experimentation," computational modeling, technological advances, ideas, and theory (Moodie *et al.*, 2010). The computationally organizing, analyzing, and predicting complex data from contemporary molecular and biochemical techniques is referred to as "bioinformatics" (Rashidi and Buehler, 2005). Key elements of bioinformatics are knowledge management and expansion. Considerable attention is paid to technologies (Table 13.1) and computational techniques for analyzing the aforesaid "omics."

Genomics

Genomics is an integrated study of functional and structural property of the heredity unit of every living cell. It uses modern techniques such as recombinant DNA technology, DNA sequencing, and bioinformatics, and it helps in uncovering the fine-scale structure and combined action of different parts of the genome inside a cell. Genomics also includes complex allelic interactions inside a genome and interactions between different loci of it. Instead of focusing on a single gene, genomics looks into the combined effect of every gene and every part of the genome as a genetic network.

Networks and robustness

Network theory has been applied successfully to a varied range of real-world complex networks. The idea of networks in biology is to model the intricate design of living systems among different entities such as genes, proteins, or metabolites. The application of network approaches needs proper identification of components of interest from the concerned networks, termed "nodes," and also the various interactions among these components that form the edges (or arcs) in the network. To cite a couple of examples, (i) in protein–protein interaction networks the nodes are proteins, and interaction among two proteins represent an edge between them; (ii) networks of brain neurons represent each neuron as node, and an edge is conceived between two neurons if they are connected by a synapse. Networks serve as an important tool in the search for "emergent properties" in complex systems. They offer a marked and fresh departure from the old reductionist approach by means of visualizing and analyzing a system as one aggregated functional unit instead of merely looking into the individual components of the system. In the process, network approaches might lend topological insights into an understanding of the system. The comprehension of the mechanism of underlying system topology is an important consideration not only in transport or technical systems but also in biological networks. The small-world property seen in metabolic networks and neuronal network of *Caenorhabditis elegans* highlights the efficiency of signal transmission in these networks. A gene regulatory network consists of interactions among various genes via RNAs or proteins that regulate them. Some proteins, called transcription factors, bind to the promoter region of other genes and regulate it, thus acting as a principal component of gene regulatory network. With respect to the gene regulatory network, the regulation architecture will be different for every part of the plant body. The architecture might also change with different phases of development. These networks are highly robust, that is, they are resistant to external perturbations. Signaling network of effector triggered immunity (ETI) in plants is an example of such robust network (Tsuda *et al.*, 2009). For the functional adaptivity and evolutionary change, network robustness is highly important (Aldana *et al.*, 2007).

In *Arabidopsis thaliana*, the size of stem cell in plant meristem is positively controlled by stem cell-promoting signaling pathway and is itself being negatively regulated by stem cell-restricting pathway.

Techniques of DNA and genome analysis

There are various techniques used for genomic analysis: agarose gel electrophoresis, $C_o t$ analysis, DNA fingerprinting, polymerase chain reaction (PCR), DNA microarrays, etc. Using these techniques, various types of data sets are prepared, which helps in theoretical analysis.

Agarose gel electrophoresis

This electrophoresis technique separates DNA fragments by size (Figure 13.2). These DNA fragments of specific molecular weights are obtained from restriction digestion of genomic DNA or by PCR with specific primer. An electric field is used to separate negatively charged DNA fragments. The DNA will migrate via size at a rate proportional to its charge to mass ration. DNA brands are visualized with ethidium bromide and fragment sizes are ascertained by comparison to a set of known ladders (Figure 13.2). Those isolated fragments are used for PCR, sequencing, etc.

$C_o t$ analysis

This procedure involves DNA renaturation kinetics in 0.18 M sodium phosphate. DNA molecules are first heated to 100°C to separate its two strands. They can be renatured following slow cooling at 60°C. The 260 nm absorbance peak is altered during renaturation, which allows measurement of DNA base pairing by denaturation/renaturation kinetics as described by M. Waring and R.J. Britten in 1966. This procedure permits determination of genome size and the fraction of single-copy DNA in a genome. Low copy and repetitive sequence quantification

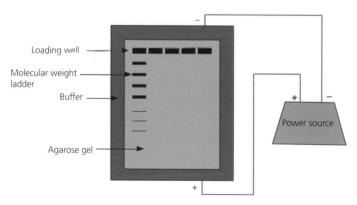

Figure 13.2 Agarose gel electrophoresis.

by C_0t-based sequence analysis can be efficiently used for clone generation. C_0t analysis is also used in hydroxyapatite chromatography for preparing genomic libraries (Peterson *et al.*, 2002).

DNA fingerprinting and PCR

Chemical hybridization-based fingerprinting involves fragmenting genomic DNA with a restriction enzyme and electrophoretic separation of resulting DNA fragments according to size. This is followed by the detection of polymorphic multilocus banding pattern by gel dying hybridization with a probe, a labeled complementary DNA sequence given by K. Weising and his coworkers in 1995. This technique is used in various cases in plant biology to check genome similarity between different plants. In addition to this, PCR is also used. PCR is mainly used to amplify DNA *in vitro*. It is also used as part of DNA fingerprinting. A PCR protocol is presented in Figure 13.3.

PCR reaction mixture consists of primers, thermostable DNA polymerase, target DNA, buffer, and dNTPs. PCR is performed in a thermocycler. Each reaction cycle includes steps such as denaturation, annealing, and extension. First, the mixture is heated to separate the target DNA into single strands and is subsequently cooled to allow oligonucleotide pieces (primers) to get attached to their complementary DNA sequences. DNA polymerase adds dNTPs to the growing strand and DNA is thus extended. Single primer amplification reactions together with minisatellite and arbitrarily sequence primer are very useful tools for determining genetic diversity by genome comparison in plants (Ranade and Farooqui, 2002).

DNA mapping

There are three types of DNA maps: cytological, genetic, and physical. The cytological map originates from organisms for which genetic chromosome maps exist. Every gene focused on a genetic map is correlated with a cytological map

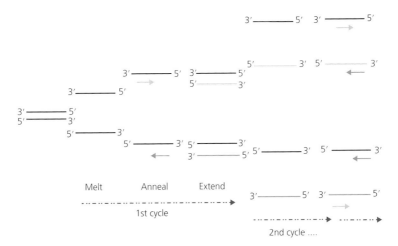

Figure 13.3 PCR protocol.

to an existing chromosome. Cytological maps are generated by analyzing chromosomal aberrations. Maize and tomato are two of the most studied plants whose maps were clearly understood using several hundred markers. Genetic mapping uses genetic techniques to yield maps depicting the position of genes and other features of a genome. Various chromosomal mapping such as tDNA insert mapping in transgenic *Petunia* sp. following *in situ* hybridization is well demonstrated. Physical mapping uses molecular biological methods to examine DNA molecules directly to produce maps, which illustrates the position of a sequence. Some major rDNA loci have also been physically mapped by *in situ* hybridization in *Brassica* species.

DNA microarrays

DNA microarrays enable studying genes that are active and inactive in different cell types and in different states. These microarrays are generated by machines that arrange miniscule quantities of hundreds of thousands of gene sequences on a single microscope slide. The steps for producing a DNA microarray are presented in Figure 13.4.

In DNA microarray, probes of the desired gene sequences are synthesized and attached to different types of solid platforms. These platforms are flooded by labeled test samples. Desired test samples when attached with the probes can be monitored after washing. Response to various stress conditions and different developmental stages are regulated by various intercellular networks that consist of DNA, RNA, and protein. These regulatory networks can be extensively studied by DNA microarray techniques that tell us about different gene expression levels at different stages (Pu and Brady, 2010). In medicinal plant study, the host gene expression profile upon treatment of a particular substrate helps to better understand the mechanism of action of that substrate (Kumar *et al.*, 2012).

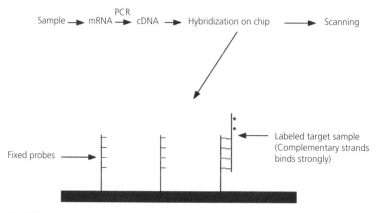

Figure 13.4 DNA microarrays.

DNA sequencing

DNA sequencing is the basis of various techniques used in plant systems biology, DNA microarray, mapping, and C_0t analysis. The two classical DNA sequencing techniques—the chain termination and the chemical degradation—are briefly discussed here. The former has been more widely employed.

Chain termination sequencing: This procedure, developed by F. Sanger, S. Nicklen, and A.R. Coulson in 1977, centers around the synthesis of new DNA strands, which are complementary to a single-stranded template (Figure 13.5).

DNA synthesis via DNA polymerase does not proceed indefinitely because a small amount of dideoxynucleotide is added to the reaction mixture. These nucleotides lack a 3′ hydroxyl group required for connection to the next nucleotide. New chains of varying length result whose masses can be quantified by matrix-assisted laser desorption ionization time-of-flight (MALDI-TOF) mass spectrometry (MS). This method has been adapted to develop automated sequencing machines, for example, cycle sequencing.

Sequencing by chemical degradation: This method, developed by A.M. Maxam and Gilbert (1977), involves P^{32} labeling at the 3′-end of DNA. After the separation of two strands, chemical degradation analysis of the 3′-end moiety is done. Analysis is done by electrophoresis and autoradiography methods. The steps are similar to the chain termination method given by T.A. Brown in 1994. The steps are (i) treatment of DNA by dimethyl sulfate, leading to the attachment of a methyl (CH_3) group to the purine nucleotides; (ii) removal of the modified purine ring following DNA cleavage at the phosphodiester bond after piperidine treatment; (iii) fragmentation of DNA molecules, with labeling of some molecules but not all; and (iv) electrophoresis of DNA followed by autoradiography.

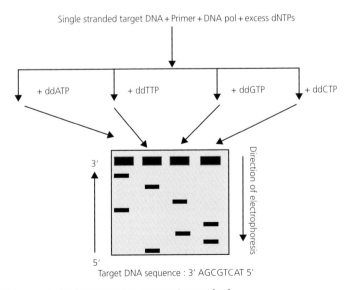

Figure 13.5 Sanger's chain termination sequencing method.

Automated versions of the Sanger method belong to the "first-generation sequencing techniques." After that, next-generation sequencing techniques were evolved. These techniques are based on the fragmentation of the whole genome, followed by parallel sequencing and reassembling of small sequenced fragments. First, the whole genome is "fragmented" into small parts. All parts are then sequenced in parallel. At this stage, various sequencing methods vary in terms of their template preparation and sequencing process. Consequently, using computation tools these small parts are then "assembled" back while checking the overlapping region. Next-generation sequencing techniques have been done on different types of platforms such as massively parallel signature sequencing (MPSS); sequencing by oligonucleotide ligation and detection (SOLiD sequencing); ion torrent semiconductor sequencing (often used interchangeably with the following terms: ion sequencing, pH-mediated sequencing, silicon sequencing, or semiconductor sequencing); DNA nanoball sequencing; and Helicos single-molecule fluorescent sequencing (Metzker, 2010).

Restriction fragment length polymorphisms

Restriction fragment length polymorphisms (RFLPs), which originate in base sequence alterations or DNA rearrangements, are naturally occurring, simply inherited, Mendelian traits. They are restriction fragments from a given chromosomal locus that vary in either size or length. These RFLPs are stable conditions of the DNA itself that can be utilized to generate genetic maps (Kochert *et al.*, 1991). This technique involves extracting double-standard DNA from a sample followed by DNA digestion by sequence-specific enzymes. Agarose gel electrophoresis separates the DNA fragments based on their size and molecular weights.

The RFLPs have a myriad of uses in studying genomes including DNA fingerprinting, employment as phenetic characters, plant breeding, tagging genes, introgression, analysis of quantitative traits, and cloning genes. Transfer of novel gene between related species and analysis of complex phylogenetic characters are getting easier with RFLP analysis.

Lipidomics

Lipidomics refers to systems-level study of lipids and their interaction with other biomolecules including proteins and other metabolites. It is often considered as a subset to "metabolome," which further comprises sugars, nucleic acids, and proteins or amino acids (Wenk, 2005). The functional specificity of lipids in comparison to other metabolites led to its study as a distinct field of science. Lipids are broadly defined as hydrophobic or amphiphilic small-molecule comprising fats, fat-soluble vitamins (i.e., vitamins A, D, E, and K), waxes, monoglycerides, diglycerides, and phospholipids, among others (Fahy *et al.*, 2009). Lipids are a key to many physiological activities varying from their participation in signaling pathways to their function as sources for energy storage. They also serve as structural components of cell membranes. Lipids constitute distinct subcellular membrane compartments of cells that differ in their membrane partitioning properties with their unique membrane organizing properties.

The study of lipidomics involves identification of lipid structure, lipid profiling, and understanding of the functional aspects. The interactions of lipids among themselves and with other proteins/metabolites are also of interest to the investigators in lipidomics. Such studies compare the changes occurred within the cells upon external perturbations to wild-type cell activity to gain major insights in underlying biological mechanism together with the knowledge about substrates and products of genetically altered enzymes (Welti *et al.*, 2002). In plant lipidomics, the organic extract of plant materials without chemical modifications is used to gain quantitative information of distinct lipid species. Mostly, membrane lipids are perfect candidates for such comprehensive and direct analyses. The comparison between *Arabidopsis*, wild-type, and phospholipase D alpha (PLD-alpha)-deficient plant unravels the role played by PLD-alpha genes under stress-induced lipid changes. The profiling of membrane lipids in this experiment shed light on the wild-type plant behavior/response, undergoing freezing, cold stresses, etc.

A powerful tool of lipid profiling is coupled electrospray ionization and quadruple MS. In addition, gas–liquid chromatography is useful. Scan modes for distinctive plant chloroplast lipids have been defined, making it possible to have rapid and quantitative analysis of major classes of lipids found in plant chloroplasts via electrospray ionization tandem MS (Welti *et al.*, 2003). Sphingolipids is another important class of lipids known to play an important role in various cellular processes, including membrane stability, stress response, cell signaling, and apoptosis, and it forms structural components of endomembranes of cells. Plant sphingolipids differ from their counterparts in animals and fungi, in their structural features. To have better understanding of other biological functions that may be carried out by sphingolipids in plants and signaling molecules derived from it, there is the need to have both qualitative and quantitative knowledge of their total content in plants. Using high-performance liquid chromatography (HPLC) technique, major classes of these lipids from leaves of *A. thaliana* have been identified (Markham *et al.*, 2006). Reversed-phase HPLC coupled with electrospray ionization and quadruple MS has helped in measuring 168 sphingolipids from *A. thaliana* using a single-solvent system (Markham and Jaworski, 2007). Two techniques of lipid profiling, namely, gas chromatography and HPLC, are briefly described.

Gas chromatography

The basic principle of chromatography is to separate out different ingredients based on their interaction with mobile phase and stationary phase. In gas chromatography, the sample is vaporized without decomposition. Here, the mobile phase used is basically an inert gas such as helium and the stationary phase is basically a thin film of active liquid or polymer on a solid support. The stationary phase kept in a metal cylinder is called "column." The substance passes through the column with the inert gas. Inside the column, different constituents of the sample interact with the stationary phase; and on the basis of their different partition coefficients, they are separated from one another.

The detector used here are flame ionization detector (FID) and the thermal conductivity detector (TCD). Thermal desorption spectrometer (TDS) is more widely used and are of high sensitivity among several detectors.

High-performance liquid chromatography

HPLC is also a chromatographic technique. In normal liquid chromatography, liquid sample is passed over the solid, stationary phase. In HPLC, though the basic principle is same, the optimum pressure (50–350 bar) is much higher than normal liquid chromatography. It also uses smaller size of packing material. There are two types of HPLC: normal phase and reversed phase. In normal phase HPLC, the column is filled with silica particle and nonpolar solvent is used; whereas in reversed phase HPLC, silica is modified by attaching a long hydrocarbon chain and a polar solvent is used.

Metabolomics

Metabolomics mainly deals with the analysis of various low-molecular-weight metabolites in a biological fluid (i.e., urine or serum in humans) to understand the dynamic changes in metabolism or to monitor the effect of some external stimuli (Clarke and Haselden, 2008). In other words, it can measure the change in terms of metabolite profiling for any type of biological perturbations (Lindon et al., 2007). In a sense, it acts as a bridge between genotypes and phenotypes (Fiehn, 2002). Various experimental techniques are available for studying metabolomics; the two most important of them are nuclear magnetic resonance (NMR) and MS. In the following, we discuss these techniques in greater detail.

One the one hand, proton NMR or H-NMR is one of the spectroscopic techniques where H nuclei are used as a reference for determining the structure of unknown molecule via NMR. This technique is based on the magnetic property of the nucleus of few atoms such as hydrogen and carbon. Deuterated water (D_2O) is used to reduce the background noise. NMR is extensively used in biological sciences because of its compatible application in wide-ranged biological fluids, without any special optimization technique needed for any analyte (Lenz et al., 2005). It is very useful for metabolite analysis.

On the other hand, MS analysis produces the spectra of the elemental composition of the molecule. In this technique, first the charged components of the molecules are separated by the mass-to-charge ratio. After that, the components are analyzed by plotting their relative abundance as a function to their mass-to-charge ratio. The fragmentation pattern of those components is one of the characteristics of this technique. MS is also very helpful to analyze the metabolites from particular biological fluids.

Every technique has its own advantages and disadvantages. Both these techniques are used either in "open-" (total component) or "close-" (specific component analysis) type analysis of metabolites. But MS analysis is used mainly in case of "'close" analysis. NMR technique is very useful for structural analysis and is relatively faster to perform. MS is generally attached with liquid or gas chromatography and is comparatively more sensitive for small component analysis than NMR (Clarke and Haselden, 2008).

In some studies, both NMR and MS are used simultaneously. This is so because some metabolites such as *S*-methyl methionine and the dipeptide alanyl-glycine can only be detected via NMR, whereas MS can do justice to sensitive quantification of many metabolites. For example, one recent study includes the total metabolic profiling of rice plant in some adverse condition such as natural disasters (Barding *et al.*, 2013). H-NMR is widely used in the estimation of primary and secondary metabolites in plants such as Opium (poppy). This plant can produce different types of bioactive benzylisoquinoline alkaloids and could become a model system for studying plant alkaloid metabolism by metabolic profile analysis.

An overview of data analyzing techniques

With the help of multivariate statistics, applied mathematics, and computer science modeling, analysis of complicated chemical systems is now possible in a data-driven manner and goes by the name of chemometrics. In metabolomics, it is used to characterize and model the metabolic pathways. For this purpose, various types of processes such as principal component analysis (PCA), statistical experimental design (SED), partial least squares (PLS) and orthogonal-PLS (OPLS) are used (Trygg *et al.*, 2007). In metabolomics, data from gas chromatography-mass spectrometry (GC-MS), liquid chromatography-nuclear magnetic resonance (LC-NMR), or liquid chromatography-mass spectrometry (LC-MS) techniques are used. In most of the cases, PCA is used as the starting point for these multivariate data analysis. This generates sufficient information, although additional information is analyzed with some new techniques such as PLS discriminant analysis (PLS-DA) and OPLS-DA. PLS is mostly used in dynamic data analysis.

Principal component analysis

PCA is used to analyze a multivariate data where each observation might be associated with various attributes (variables). Thus, the resulting space is a multidimensional hyperspace that is difficult to visualize. PCA helps in the reduction of dimensionality by finding a new set of variables that are uncorrelated, called "principal components." Each of these components is a linear combination of original variables and at the same time retains most of the sample's valuable information. The information is described as the variance present in the observed sample (Jolliffe, 2002). The principal components are so obtained such that the first PC contains most of the variation present in the original data. The subsequent components are ordered in decreasing order of variation of the information they hold. The eigenvectors of correlation matrix of sample data are used for the determination of principal components.

Partial least squares method

PLS method mainly defines quantitative relationship between two data sets "A" and "B"; say one comprising the chromatographic data of a set of calibration sample and comprising concentrations of endogenous metabolites by projecting the two data sets to new spaces (Trygg et al., 2007).

OPLS method

In order to facilitate model interpretation and simultaneously smoothen the way for execution of models onto the sample data sets, the data analysis techniques are further modified. Here in OPLS, the idea is to separate the systematic variation of "A" into two—differentiating the predictive information (linearly related to "B") and uncorrelated information (orthogonal to "B") (Trygg et al., 2007).

Proteomics

Since proteins are essential components for various cellular functions within an organism, their structural and functional analyses are indeed necessary. Proteomics deals with the study of protein structure and function on a large scale. The term "proteome" refers to the protein complement expressed by the genome. The proteome can consist of hundreds or thousands of proteins due to posttranslational modifications. The order of complexity is much larger in the study of the proteome as compared to genomics. Distinct sets of genes are expressed within different types of cells adding to huge protein sum. In fact, at a given instant of time, two similar types of cells may have different levels of protein expression subjected to the varying environmental conditions. This underlying fact amounts to the complexity of proteomics. Nevertheless, proteomics enables a depiction of the dynamical state of a cell presenting an advantage over genomics, which presents a static snapshot of state of cell expressing its genes in various ways. Thus, the study of proteins gives a more realistic view of a cell's state, and hence an organism's physiological activity. The growing interest in drug discovery further elicits the need for protein characterization in search for new drug targets and has, henceforth, led to the science of proteomics since it focuses on not only genome-encoded events such as protein translation but also non-genome-encoded events that include posttranslational modifications and diverse macromolecular interactions, for example, between proteins, nucleic acids, and lipids, instead of just DNA.

Elementary proteomics deals with protein identification, the process of which starts from protein extraction followed by peptide/protein separation, finally culminating in data integration. The study of proteomics is being undertaken using various advanced analytical tools such as two-dimensional (2D) gel electrophoresis, peptide mass fingerprinting by MALDI-TOF, and electrospray ionization MS/ESI-MS. Advanced techniques such as MS have been widely applied for protein identification or characterization. NADH dehydrogenase complex (complex I) of the oxidative phosphorylation (OXPHOS) system, located at the inner mitochondrial membrane, has been shown to have a unique structure in plants (Dudkina *et al.*, 2005). Recently, its protein constituents from *A. thaliana* have been identified using MS after the subunits are resolved using SDS/SDS-PAGE (sodium dodecyl sulfate polyacrylamide gel electrophoresis) (Peters *et al.*, 2013). Protein profiling of a plant species undergoing stressed conditions necessitates the use of refined molecular techniques and sheds new insight into stress signaling mechanism in plants caused either by abiotic or intracellular stress. The aim of such studies is to develop genetically engineered plants, which are able to tolerate significant stress conditions (Hossain *et al.*, 2012).

Another crucial aspect of plant study includes plant–insect interactions. Proteomics leads to a better understanding of defense mechanism of plants against herbivores. The use of high-throughput techniques, such as 2D gel electrophoresis and MALDI-TOF, helps in the identification of novel genes involved in defensive response of rice to insect infestation (Rakwal and Komatsu, 2000; Sangha *et al.*, 2013).

The essential experimental techniques, namely, gel electrophoresis and blotting, required in proteomics analysis, are described in the following text.

Gel electrophoresis

Electrophoresis separates molecules such as DNA, RNA, or proteins on agarose or polyacrylamide based on their size and charge in an electric field. For protein molecules, native polyacrylamide gel is normally used for molecular charge-based separation. But in denaturing SDS-PAGE, the charge interference of the molecule can be neglected as the huge negative surface charge of the molecule is imparted by SDS. Electrophoresis can be classified into two— 2D gel and 3D gel.

2D gel electrophoresis

In 2D gel electrophoresis, protein molecules are separated based on two different properties. First, they are separated according to their native size. Second, they are separated according to the isoelectric points in a different axis. As a result, different protein molecules or molecular fragment having same mass are separated.

Separation based on the isoelectric point is called "isoelectric focusing." An isoelectric point is basically expressed as one "pH value" of that protein, when the whole molecule becomes neutral. So despite being in an electric field, a protein molecule cannot move toward opposite charge. This is why two proteins with different pH value are separated from one another. So a pH gradient is required in the axis of second separation.

3D gel electrophoresis

In 3D gel electrophoresis, protein molecules are separated in a 3D manner. Here, the gel is used as a cube or a solid block of polyacrylamide. On the top of the gel, protein molecules are loaded in a 2D manner such as a microtiter plate. A laser beam plane illuminates fluorescent label protein in different planes. A digital camera below the gel records the protein bands in different planes of the gel. Up to 1536 samples can be loaded at a time in the gel. The temperature is maintained throughout the gel.

Blotting techniques

Blotting is one of the analytical techniques where samples from gel are transferred to the membrane for detection. In the case of protein analysis, Western blot technique is used. From the polyacrylamide gel, the protein samples are transferred to nitrocellulose or polyvinylidene difluoride membrane. On the membrane, labeled antibodies detect specific proteins. This technique is modified to enzyme-linked immunosorbent assay where an enzyme is linked with an antibody for detection.

Transcriptomics

Transcriptomics basically deals with the total amount of RNA molecules (mRNA, tRNA, and rRNA) inside the cell. The study of transcriptomics helps in determining the gene expression profile of different cells at different time points. It provides a deeper understanding over changes in gene expressions brought about by various diseases, treatments, and environmental fluctuations. Transcriptome analysis of different plant cells helps in understanding different signaling pathways and molecular mechanisms underlying various cellular processes. Phytoremediation, that is, remediation of contaminated water or soil using plants, necessitates the identification of metabolic pathways involved. Global techniques such as transcriptomics have led to important insights into the mechanism of detoxification process employed by the plants that further help in the identification of reliable markers for fast remedial actions (Coleman et al., 2005). The polyploidy nature of crops adds to the difficulties in the identification of required markers and their genome sequencing. Techniques have been developed to quantify gene expressions and to correlate trait variation with them and also with sequence variation of transcripts. Applying this method, the so-called associative transcriptomics, on *Brassica napus*, a tetraploid, reveals markers linked to quantitative trait loci for glucosinolate content of seeds (Harper *et al.*, 2012). Transcriptome analysis of cultivated tomato and its comparative analysis with five related species have recently revealed hundreds of genes with changes in gene expression levels that occur as a result of natural selection (Koeniga *et al.*, 2013). The evolved genes have been related to environmental response and stress tolerance. To analyze the effect of different genome organization on plant adaptation to stress, three different wheat cultivars, namely, bread wheat cultivar Chinese Spring (CS), a Chinese Spring terminal deletion line (CS_5AL-10), and the durum wheat cultivar Creso were subjected to mild and severe drought stress at grain-filling stage (Aprile *et al.*, 2009). Transcriptome profiling provides evidences of differential responses along the wheat genome under these stressed conditions with many of the clusters exhibiting different levels of expressions. Also by serial analysis of gene expression (SAGE), we can compare the expression level of various genes by mRNA tagging.

Nowadays, transcriptome data sets of different organisms are being prepared by different sequencing techniques. DNA microarray data is also helpful to prepare transcriptome data sets. In *A. thaliana*, transcriptome analysis of pollen grains showed over-repression of DNA repair, ubiquitin-mediated proteolysis, and cell cycle progression (Day *et al.*, 2008). Transcriptome profiling is basically done using RNA sequencing and DNA microarray techniques.

RNA sequencing

RNA sequencing refers to the sequencing of total ribonucleic acid content in a cell. It is also known as "whole-transcriptome shotgun sequencing." In comparison to DNA sequencing, RNA sequencing data gives more insights into the gene regulation and protein translation. With the help of next-generation sequencing or high-throughput sequencing, RNA sequencing has become a powerful tool for transcriptomic analysis.

RNA sequencing has been done on different types of platforms such as Illumina Genome Analyzer (for sequencing mammalian transcriptomes), ABI SOLiD sequencing (for sequencing stem cell transcriptomes), Roche's 454 sequencing (for sequencing single nucleotide polymorphisms or SNPs in maize). Depending on the different types of platforms used, the methods could be different. But for mRNA analysis, mainly the poly (T) oligonucleotides are used as a template for the poly (A) tail of mRNA. This helps in separating coding RNA strand from the total RNA pool. Those primers are basically attached with magnetic beads. After that, reverse transcription is done.

Biological database

Various research areas such as genomics, proteomics, transcriptomics, and microarray, produce huge amount of experimental data. These data sets are stored in different formats and are further processed by means of computational analyses of different biological process. These include gene expression profile, structural data, interaction profile, etc. Biological phenomena such as evolution, protein structure/function, interactions, and expression profiles are computationally analyzed and interpreted using these data sets. Various biological databases available are nucleotide sequence databases (DDBJ, NCBI, etc.); genome databases (RegulonDB, Ensembl, etc.); protein sequence databases (Uniport, PDB, DIP, etc.), protein–protein interaction databases (Biogrid, STRING, etc.); and metabolic pathway databases (KEGG, MANET, etc.).

Synthetic biology

Synthetic biology deals with engineering biomolecular systems. Studies in this area of research are broadly undertaken by two kinds of approaches. One of the approaches considers developing synthetic or unnatural systems by assembling together naturally occurring components of biology, whereas the other approach focuses on reproducing behaviors of natural systems from unnatural molecules (Benner and Sismour, 2005). The first approach focuses on amalgamation of living system components that can be interchanged to result into system that may or may not have resemblance to living systems, that is, unnatural in their formation. The other approach probes into building of artificial circuits that are able to perform when injected within the living systems, or replicates the emergent properties from natural biology, recreating artificial life. The invention of two artificial circuits namely, toggle switch and oscillator, further instigates the interest of scientists toward this developing field. The development of synthetic biology needs efforts from nonbiologists (i.e., physicists, chemists, and engineers) and biologists to achieve its goals of building new/artificial biological functions.

The genetic toggle switch is an example of artificial genetic circuit based on mathematical concept of "bistability," which was later found to be a naturally occurring phenomenon in many biological processes such as cell cycle, cellular differentiation, and apoptosis (Tyson *et al.*, 2003). Bistability refers to a condition in which the systems are provided with two steady stable states. There are two kinds of response of a control system—mutual activation circuits and mutual inhibition circuits—as observed in graphs shown in Figures 13.6b and 13.7b, respectively. The *y*-axis represents response of control system as a function of inducer, *I*, and the observed behavior is known as bistability (Tyson *et al.*, 2003).

Figure 13.6a is an example of *mutual activation system* where A activates protein B, which in turn enhances the synthesis of A. The system undergoes an abrupt and irreversible change as the signal/inducer magnitude is increased, greater than the critical value, I_c (Figure 13.6b). If *I* values are decreased, the systems response remains high, an irreversible process. The control system exhibits bistability for *I* values between 0 and I_c, that is, two stable steady states marked by solid lines separated by an unstable steady state, dashed line.

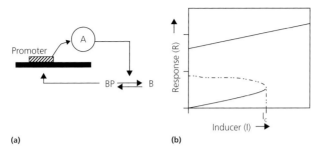

Figure 13.6 (a) Mutual activation and (b) response curve for a mutual activation circuit.

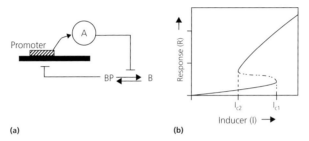

Figure 13.7 (a) Mutual inhibition circuit and (b) response curve for a mutual inhibition circuit.

The *mutual inhibition system* (Figure 13.7a), where A inhibits the activity of B while B promotes the degradation of A, also offers bistability, but the resultant process in not an irreversible one (Figure 13.7b).

Once the signal/inducer value is increased beyond I_{c1}, the system response will remains high (on-state) even if the signal is decreased below I_{c1}. It falls back to off-state only after the value of the signal is decreased below the critical value I_{c2} that is different from I_{c1}. For values of I between I_{c2} and I_{c1}, the system can have either high or low response depending upon the change in the signal. The toggle switch is a two-way, discontinuous switch represented by a mutual inhibition system (described earlier). The genetic toggle switch was synthesized in *Escherichia coli* exhibiting bistability conditions and paved the way for its applications in gene therapy and biotechnology (Gardner *et al.*, 2000). It consists of two promoters and two repressors, each promoter being inhibited by the repressor transcribed by the other promoter. Bistability ensures the tolerance of system response to inherent noise in gene expression with two critical values. Moreover, it has shown the possibility of constructing synthetic gene circuits with applications in living systems.

Another synthetic design of genetic network consists of three transcriptional repressors, each inhibiting the activity of the next repressor forming a complete cycle namely a negative feedback loop (Figure 13.8), conferring oscillatory behavior to the system. This synthetic circuit, called *repressilator*, has been constructed in *E. coli*, and the oscillations in the levels of three

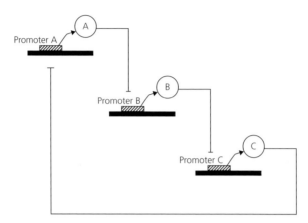

Figure 13.8 A repressilatory circuit consisting three products A, B, and C. Each protein inhibits the activity of the next gene, thus resulting in a negative feedback loop.

repressor proteins are observed (Elowitz and Leibler, 2000). Such rational network designs have led the foundation for more such synthetic circuits constructed by assembling together individual natural components having new functional properties.

Considering tremendous potential of synthetic biology, it is being applied extensively in the biomedical field (Ruder *et al.*, 2011). Synthetic designs are being developed for treating bacterial infections. Such studies include engineered bacteriophage that destroys the bacterial cells present in bacterial biofilms by degrading the biofilms (Lu and Collins, 2007). Biofilms act as a protection layer for bacteria that resist host immune defense as well as antibiotics. Another study focuses on increasing the efficacy of existing antibiotics via engineering bacteriophage (Lu and Collins, 2009).

Although, the field of synthetic biology has gained success by developing new biotechnological tools such as synthetic pathways and genetic circuits, its application to plant science is still limited (Zurbriggen *et al.*, 2012). The genetic engineering of plants till date is based upon reductionist approach, that is, manipulation of components such as enzymes and transcriptional regulators. There is the need to synthesize new biological systems with desired properties from well-characterized biological modules. This will substantially contribute toward the production of better quality plants, possessing high stress tolerance as well as effective disease and pest control properties.

Acknowledgments

We thank Prof. Devrani Mitra of Presidency University, Kolkata, for critical comments and for a careful reading of this chapter.

References

Aldana, M., Ballezaa, E., Kauffman, S. and Resendiz, O. (2007) Robustness and evolvability in genetic regulatory networks. *Journal of Theoretical Biology*, **245** (3), 433–448.

Aprile, A., Mastrangelo, A.M., Leonard, A.M.D. *et al.* (2009) Transcriptional profiling in response to terminal drought stress reveals differential responses along the wheat genome. *BMC Genomics*, **10** (279). doi:10.1186/1471-2164-10-279.

Barding, G.A. Jr., Béni, S., Fukao, T. *et al.* (2013) Comparison of GC-MS and NMR for metabolite profiling of rice subjected to submergence stress. *Journal of Proteome Research*, **12**, 898–909.

Benner, S.A. and Sismour, A.M. (2005) Synthetic biology. *Nature Reviews Genetics*, **6**, 533–543.

Clarke, C.J. and Haselden, J.N. (2008) Metabolic profiling as a tool for understanding mechanisms of toxicity. *Toxicologic Pathology*, **36**, 140–147.

Coleman, J.O., Haslam, R.P. and Downie, A.L. (2005) Transcriptomics and proteomics: tools for optimising phytoremediation activities. *Zeitschrift für Naturforschung*, **60** (7–8), 544–548.

Day, R.C., Herridge, R.P., Ambrose, B.A. and Macknight, R.C. (2008) Transcriptome analysis of proliferating *Arabidopsis* endosperm reveals biological implications for the control of syncytial division, cytokinin signaling, and gene expression regulation. *Plant Physiology*, **148**, 1964–1984.

Dudkina, N.V., Eube, H., Keegstra, W., Boekema, E.J. and Braun, H.P. (2005) Structure of a mitochondrial supercomplex formed by respiratory-chain complexes I and III. *Proceedings of the National Academy of Sciences of the United States of America*, **102** (9), 3225–3229.

Elowitz, M.B. and Leibler, S. (2000) A synthetic oscillatory network of transcriptional regulators. *Nature*, **403**, 335–338.

Fahy, E., Subramaniam, S., Murphy, R.C. *et al.* (2009) Update of the LIPID MAPS comprehensive classification system for lipids. *Journal of Lipid Research*, **50** (Suppl.), S9–S14.

Fiehn, O. (2002) Metabolomics—the link between genotypes and phenotypes. *Plant Molecular Biology*, **48**, 155–171.

Gardner, T.S., Cantor, C.R. and Collins, J.J. (2000) Construction of a genetic toggle switch in *Escherichia coli. Nature*, **403**, 339–342.

Harper, A.L., Trick, M., Higgins, J. *et al.* (2012) Associative transcriptomics of traits in the polyploid crop species *Brassica napus. Nature Biotechnology*, **30**, 798–802.

Hossain, Z., Nouri, M.Z. and Komatsu, S. (2012) Plant cell organelle proteomics in response to abiotic stress. *Journal of Proteome Research*, **11** (1), 37–48.

Jolliffe, I.T. (2002) *Principal Component Analysis*, Springer-Verlag, New York/Secaucus, NJ.

Kitano, H. (2002) Systems biology: a brief overview. *Science*, **295** (5560), 1662–1664.

Kochert, G., Halward, T., Branch, W.O. and Simpson, C.E. (1991) RFLP variability in peanut (Arachis hypogaea L.) cultivars and wild species. *Theoretical and Applied Genetics*, **81**, 565–570.

Koeniga, D., Jiménez-Gómeza, J.M., Kimura, S. *et al.* (2013) Comparative transcriptomics reveals patterns of selection in domesticated and wild tomato. *Proceedings of the National Academy of Sciences of the United States of America*, **110** (28), E2655–62.

Kumar, A., Asthana, M., Sharma, S. *et al.* (2012) Importance of using DNA microarray in studying medicinal plant. *Webmed Central Molecular Biology*, **3** (12), WMC003876.

Lenz, E.M., Bright, J., Knight, R. *et al.* (2005) Metabonomics with ^{1}H-NMR spectroscopy and liquid chromatography-mass spectrometry applied to the investigation of metabolic changes caused by gentamicin-induced nephrotoxicity in the rat. *Biomarkers*, **10** (15), 173–187.

Lindon, J.C., Holmes, E. and Nicholson, J.K. (2007) Metabonomics in pharmaceutical R&D. *FEBS Journal*, **274**, 1140–1151.

Lu, T.K. and Collins, J.J. (2007) Dispersing biofilms with engineered enzymatic bacteriophage, *Proceedings of the National Academy of Sciences of the United States of America*, **104** (27), 11197–11202.

Lu, T.K. and Collins, J.J. (2009) Engineered bacteriophage targeting gene networks as adjuvants for antibiotic therapy. *Proceedings of the National Academy of Sciences of the United States of America*, **106** (12), 4629–4634.

Markham, J.E. and Jaworski, J.G. (2007) Rapid measurement of sphingolipids from *Arabidopsis thaliana* by reversed-phase high-performance liquid chromatography coupled to electrospray ionization tandem mass spectrometry. *Rapid Communication Mass Spectrometry*, **21**, 1304–1314.

Markham, J.E., Li, J., Cahoon, E.B. and Jaworski, J.G. (2006) Separation and identification of major plant sphingolipid classes from leaves. *The Journal of Biological Chemistry*, **281** (32), 22684–22694.

Maxam, A.M. and Gilbert, W. (1977) A new method for sequencing DNA. *Proceedings of the National Academy of Sciences of the United States of America*, **74**(2), 560–564.

Metzker, M. L. (2010) Sequencing technologies—the next generation. *Nature Reviews Genetics*, **11** (1), 31–46.

Moodie, S.L., Novère NL, Sorokin, A. *et al.* (2010) The systems biology graphical notation: process description language level 1. *Nature Precedings V1*, posted 4.

Peters, K., Belt, K. and Braun, H.P. (2013) 3D gel map of *Arabidopsis* complex I. *Frontiers in Plant Science*, **4**, 153; doi:10.3389/fpls.2013.00153.

Peterson, D.G., Schulze, S.R., Sciara, E.B. *et al.* (2002) Integration of C_0t analysis, DNA cloning, and high-throughput sequencing facilitates genome characterization and gene discovery. *Genome Research*, **12**, 795–807.

Pu, L. and Brady, S. (2010) Systems biology update: cell type-specific transcriptional regulatory networks. *Plant Physiology*, **152**, 411–419.

Rakwal, R. and Komatsu, S. (2000) Role of jasmonate in the rice (*Oryza sativa* L.) self defense mechanism using proteome analysis. *Electrophoresis*, **21**, 2492–2500.

Ranade, S.A. and Farooqui, N. (2002) Assessment of profile variations amongst provenances of neem using single-primer-amplification reaction (SPAR) techniques. *Molecular Biology Today*, **3**, 1–10.

Rashidi, H. and Buehler, L.K. (2005) *Bioinformatics Basics Applications in Biological Science and Medicine*, CRC Press, Boca Raton, FL.

Ruder, W.C., Lu, T. and Collins, J.J. (2011) Synthetic biology moving into the clinic. *Science*, **333**, 1248–1252.

Sangha, J.S., Yolanda, H.C., Kaur, J. *et al.* (2013) Proteome analysis of rice (*Oryza sativa* L.) mutants reveals differentially induced proteins during brown planthopper (*Nilaparvata lugens*) infestation. *International Journal of Molecular Sciences*, **14**, 3921–3945.

Trygg, J., Holmes, E. and Lundstedt, T. (2007) Chemometrics in metabonomics. *Journal of Proteome Research*, **6**, 469–479.

Tsuda, K., Sato, M., Stoddard, T. *et al.* (2009) Network properties of robust immunity in plants. *PLoS Genetics*, **5** (12), e1000772.

Tyson, J.J., Chen, K.C. and Novak, B. (2003) Sniffers, buzzers, toggles and blinkers: dynamics of regulatory and signaling pathways in the cell. *Current Opinion in Cell Biology*, **15** (2), 221–231.

Welti, R., Li, W., Li, M. *et al.* (2002) Profiling membrane lipids in plant stress responses —role of phospholipase d in freezing-induced lipid changes in *Arabidopsis*. *The Journal of Biological Chemistry*, **277** (35), 31994–32002.

Welti, R., Wang, X. and Williams, T.D. (2003) Electrospray ionization tandem mass spectrometry scan modes for plant chloroplast lipids. *Analytical Biochemistry*, **314**, 149–152.

Wenk, M.R. (2005) The emerging field of lipidomics. *Nature Reviews Drug Discovery*, **4** (7), 594–610.

Zurbriggen, M.D., Moor, A. and Weber, W. (2012) Plant and bacterial systems biology as platform for plant synthetic bio(techno)logy. *Journal of Biotechnology*, **160**, 80–90.

Further reading

Anand, R.K. and Chiu, D.T. (2012) Analytical tools for characterizing heterogeneity in organelle content. *Current Opinion in Chemical Biology*, **16**, 391–399.

Grewal, R.K., Mitra, D. and Roy, S. (2015) Mapping networks of light-dark transition in LOV photoreceptors. *Bioinformatics [Oxford]*, **31**, 3608–3616.

Heinig, U., Gutensohn, M., Dudareva, N. and Aharoni, A. (2013) The challenges of cellular compartmentalization in plant metabolic engineering. *Current Opinion in Biotechnology*, **24**, 239–246.

Kalluri, U.C., Yin, H., Yang, X. and Davison, B.H. (2014) Systems and synthetic biology approaches to alter plant cell walls and reduce biomass recalcitrance. *Plant Biotechnology Journal*, **12** (9), 1207–1216.

Libault, M. and Chen, S. (2016) Plant single cell type systems biology. *Frontiers in Plant Science*, **7**, 35, doi: 10.3389/fpls.2016.00035.

Lucas, M., Laplaze, L. and Bennett, M.J. (2011) Plant systems biology: network matters. *Plant Cell and Environment*, **34** (4), 535–553.

Roy, S. (2014) Perspectives in systems biology. *Systems and Synthetic Biology*, **8**, 1–2.

Satori, C.P., Kostal, V. and Arriaga, E.A. (2012) Review on recent advances in the analysis of isolated organelles. *Analytica Chimica Acta*, **753**, 8–18.

Taylor, N.L., Ströher, E., and Millar, A.H. (2014) Arabidopsis organelle isolation and characterization. *Methods in Molecular Biology*, **1062**, 551–572.

APPENDIX A

Four major compounds in plant cells and tissues[a]

Class	Structure/function	References
Carbohydrates Monosaccharide	Simple sugars consisting of two families, aldoses and ketoses distinguished by the carbaryl oxygen; there is a variety of monosaccharide derivatives such as sugar alcohols, sugar acids and amino sugars	Bols (1996) Lehmann (1998) Robyt (1998)
Sugar alcohols Examples are glacitol, sorbitol, xylitol	Polyhydroxy compounds compounded with carbaryl group of either the aldehyde or hexane group	Robyt (1998)
Sugar acids	'Acidic' oxidation of aldehyde group of aldose carbohydrate 'anonic' acid oxidation of primary alcohol group, anonic acids are important in plant cell wall structure	Robyt (1998)
Amino sugars	Replacement of a sugar OH with an amino group	Robyt (1998)
Oligosaccharides	Saccharides composed of 2–10 monosaccharides joined in glycosidic exchange; diverse functions as in the receptors in recognition	Dey (1990) Avigad (1982)
Polysaccharides	Higher-molecular-weight saccharides such as starch, fructans, hemicelluloses, cellulose and pectin (see Chapter 11 on photosynthesis and Chapter 9 on cell walls in the present volume)	Carpita and Gibeaut (1993) Albersheim et al. (2010)
Lipids	Long-chain organic acids consisting of 4–24C atoms	Gur et al. (2002) Williams et al. (2010) Azimova et al. (2012) Tevini and Lichetenthaler (1977)

[a]The functions of these compounds are numerous and are discussed throughout the present volume.

Plant Cells and their Organelles, First Edition. Edited by William V. Dashek and Gurbachan S. Miglani.
© 2017 John Wiley & Sons, Ltd. Published 2017 by John Wiley & Sons, Ltd.

Class	Structure/function	References
Fatty acids		
Saturated	Hydrocarbon chain contains only single carbon bonds	Mostofsky and Yehuchu (1996)
Unsaturated	Hydrocarbon chain contains one or more double bonds	
Triacylglycerol	Esters of glycerol with three fatty acids	Harwood (1980)
Glycolipids	Chloroplast lipids such as mono- and di-galactosyldiacylglycerol and sulpholipids	Benning (1998)
Phospholipids	Polar lipids that contain one or more polar heads besides their hydrocarbon tails	Cevc (1993)
Sphingolipids	Polar head and two non-polar tails lacking glycerol	
Steroles	Polar head and a remaining non-polar variety	Lukas (1994)
Proteins	Major group of macromolecules comprising amino acids in peptide linkage	Howard and Brown (2002)
Amino acids	Organic molecules containing a carboxyl (COOH) and amino (NH_2) groups attached to a single carbon atom	Smith (1998) Bender (1998)
Peptides	Two to ten amino acids joined in peptide linkage	Bodansky (1993) Barett and Elmore (1998)
Polar		
Non-polar		
Polypeptides	Many amino acids in peptide linkage	Wieland and Bodansky (1991)
Proteins	Complex polypeptides exhibiting primary, secondary, tertiary and quaternary structures; there are many types of proteins and their function are discussed throughout this volume	Bender (2012) Smith and Gestwicki (2012)
Nucleic acids	Covalently linked nucleotides	Brichell and Darling (1995) Boulter (2012) See Chapter 8 (present volume)
DNA and RNA		
Purines	Six-member pyrimidine ring fused to a five-membered imidazole ring; three or four nitrogen atoms	Lister (1996)
Pyrimidines	Six-membered ring compounds containing two nitrogen atoms	Brown (1994)
Nucleosides	Composed of purine or pyrimidine plus a sugar	Townsends (1988)
Nucleotides	Composed of a heterocyclic nitrogenous base, a pentose and a phosphoryl group	Townsends (1998)
Polynucleotides	Consist of many nucleotides	Boulter (2012)

Chemicals other than carbohydrates, lipids, proteins and nucleic acids important in cell biology[a]

Chemical class	Structure and function	References
Alkaloids	Basic substances containing one or more nitrogen atoms	Rahman (1990)
Classes	Growth regulators	Jackson *et al.* (1994)
Pyridine	Insect repellents or attractants	
Tropone	Maintain ionic balance	
Isoquinoline	Nitrogen storage reservoirs	Roberts and Wink (1998)
Flavonoids	Secondary compounds; compounds with two benzene rings separated by a propane unit; phenolic pigment compounds	Shirley (1996)
Examples		
Anthocyanins	Flower coloring	Clark and Titus (1997)
Aurones	Alleopathy	Packer (2001)
Flavonols	Effective in mammalian systems against inflammation, heart disease and cancer	Busling and Matley (2002)
Catechins	Auxin transport inhibitor	Lepiniec *et al.* (2006)
Chalcones	Dormancy or viability	Havsteen (2007)
Isoflavonoids	UV protection	Buer (2010)
Hormones		
Abscisic acid	A sesquiterpenoid; affects induction and maintenance of dormancy; induces gene transcription for protease inhibitors in response to wounding; induces seeds to synthesize storage proteins; inhibits shoot growth; inhibits GA-induced amylase synthesis; stimulates stomatal closure	Walton and Li (1995) Arteca (1996)
Auxins	Indole-compounds; non-indole synthetic varieties Apical auxin suppresses lateral bud growth; can inhibit or promote leaf and fruit abscission; delay leaf senescence; delay fruit ripening; can enhance ethylene productions; induce cambium cell division; may induce fruit set; may affect phloem transport; promote flowering in bromeliads; promote femaleness in defensive plants; stimulate cell elongation; stimulate phloem and xylem differentiation; stimulate rooting	Bandurski *et al.* (1995) Bartel (1997) Zizimalova and Napier (2003)
Brassinosteroids	Steroidal compounds; promote plant growth; affect plasmodesmata organization and transport; decrease fruit drop; enhance xylem differentiation; increase RNA and DNA polymerase; inhibit root growth and development; promote ethylene biosynthesis; promote germination; can promote shoot elongation	Arteca (1995) Culte *et al.* (1993) Khripach *et al.* (1999)

[a]Chemical formulae for these compounds occur in Dashek, W.V. and Harrison, M. (2006) *Plant Cell Biology*, Science Publishers, Enfield, NH.

Chemical class	Structure and function	References
Cytokinins	Naturally occurring purines related to adenine; synthetic purines; enhanced leaf expansion; may affect stomatal opening; promote chlorophyll synthesis; stimulate short initiation/lipid formation in tissue culture; stimulate growth of lateral bud release of apical dominance	Haberer and Kieber (2007) Kakimots (2003) Lightfoot *et al.* (1997) McGaw and Burch (1995) Mok and Mok (1994)
Ethylene	A gaseous hormone; enhances leaf and fruit abscission; induces fruit ripping; induces femaleness in dioecious flowers; may be important in oderations root formation; promotes flower opening; stimulates flower and leaf senescence; stimulates root and shoot growth and differentiation	Abeles *et al.* (1992) Bleecher and Kende (2000) Chang *et al.* (1997) Kende (1993)
Gibberellins	Acid compounds; diterpenes; break seed germination; can cause parthenocarpic (seedless) fruit development; enhance stem elongation; induce maleness in dioecious flowers, may delay senescence or promote flowers in long dry plants; stimulation of α-amylase in germinating seeds	Hedden (1999) Rademacher (2000) Sponsel (1995a, b) Takahashi *et al.* (1991)
Jasmonates	Jasmonic acid is synthesized from biolinelic acid; promote flowering; stimulate plant pathogenesis	
Organic acids	Weakly acidic acids; components of citric acid cycle	Dashek and Michaels (1992) Duke (1992) Fox and Powell (2001) Szmigielska *et al.* (2002) Rivasseau *et al.* (2006)
Phenols	Hydroxylated ring structure; simple monomeric components of polymeric polyphenols; important in human health	Tyman (1996) Gross *et al.* (1999) Pobłocka-Olech *et al.* (2010)
Phenylpropanoids	Plant compounds with an attached 3C side chain; caffeic acid activity	Seigler (1998) Kurkin (2003)
Phytoalexins	Many are phenolic phenylpropanoids; some are isopropanoids; few are polyacetylenes; antimicrobial substances; components of the hypersensitive response	Kuc (1994) Daniel and Purkayastha (1995) Yu (1997) Fatima *et al.* (2008) Fatima *et al.* (2008)
Quinones	Cyclic ketones; function in electron transport	Martinez and Benito (2005) Costa *et al.* (2009) Zafar (1994)
Terpenes	Organic compounds derived from isoprene units; usually lipid-soluble; some have hormone activity; others are pigments	
Vitamins	Organic compounds which are either water- or fat-soluble; coenzymes	Connolly *et al.*(1991) Hambone and Tomos-Barberan (1991)

References

Abeles, F.B., Morgan, P.W. and Saltviet, M.E. (1992) *Ethylene in Plant Biology*, Academic Press, San Diego, CA.

Albersheim, P., Danvill, A., Roberts, K., Secleroff, R. and Staehelin, A. (2010) *Plant Cell Walls*, Garland Publishing, New York, NY.

Arteca, R.N. (1995) Jasmonates, salicylic acid and brassinosteroids. In: *Plant Hormones: Physiology, Biochemistry and Molecular Biology*, 2nd Edition, (Davies, P.J., ed.) Kluwer Academic Publ., Boston, MA, pp. 206–213.

Arteca, R.N. (1996) *Plant Growth Substances: Principles and Applications*, Kluwer Academic Publ., Dordrecht, the Netherlands.

Avigad, G. (1982) Sucrose and other disaccharides. In: *Encyclopedia of Plant Physiology, New Series Plant Carbohydrates I. Intracellular Carbohydrates*. Vol. **13A**, (Loewus, F.A. and Tanner, W. eds) pp. 216–247, Springer-Verlag, Berlin, Germany.

Azinova, S.S., Glushenhova, A.I. and Vinogradawa, V.I. (2012) *Lipids, Lipophilic Components and Essential Oils from Plant Sources*, Springer, New York, NY.

Bandurski, R.S., Cohen, J.D., Slovin, J.P. and Reinecke, D.M. (1995) Auxin biosynthesis and metabolism. In: *Plant Hormones*, (Davies, P.J., ed.) pp. 39–65. Kluwer Academic Publ., Dordrecht, the Netherlands.

Barett, G.C. and Elmore, D.T. (1998) *Amino Acids and Peptides*, Cambridge University Press, Cambridge, England.

Bartel, B. (1997) Auxin biosynthesis. *Annual Review of Plant Physiology and Plant Molecular Biology*, **48**, 51–66.

Bender, D.A. (1998) *Amino Acid Metabolism*, CRC Press, Boca Raton, FL.

Bender, D.A. (2012). The Metabolism of surplus amino acids. *British Journal of Nutrition*, **108**, S113–S121.

Benning, C. (1998) Biosynthesis and function of the sulfolipid sulfoquinovosyl diacylglycerol. *Annual Review of Plant Biology*, **49** (1), 53–75.

Bleecher, A.B. and Kende, H. (2000) Ethylene: a gaseous signal molecule in plants. *Annual Review of Cell Development and Biology*, **16**, 1–18.

Bodansky, M. (1993) *Peptide Chemistry: A Practical Textbook*, Springer-Verlag, Berlin, Germany.

Bols, M. (1996) *Carbohydrate Building Blocks*, John Wiley & Sons, Inc., New York, NY.

Boulter, D. (2012) *Nucleic Acids and Proteins in Plants I Structure, Biochemistry and Physiology of Proteins*, Springer, Berlin, Germany.

Brichell, F. and Darling, D. (1995) *Nucleic Acid XXX – The Basics*, Oxford University Press, Oxford, England.

Brown, D.J. (1994) *The Pyrimidines*, John Wiley & Sons, New York, NY.

Buer, C.S., Imin, N. and Djordjevic, M.A. (2010) Flavonoids: new roles for old molecules. *Journal of Integrative Plant Biology*, **52**, 98–111.

Busling, B.S. and Matley, J.A. (2002) *Flavonoids in Cell Function*, Plenum Publ., Kluwer Academic Publ., New York, NY.

Carpita, N.C. and Gibeaut, D.M. (1993) Structural models of primary cell walls in flowering plants: consistency of molecular structure with the physical properties of the walls during growth. *The Plant Journal*, **3** (1), 1–30.

Cevc, G. (1993) *Phospholipids Handbook*, Marcel Dekker, New York, NY.

Chang, C., Grievson, D., Kanellis, A.K. and Kende, H. (1997) *Biology and Biochemistry of the Plant Hormone Ethylene*, Kluwer Academic Publ. Boston, MA.

Clark, W.D. and Titus, G.P. (1997) Applications in flavonoid research. In: *Methods in Plant Biochemistry and Molecular Biology*, (Dashek, W.V., ed.), pp. 217–227. CRC Press, Boca Raton, FL.

Connolly, S.J., Laupacis, A., Gent, M., Roberts, R.S., Cairns, J.A. and Joyner, C. (1991) Canadian atrial fibrillation anticoagulation (CAFA) study. *Journal of the American College of Cardiology*, **18** (2), 349–355.

Costa, A., Jan, E., Sarnow, P. and Schneider, D. (2009) The IMD pathway is involved in antiviral immune responses in *Drosophila*, *PLoS ONE*, **4**:e7436. doi: 10.1371/journal.pone.0007436.

Culte, H.G., Yokota, T. and Adam, G. (1993) *Brassinosteroids: Chemistry, Bioactivity and Applications*, American Chemical Society, Washington, DC.

Daniel, M. and Purkayastha, R.P. (1995) *Handbook of Phytralexin Metabolism and Action*, Marcel Dekker, New York, NY.

Dashek, W.V. and Michaels, J. (1992) Assay and purification of enzymes – oxalate decarboxylase. In: *Methods in Plant Biochemistry and Molecular Biology*, (Dashek, W.V., ed.) pp. 49–71. CRC Press, Boca Raton, FL.

Dey, P.M. (1990) Oligosaccharides. In: Dey, P.M. (ed.) *Carbohydrate Methods in Plant Biochemistry*, Academic Press, San Diego, CA.

Duke, J.A. (1992) *Handbook of Biologically Active Phytochemicals and Their Activities*, CRC Press, Boca Raton, FL.

Fatima, B., Arman, M. and Iqbal, S. (2008) HPLC separation of phytoalexins from *Phaseolus vulgaris* treated with elicitor from *Colletotrichum lindemuthianum*. *Asian Journal of Scientific Research*, **1**, 160–165.

Fox, R.B. and Powell, W.H. (2001) *Nomenclature of Organic Compounds: Principles and Practice*. Oxford University Press, Oxford, England.

Gross, G.C., Hemingway, R.W. and Yoshida, J. (1999) *Plant Polyphenols 2: Chemistry, Biology, Pharmacology, Ecology*, Kluwer Academic Publs., New York, NY.

Gurr, N.J., Harword, J.L. and Frayn, K.N. (2002) *Lipid Biochemistry*, Blackwell Science, Malden, MA.

Haberer, G. and Kieber, J.J. (2007) Cytokinin: new insights into a classic phytohormone. *Plant Physiology*, **128**, 354–362.

Hambone, J.B. and Tomos-Barberan, F.A. (1991) *Ecological Chemistry and Biochemistry of Plant Terpenoids*, Claredon Press, Oxford, England.

Harwood, J.L. (1980) Plant acyl lipids: structure, distribution and analysis. In: *Biochemistry of Plants*, vol. **4**, (Stumpf, P.K. and Conn, E.E., eds.) pp. 1–55. Academic Press, New York, NY.

Havsteen, B.H. (2007) The biochemistry and medical significance of the flavonoids. *Pharmacology & Therapeutics*, **96** (2–3), 67–202.

Hedden, P. (1999) Recent advances in gibberellins biosynthesis. *Journal of Experimental Botany*, **50**, 553–563.

Howard, G.C. and Brown, E. (2002) *Modern Protein Chemistry. Practical Aspects*, CRC Press, Boca Raton, FL.

Jackson, M.B., Belcher, A.R. and Brain, P. (1994) Measuring shortcomings in tissue culture aeration and their consequences for explant development. In: *Physiology, Growth and Development of Plants in Culture*, (Lumsden, P.J., Nicholas, J.R. and Davies, W.J., eds.) pp. 191–203. Springer, Dordrecht, the Netherlands.

Kakimots, T. (2003) Biosynthesis of cytokinins. *Journal of Plant Research*, **116**, 223–239.

Kende, J. (1993) Ethylene biosynthesis. *Annual Review of Plant Physiology and Plant Molecular Biology*, **44**, 283–307.

Khripach, V.A., Zhabinski, V.N. and de Groot, A.E. (1999) *Brassinosteroids: A New Class of Plant Hormones*, Academic Press, San Diego, CA.

Kuc, J. (1994) Relevance of phytoalexins – a critical review. *Acta Horticulture*, **381**, 529–539.

Kurkin, V.A. (2003) Phenylpropanoids from medical plants: distribution, classification, structural analysis and biological activity. *Chemistry of Natural Compounds*, **39**,123–153.

Lehmann, J. (1998) *Chemistry of Carbohydrates Structure and Biology*, Koenlenhydrate Chemis and Biologi, Stuttard, Germany.

Lepiniec, L., Debeaujon, I., Routaboul, J.M. *et al.* (2006) Genetics and biochemistry of seed flavonoids. *Annual Review of Plant Biology*, **57**, 405–430.

Lightfoot, D.A., McDaniel, K.L., Ellis, J.K., Hammerton, R.H. and Nicander, B. (1997) Methods for analysis of cytokinin content, metabolism and response. In: *Methods in Plant Biochemistry and Molecular Biology*, Dashek, W.V. (ed.), CRC Press, Boca Raton, FL.

Lister, J.H. (1996) *The Purines Supplement*, John Wiley & Sons, New York, NY.

Lukas, S.E. (1994) *Steroids*, Enslow Publs., Springfield, NI.

Martinez, M.J.A. and Benito, P.B. (2005) Biological activity of quinines. *Studies in Natural Products Chemistry*, **30**, 303–366.

McGaw, B.A. and Burch, L.R. (1995) Cytokinin biosynthesis and metabolism. In: *Plant Hormones*, (Davies, P.J., ed.) Kluwer Academic Publs., Amsterdam, the Netherlands.

Mok, D.W.S. and Mok, M.C. (1994) *Cytokinins*, CRC Press, Boca Raton, FL.

Mostofsky, D.I. and Yehuchu, S. (1996) *Fatty Acids Biochemistry and Behavior*, Humana Press, Totowa, NJ.

Packer, L. (2001) *Flavanoids and Other Polyphenols*, Academic Press, San Diego, CA.

Pobłocka-Olech, L., Krauze-Baranowska, M., Głód, D., Kawiak, A. and Łojkowska, E. (2010) Chromatographic analysis of simple phenols in some species from the genus *Salix*. *Phytochemistry Analysis*, **21**, 463–469.

Rademacher, W. (2000) Growth retardants: effects on gibberellin biosynthesis and other metabolic pathways. *Annual Review of Plant Physiology and Plant Molecular Biology*, **51**, 501–531.

Rahman, A. (1990) *Diterpenoid and Steroidal Alkaloids*. Elsevier Health Science, Amsterdam, the Netherlands.

Rivasseau, C., Boisson, A.M., Mongélard, G., Couram, G., Bastien, O. and Bligny, R. (2006) Rapid analysis of organic acids in plant extracts by capillary electrophoresis with indirect UV detection: directed metabolic analyses during metal stress. *Journal of Chromatography A*, **1129** (2), 283–290.

Roberts, M.F. and Wink, M. (1998) *Alkaloids Biochemistry Ecology and Molecular Applications*, Plenum Press, New York, NY.

Robyt, J.F. (1998) *Essentials of Carbohydrate Chemistry*, Springer, New York, NY.

Seigler, D.S. (1998) Phenylpropanoids. In: *Plant Secondary Metabolism*, (Seigler, D.S., ed.) pp 106–129. Kluwer Academic Publs., Boston, MA.

Shirley, B.W. (1996) Flavonoid biosynthesis new functions for an old pathway. *Trends in Plant Science*, **1**, 377–387.

Smith, D.M. (1998) Protein separation and characterization. In: *Food Analysis*, pp. 261–282, (Nielson, S.S., ed.) Aspen Publishers, Gaithersburg, MD.

Smith, M.C. and Gestwicki, J.E. (2012) Features of protein-protein interactions that translate into potent inhibitors: topology, surface area and affinity. *Expert Review of Molecular Medicine*, **14**, e16, doi: 10.1017/crm.2012.10.p.

Sponsel, V.M. (1995a) The biosynthesis and metabolism of gibberellins in higher plants. In: Davies, P.J. (ed.) *Plant Hormones*, Kluwer Academic Publs., Amsterdam, the Netherlands.

Sponsel, V.M. (1995b) The biosynthesis and metabolism of gibberellins in higher plants. In: Davies, P.J. (ed.) *Plant Hormones: Physiology, Biochemistry and Molecular Biology*, 2nd Edition, pp 66–97. Kluwer, Norwell, MA.

Szmigielska, A.M., Van Rees, K.C.J. and Cieslinski, G. (2002) Gas chromatographic analysis of low molecular weight organic acids in roots and shoots of durum wheat plants. *Communications in Soil Science Plant Analysis*, **33**, 1415–1423.

Takahashi, N., Phinney, B.O. and MacMillan, J. (1991) *Gibberellins*, Springer-Verlag, Berlin, Germany.

Tevini, M. and Lichtenthaler, H.K. (1977) *Lipids and Lipid Polymers in the Higher Plants*, Springer, Berlin, Germany.

Townsends, L.B. (1988) *Chemistry of Nucleosides and Nucleotides*, Phlegm Press, New York, NY.

Tyman, J.H.P. (1996) *Synthetic and Natural* Phenols, 1st Edition. Elsevier Science, Amsterdam, the Netherlands.

Walton, D.C. and Li, Y. (1995) Abscisic acid biosynthesis and metabolism. In: *Plant Hormones: Physiology, Biochemistry and Molecular Biology*, 2nd Edition, (Davies, P.J., ed.) pp 140–157. Kluwer, Norwell, MA.

Wieland, T. and Bodansky, M. (1991) *The World of Peptides. A Brief History of Peptide Chemistry*, Springer, Berlin, Germany.

William, J.P., Khan, M.L. and Len, N.W. (2010) *Physiology, Biochemistry, and Molecular Biology of Plant Lipids*, Springer, Berlin, Germany.

Yu, L. (1997) The solution and assay of elicitins. In: *Methods in Plant Biochemistry and Molecular Biology*, (Dashek, W.V., ed.) pp. 265–279. CRC Press, Boca Raton, FL.

Zafar, R. (1994) *Medicinal Plants of India*, pp. 48–61. CBS Publs. and Distri., Delhi, India.

Zizimalova, E. and Napier, R.M. (2003) Points of regulation for auxin action. *Plant Cell Report*, **21**, 625–634.

APPENDIX B

Classification of enzymes

Enzyme structures are classified by their E.C. number (using the May 2000 v 26.0 release of the ENZYME Data Bank)

Enzyme classification (E.C.)[a], [b]	Reaction catalyzed
E.C.1. Oxidoreductases [1064 PBD entries] $A - B + C = A = B - C$	Electron transfer
E.C.2. Transferases [1435 PDB entries] $A - B + C = A + B - C$	Group transfer
E.C.3. Hydrolases [3173 PDB entries] $A - B + H_2O \leftrightarrows A - H + B - OH$	
E.C.4. Lyases [419 PDB entries] $\dfrac{XY}{AB} = A = B + X - Y$	Group additional removal to/from double bond
E.C.5. Isomerases [309 PDB entries] $\dfrac{XY}{AB} = \dfrac{YY}{AB}$	Transfer of groups in molecules resulting in isomeric forms
E.C.6. Ligases [162 PDB entries] $A - B = A - B$	Generation of C–C, S–S, G–O, and C–N bonds condensation and ATP cleavage

The enzyme structures can also be accessed via the PDBsum database:
- PDBsum – Summaries and structural analyses of PDB data files

[a] The student is referred to the Nomenclature Committee of the International Union of Biochemistry and Molecular Biology for discussion of enzyme nomenclature
[b] From Dashek and Michaels (1997) and Laskowski (2001): http://www.biochem.ucl.oc.uk/bsm/enzymes

Plant Cells and their Organelles, First Edition. Edited by William V. Dashek and Gurbachan S. Miglani.
© 2017 John Wiley & Sons, Ltd. Published 2017 by John Wiley & Sons, Ltd.

Summary of the laws of thermodynamics and related terms and concepts

---

Laws of thermodynamics

First law – total energy of the universe remains constant

Second law – universe constantly changes and becomes more disordered or systems change spontaneously from states of lower probability (more ordered) to states of higher probability

Terms and concepts

Free energy (G) – criterion for an increase in disorder

ΔG change in free energy, i.e. a measure of the extent of disorder

ΔG release of free energy, reaction occur spontaneously, creation of disorder, energetically favourable

$+\Delta G$ occurs only if coupled to a second reaction with a $-\Delta G$; creates order but energetically unfavourable

Entropy = degree of disorder of a state (S)

$\Delta S = R \ln pB/pA$

Change in free energy that occurs when A $\rightarrow$ B, i.e. one mole of A is converted into one mole of B. pA and pB = probabilities of two states. Note: Reaction with a large increase in S occurs spontaneously

R = gas constant

Enthalpy (H) or $E + PV$ change in free energy ΔH = heat absorbed

Free energy change is a direct measure of entropy change in the universe

H = enthalpy V = volume

G = Gibbs free energy E = energy

T = absolute temperature S = entropy

P = pressure

Summary of the effects of inhibitors on Lineweaver–Burk plots $1/V$ vs $1/[S]$

	Slope	Intercept on ordinate
No inhibitor	$\dfrac{K_m}{V_{max}}$	$\dfrac{1}{V_{max}}$
Competitive	$\dfrac{K_m}{V_{max}}\left(1+\dfrac{[I]}{K_1}\right)$	$\dfrac{1}{V_{max}}$
Uncompetitive	$\dfrac{K_m}{V_{max}}$	$\dfrac{1}{V_{max}}\left(1+\dfrac{[I]}{K_1}\right)$
Noncompetitive	$\dfrac{K_m}{V_{max}}\left(1+\dfrac{[I]}{K_1}\right)$	$\dfrac{1}{V_{max}}\left(1+\dfrac{[I]}{K_1}\right)$

References

Dashek, D.V. and Michaelis, I.A. (1997) Assay and purification of enzymes – oxalate decarboxylase. In: *Methods in Plant Biochemistry and Molecular Biology*, (Dashek, W.V., ed), pp. 49–71. CRC Press, Boca Raton, FL.

Laskowski, R.A. (2001) PDBsum: summaries and analyses of PDB structures. *Nucleic Acids Research*, **29**, 221–222.

APPENDIX C

Summary of type I and type II errors*,†

Statistical choice	Null hypothesis positive H_0	(H_0) Decision negative H_0
Negate H_0	Type I error‡	Affirm
	P (type I error) $= \alpha$	
	More serious than a type II error	Decision
	Error of the first kind	
H_0 not negated, H_1	Affirm	Type II error§
	Decision	P (type II error) $= \beta$
		Frequently due to a small sample size
		Error of the second kind

Note: This table is a composite from multiple readings, e.g. http://www.cas.lancs.ac.uk/glossary_v1.1/main.html and other websites.

* Type I and II errors are inversely related.

† The student should consult statistics texts (see References for discussions of p values and confidence levels (Rohlf, 1994; Zar, 1998; Rosner, 2010; Looney, 2002).

‡ Type I errors occur when the null hypothesis is true but rejected.

§ Type II errors result when the null hypothesis is false but not rejected. Power of a hypothesis test $= 1 - P$ (Type II error) $= 1 - d$.

Plant Cells and their Organelles, First Edition. Edited by William V. Dashek and Gurbachan S. Miglani.
© 2017 John Wiley & Sons, Ltd. Published 2017 by John Wiley & Sons, Ltd.

Summary of statistical test choices*[,†]

Variables	Other parameters	Recommended choice	Description
One	One group	$\bar{x} \pm$ SD	$\bar{x}$ = sum of all the numbers of items in the set, SD – square root of the variance
	Two groups	T-test[‡]	A parametric statistical test requiring normality assumptions
	Three groups	ANOVA	An acronym of a category of test termed *analysis of variance*
Two	Both continuous	Correlation	Strength of the relation between variables
	One continuous One discrete	ANOVA	See above
	Both categorical	Chi-square	Nonparametric alternative to the *t*-test
Three or more	One group	Multiple regression	Serves to make predictions regarding multiple values
		Factor analysis	Used to discover simple patterns in the pattern of relationships of any variable
		ANOVA repeated measures	Used for repeated measures of variables
		Analysis of covariance	Combines features of simple linear regression with one-way analysis of variance
		Multivariate ANOVA	A set of tests used when there are three or more independent variables with two or more treatments per variable
		Discriminant function	Employed to determine which variable discriminates between two or more naturally occurring groups

Note: Experiments may be those that lead to predictions or physical experiments with observations. The latter require controls, statistically significant sample sizes (*N* numbers) and repetition. The significance of the scientific method is its predictive power. Adapted from: http://maps.unomaha. edu/Maher/geo117/scimethod.html

* Modified from T. Lee Willoughby – http://research.med.umkc.edu/tlwbiostats/choosetest.html

† Descriptions of these tests can be found in Fry (1993a, b), Iles (1993a, b), Dytham (1999), Wardlaw (2003), Myers and Well (2003), and Maxwell and Delaney (2004).

‡ See http://www.cas.lancs.ac.uk/glossary_v1.1/main.html

Statistical software packages*

Package	Usefulness
GLIM Version 3.77[†]	ANOVA and linear and multiple regression
Minitab release 7.1 and 8.2[‡]	Comprehensive statistical package ANOVA, linear bivariate and multiple regression
SAS release 6.03[†]	Comprehensive statistical package, ANOVA, linear and multiple regression, cluster analysis and others
SAS, Sigma plot and Deltagraph[†]	Nonlinear curve fitting
SPSS/PC + version 3.0[§]	Comprehensive statistical package, ANOVA, linear and multiple regression, cluster analysis and others
Stratagraphics version 2.1[†]	Comprehensive statistical package, ANOVA, bivariate linear and nonlinear regression, multiple regression, cluster analysis and others
Delta Graph, Table Curve, Sigma Plot, Origin, Igor and Prism**	Curve fitting of complex functions
Matlab, Lab View**	Data analysis programs

*Harvard Graphics and Statistica are well-established software packages with spreadsheets.
[†] Adapted from: Appendix A, Fry (1993b).
[‡] Wardlaw (2003). Provides extensive laboratory examples which use Minitab.
[§] SPSS is highlighted by Campbell (1989).
**Young (2001).

References

Campbell, R.C. (1989) *Statistics for Biologists*, Cambridge University Press, Cambridge, UK.

Dytham, C. (1999) *Choosing and Using Statistics: A Biologists Guide*, Blackwell Science, Malden, MA.

Fry, J.C. (1993a) One way analysis of variance. In: *Biological Data Analysis. A Practical Approach*, (Fry, J.C., ed.), pp. 1–39. IRL Press, Oxford, UK.

Fry, J.C. (1993b) Biovariate regression. In: *Biological Data Analysis. A Practical Approach*, (Fry, J.C., ed.), pp. 81–125. IRL Press, Oxford, UK.

Iles, T.C. (1993a) Crossed and hierarchical analysis of variance. In: *Biological Data Analysis. A Practical Approach*, (Fry, J.C., ed.), pp. 41–80. IRL Press, Oxford, UK.

Iles, T.C. (1993b) Multiple regression. In: *Biological Data Analysis. A Practical Approach*, (Fry, J.C., ed.), pp. 127–172. IRL Press, Oxford, UK.

Looney, S.W. (2002) *Biostatistical Methods*, Humana Press, Totowa, NJ.

Maxwell, S.E. and Delaney, H.D. (2004) *Designing Experiments and Analyzing Data: A Model Comparison Perspective*, Lawrence Erlbaum Association, Mohawk, NJ.

Myers, J. and Well, A.D. (2003) *Research Design and Statistical Analysis*, Lawrence Erlbaum Association, Mohawk, NJ.

Rohlf, F.J. (1994) *Biometry: The Principles and Practice of Statistics in Biological Research*, W.H. Freeman, New York, NY.

Rosner, B. (2010) *Fundamentals of Biostatistics*, Daxbury Press, Independence, KY.

Wardlaw, A.C. (2003) *Practical Statistics for Biologists*, John Wiley & Sons, New York, NY.

Young, S.S. (2001) *Computerized Data Acquisition and Analysis for the Life Sciences*, Cambridge University Press, Cambridge, UK.

Zar, J.H. (1998) *Biostatistical Analysis*, Prentice Hall, Upper Saddle River, NJ.

Index

Plant Cells and their Organelles, First Edition. Edited by William V. Dashek and Gurbachan S. Miglani.
© 2017 John Wiley & Sons, Ltd. Published 2017 by John Wiley & Sons, Ltd.